Electronic Measurement Systems

Theory and Practice

Second Edition

Electronic Measurement Systems

Theory and Practice

Second Edition

Anton F P van Putten

Eindhoven and Middlesex Universities

Institute of Physics Publishing
Bristol and Philadelphia

First edition published 1988 by Prentice Hall International (UK) Ltd
Second edition published 1996 by IOP Publishing Ltd

British Library Cataloguing-in-Publication Data

A catalogue record for this book is available from the British Library.

ISBN 0 7503 0339 5 (hbk)
    0 7503 0340 9 (pbk)

Library of Congress Cataloging-in-Publication Data

Putten, Anton F. P. van, 1939–
    Electronic measurement systems: theory and practice/Anton F. P. van Putten—2nd ed.
      p.   cm.
    Includes bibliographical references and index.
    ISBN 0-7503-0339-5 (hbk: alk. paper).—ISBN 0-7503-0340-9 (pbk: alk. paper)
    1. Electronic measurements. 2. Electronic instruments.
    I. Title.
    TK7878.P87   1996
    681'.2—dc20
                                  96-29304
                                    CIP

Published by Institute of Physics Publishing, wholly owned by The Institute of Physics, London

Institute of Physics Publishing, Techno House, Redcliffe Way, Bristol BS1 6NX, UK

US Editorial Office: Institute of Physics Publishing, The Public Ledger Building, Suite 1035, 150 South Independence Mall West, Philadelphia, PA 19106, USA

Typeset and printed in the UK by J W Arrowsmith Ltd, Bristol

To my wife Ria

      Michael
      Maurice
      Antoinette & Johan
      Pascal

with love

Only knowledge in harmony brings progress

                                   AFPvP

CONTENTS

Preface to the First Edition   xv

Preface to the Second Edition   xix

List of Symbols   xxi

Acknowledgments   xxv

1 Introduction to Electronic Measurement Systems   1
   1.1 Information   1
   1.2 Types of measurement   4
   1.3 Why electronic measurement systems?   5
   1.4 Future trends   8
   Problems   10
   Bibliography   10

2 Types of Measuring System   11
   2.1 Introduction   11
   2.2 Functional structures   12
      2.2.1 Measuring structures for an active physical quantity   12
      2.2.2 Measuring structures for a passive physical quantity   14
   2.3 Signal structures   16
      2.3.1 Signals with a unique character   17
      2.3.2 Periodic signals   22
      2.3.3 Sampled signals   26
      2.3.4 Stochastic signals   30
   2.4 Spatial structures   32
      2.4.1 A multiple-input/output configuration   33
      2.4.2 A centralized data-acquisition system   34
      2.4.3 A decentralized data-acquisition system with a digital
           multiplexer   37
      2.4.4 A distributed multiplexer data-acquisition system   37
      2.4.5 Telemetry   38
   2.5 Automated electronic structures   39
   2.6 I/O interface   41
      2.6.1 IEEE 488   41
      2.6.2 The VXI/VME bus   42
      2.6.3 Serial and industrial networks, RS-232 (V24)   43
      2.6.4 Plug-in data-acquisition (DAQ) boards   44
   2.7 Architecture of bus systems   46

Problems   48
Bibliography   49

3 Implementations of Functions   **50**
    3.1 Introduction   50
    3.2 Tools   50
        3.2.1 Complex calculus and variables   50
        3.2.2 Laplace transform of signals   55
        3.2.3 Description of linear systems   56
    3.3 Analogue functions   67
        3.3.1 The operational amplifier   67
        3.3.2 Preamplifier with adjustable gain   69
        3.3.3 The non-inverting operational amplifier   70
        3.3.4 The differential amplifier configuration   70
        3.3.5 The comparator   71
        3.3.6 The integrator   72
        3.3.7 A constant-current source   73
        3.3.8 The logarithmic amplifier   73
        3.3.9 A current-to-voltage converter   75
        3.3.10 An analogue multiplier   75
        3.3.11 A peak detector   77
    3.4 Filters   77
        3.4.1 Types of filter   78
        3.4.2 The switched capacitor   78
    3.5 Analogue-to-digital and digital-to-analogue conversions   79
        3.5.1 Introduction   79
        3.5.2 Coding   79
    3.6 Digital-to-analogue converters   82
        3.6.1 The adder network   82
        3.6.2 The ladder network   83
    3.7 Analogue-to-digital converters   85
        3.7.1 The dual-slope integration ADC   85
        3.7.2 The sigma–delta modulator ADC   87
        3.7.3 Voltage-to-frequency converter (VFC)   89
        3.7.4 Successive approximation ADC   90
        3.7.5 Parallel converter   91
    3.8 Digital functions   91
        3.8.1 A decoder   91
        3.8.2 An encoder   92
        3.8.3 A multiplexer   92
    3.9 User-programmable logic devices   93
        3.9.1 Combinational logic   95
        3.9.2 Realizations of programmable arrays   96
        3.9.3 Sequential logic   101

3.10 Miscellaneous   102
    3.10.1 The conventional Wheatstone bridge configuration   103
    3.10.2 An integrated Wheatstone bridge   104
    3.10.3 The principle of the constant-voltage, constant-current
         bridge configuration   105
Problems   108
Bibliography   109

4 System Specifications   **110**
  4.1 Introduction   110
  4.2 Technical specifications   110
    4.2.1 Application area   111
    4.2.2 Resolution   113
    4.2.3 Accuracy and inaccuracy   113
    4.2.4 Sensitivity   116
    4.2.5 Linearity   117
    4.2.6 Offset and drift   119
    4.2.7 Rejection factor   119
    4.2.8 Dynamic range   123
    4.2.9 Reliability   126
    4.2.10 The number of measurements per unit time   126
    4.2.11 Environmental conditions   128
  4.3 Types of error   128
    4.3.1 Systematic errors   128
    4.3.2 Conditional errors   129
    4.3.3 Stochastic errors   129
  4.4 Measures for improving electronic systems   130
    4.4.1 The feedback principle   130
    4.4.2 Feedforward coupling   133
    4.4.3 The influence of feedback on input and output
         impedance   136
  4.5 A microprocessor-based measuring system   139
    4.5.1 Microprocessor fundamentals   139
Problems   150
Bibliography   152

5 Reliability   **153**
  5.1 Introduction   153
  5.2 Reliability concepts   154
    5.2.1 Definitions   154
    5.2.2 Stochastic variables, probability distributions, functions
         and reliability   155
    5.2.3 Reliability parameters   158
    5.2.4 The mean time between failure   160

5.3	Families of lifetime distributions	161
5.4	Reliability of parts, components, equipment and systems	163
	5.4.1 The ISO 9000 system	165
5.5	Availability; repairable and non-repairable systems	166
5.6	Boolean algebra and reliability	169
5.7	The reliability and MTBF of systems in series	172
5.8	The reliability and MTBF of systems in parallel	173
	5.8.1 An active on-line standby system	174
	5.8.2 A cold on-line standby system	177
	5.8.3 A parallel k out of n system	179
5.9	Methods to determine the reliability of systems	180
	5.9.1 The network reduction method	180
	5.9.2 The path-tracing technique	182
	5.9.3 The decomposition technique	182
	5.9.4 The minimal-cut-set technique	185
	5.9.5 Markov techniques	186
5.10	Design aspects	193
5.11	Causes of failure	199
5.12	Software	201
5.13	Future developments	202
	Problems	202
	Bibliography	204

6	**Transducers**	**205**
6.1	Introduction	205
6.2	Classification of transducers	206
	6.2.1 Types of energy form	207
	6.2.2 Types of energy source	208
	6.2.3 Modulating and self-generating transducers	209
6.3	State descriptions	213
	6.3.1 The steady-state description	214
	6.3.2 The dynamic-state description	216
6.4	Transducer parameters, definitions and terminology	218
	6.4.1 Transducer parameters	218
	6.4.2 Definitions and terminology	218
6.5	Characteristics of integrated circuits and transducers	221
	6.5.1 Silicon technology	224
	6.5.2 Size and linewidth	224
	6.5.3 Complexity	225
	6.5.4 Power consumption	225
	6.5.5 Reliability	226
	6.5.6 Electrons versus photons as information carrier	227
	6.5.7 Available technologies for transducers	227
	6.5.8 Price–performance ratio	230
	6.5.9 Application areas	231

6.6 Transducer effects in silicon and other compatible materials 231
    6.6.1 Transducer effects in silicon 231
    6.6.2 Transducer effects in compatible materials 231
    6.6.3 Transducer effects in optical semiconductors 232
6.7 The optical energy domain 234
    6.7.1 Physics 234
    6.7.2 Applied units in the optical domain 235
    6.7.3 The interaction of electromagnetic radiation with
        semiconductors 236
    6.7.4 The p–n junction in the optical domain 237
    6.7.5 The influence of temperature and radiation on the
        conductivity 238
    6.7.6 Radiant energy in silicon microtransducers 240
    6.7.7 Radiant energy in compatible technology 243
    6.7.8 Fibre technology 245
    6.7.9 Review of known optical effects 248
6.8 The mechanical energy domain 248
    6.8.1 Physics 248
    6.8.2 Mechanical energy in silicon microtransducers 254
    6.8.3 Mechanical energy in compatible technology 256
    6.8.4 Review of effects in the mechanical energy domain 257
6.9 The thermal energy domain 257
    6.9.1 Physics 257
    6.9.2 Thermal effects in semiconductors 260
    6.9.3 Thermal energy in silicon microtransducers 261
    6.9.4 Thermal energy in compatible technology 263
    6.9.5 Review of effects in the thermal energy domain 265
6.10 The magnetic energy domain 265
    6.10.1 Physics 266
    6.10.2 Hall effect 268
    6.10.3 Magnetic energy and silicon microtransducers 271
    6.10.4 Magnetic energy in compatible technology 273
    6.10.5 Physical effects in the magnetic energy domain 274
6.11 The chemical energy domain 275
    6.11.1 'Physics' 275
    6.11.2 Chemical energy in silicon microtransducers 278
    6.11.3 Review of chemical effects 280
6.12 Future trends in transducers 280
    6.12.1 Technological trends 280
Problems 282
Bibliography 284

7 **Offset and drift** 286
    7.1 Introduction 286
    7.2 Principles of offset voltage and current 287

7.3 Calculation of the offset voltage and current 288
   7.3.1 The bipolar transistor 289
   7.3.2 The field effect transistor 289
7.4 Offset behaviour of differential amplifier stages 291
   7.4.1 The bipolar differential input stage 291
   7.4.2 The junction FET differential input stage 293
   7.4.3 The MOSFET differential input stage 293
7.5 Offset in transducers 294
7.6 Drift behaviour of offset current and voltage 294
   7.6.1 Stochastic variations in electrical conductivity 295
   7.6.2 Aging, humidity and vibrations 295
   7.6.3 Power supply variations 295
7.7 The influence of temperature 296
7.8 Temperature drift in bipolar transistors 298
7.9 Temperature drift in FETs 300
   7.9.1 The junction field effect transistor 300
   7.9.2 The MOS field effect transistor 302
7.10 Drift in sensors 303
   7.10.1 Drift elimination by modulation 306
   7.10.2 The alternating direction method (ADM) 307
Problems 309
Bibliography 310

8 Guarding and Shielding 311
8.1 Introduction 311
8.2 The nature and causes of interference 311
   8.2.1 Sources of interference 315
8.3 Aspects of guarding 317
   8.3.1 Ground and earth 318
   8.3.2 A grounded system 319
8.4 Aspects of shielding 321
   8.4.1 A shielding cabinet 321
   8.4.2 A shielding cabinet as part of an electronic circuit 322
   8.4.3 The use of more than one shielding cabinet 323
   8.4.4 The connection of a common in a multiple-cabinet configuration 324
8.5 Interference caused by a magnetic field 328
8.6 Types of coupling mechanism 330
   8.6.1 Direct coupling 331
   8.6.2 Magnetic coupling 333
8.7 Interference caused by power supply transformers 334
8.8 The differential amplifier and guarding 338
   8.8.1 The common-mode rejection ratio 341
   8.8.2 The CMRR in shielding circumstances 342

8.9   Improvement of the common-mode rejection ratio   344
       8.9.1   The application of a floating guard   345
       8.9.2   Active guarding or the application of double-shielded
               cables   348
8.10 Worked examples   349
8.11 Electromagnetic compatibility   354
       8.11.1 New Approach, Global Approach and CE mark   354
       8.11.2 Generic emission standards   356
Problems   358
Bibliography   360

9   **Noise Calculations**   **361**
    9.1   Introduction   361
    9.2   Noise voltage, noise current and noise figure   362
          9.2.1   The noise voltage $\overline{e_n}$ 362
          9.2.2   The noise current $\overline{i_n}$ 363
          9.2.3   The noise figure NF   363
    9.3   The relationship between the noise voltage, noise current and
          noise figure   364
    9.4   Calculating total noise voltage   366
    9.5   Calculating the noise figure and the signal-to-noise ratio   370
    9.6   The noise figure myth   372
          9.6.1   The optimum source resistance   373
          9.6.2   The optimum signal-to-noise ratio   375
    9.7   Other noise calculating techniques   380
          9.7.1   Shifting noise sources   380
          9.7.2   Superposition   382
          9.7.3   Thevenin and Norton   382
          9.7.4   Noise sources in series and in parallel   383
          9.7.5   Worked examples   385
    Problems   388
    Bibliography   389

10 **Physics of Noise**   **390**
    10.1 Introduction   390
    10.2 Thermal noise or Johnson noise in conductors   390
    10.3 Shot noise   391
    10.4 Flicker noise   392
    10.5 Noise in semiconductor components   394
          10.5.1 Noise behaviour of the bipolar transistor   394
          10.5.2 Noise behaviour of the field effect transistor   396
          10.5.3 Noise behaviour of the diode   399
    10.6 Noise in sensors   401
    Bibliography   403

11   Interfacing to Sensors   **404**
   11.1   Introduction   404
   11.2   Sensors in bridge configurations   405
      11.2.1   Bridge configuration with one sensor element   405
      11.2.2   Bridge configuration with two sensor elements   406
      11.2.3   Bridge configuration with four sensor elements   407
      11.2.4   A double-bridge configuration   408
      11.2.5   Biasing bridge circuits   408
   11.3   Bridge amplifier configurations   410
      11.3.1   Circuit 1   410
      11.3.2   Circuit 2   410
      11.3.3   Circuit 3   412
      11.3.4   Circuit 4   412
      11.3.5   Circuit 5   413
      11.3.6   Circuit 6   414
      11.3.7   Circuit 7   414
   11.4   Telemetry   416
   11.5   Hybrid interfacing circuits   417
   11.6   Autozeroing   417
      11.6.1   Autozeroing in electronics   417
      11.6.2   Autozeroing in transducers   418
   Problems   418
   Bibliography   419

12   Ergonomics or Human Engineering   **420**
   12.1   Introduction   420
   12.2   Biological equipment of the human being   421
      12.2.1   The auditory window   421
      12.2.2   The visual window   421
   12.3   Some other aspects of human perception of information   423
      12.3.1   Transmission of information   423
      12.3.2   Retrieval of information   423
      12.3.3   Processing of information   424
   12.4   Perception   424
   12.5   Trends in the development of measuring systems   425
   12.6   The observer–display interaction   427
   12.7   Coding of information on displays   429
   Problems   433
   Bibliography   433

Appendix   Table of Frequently Used Physical Constants   **434**

Index   **435**

PREFACE TO THE FIRST EDITION

Electronic measurement systems have become complicated machines able to perform complex tasks in a large variety of environments. Moreover, such systems are found in every area of science and technology. The huge impact of digital electronics has changed the world of electronic measurement systems dramatically, both inside and outside the system. Multipurpose, multifunctional measurement systems which are capable of performing tasks that were impossible ten years ago are commonplace now. Another important aspect of these systems is society's increasing dependence on them: the reliability of systems has become of vital importance. Measuring systems are realizations of technical concepts able to gather, handle, modify, store, retrieve and present information. The design of a measurement system not only requires a great deal of knowledge about components and their possibilities: aspects related to the underlying philosophy of a system and its specific architecture are increasingly important. For example, a large shift to digital control and microprocessor-driven measurement systems has taken place in a large variety of applications. The design of a measurement system is a complicated process and often the designer of a system is highly specialized in a few areas, but in practice the designer is confronted with numerous questions concerning the (technical) specifications, for instance the following.

1 What degree of accuracy is required of the system?
2 What are the environmental conditions, e.g., temperature range, shock, vibrations and electromagnetic interference?
3 What is the required reliability or the expected mean time between failures?
4 Which type of transducer is needed for the physical quantity to be measured and what possibilities are available?
5 Which type of architecture should be applied?
6 What is the required sensitivity of the measurand?
7 What about noise behaviour?
8 What is the required common-mode rejection ratio?
9 What kind of ergonomic layout is best?

This book covers a large number of topics encountered in the design, application and verification of electronic measurement systems. Every system is considered to be composed of three parts: the *input* stage, the *modifier* stage and the *output* stage. Each stage can be treated separately.

In the first four chapters a general picture is presented of measurement architectures, structures, the tools used to describe systems and implementation of components. Further system specifications are discussed in detail and illustrated with many examples. The remaining chapters discuss topics of major importance to the proper functioning of a system. Not all subjects could be covered in the same detail here, mainly because of space limitations. As will be recognized, attention is devoted to subjects that are usually not found together in one book. Often it is these aspects that ultimately determine whether a system can operate within its stated specifications. For several years, the material on which this textbook is based has been used for graduate and undergraduate students in electrical and electronic engineering with a basic knowledge of electronics, but of course some topics discussed have a much broader application than to electronics systems alone. For the sake of completeness, and to make an introduction easier, some topics which may be assumed to be well known are also discussed here. The contents of the chapters may be summarized as follows.

Chapter 1 is devoted to basic ideas of measurement systems, such as what information is, and why it is preferred to handle information electronically.

Chapter 2 considers measurement architectures and structures. This includes functional structures, signal, spatial and automated structures.

Chapter 3 is devoted to the available tools, such as the Bode plot, the polar plot and the pole–zero plot and the relationships between them. Further implementations of frequently applied functions are presented. These implementations cover analogue functions, analogue-to-digital and digital-to-analogue conversion and some miscellaneous functions. Under this last heading a new integrated bridge is discussed.

Chapter 4 is devoted to general and more specific technical system specifications. All the usual technical specifications are presented with their meanings and consequences. This includes a discussion of the common-mode behaviour, linear amplification and the implications of feedback. Error definitions, classifications and calculations are also presented. After some error causes are introduced, some frequently applied improving measures are reviewed. The chapter concludes with a discussion of a microprocessor-based measuring system.

Chapter 5 provides a comprehensive introduction to reliability concepts, including the basic concepts and calculation techniques used in this area. Markov techniques are included as the most generally applied and powerful techniques for calculating reliability.

Chapter 6 is devoted to transducers or sensors and actuators for energy conversions to and from the electrical energy domain. Basic ideas, concepts and definitions are discussed in detail and a general review of most known energy conversion mechanisms is presented. The possibility of silicon and compatible technologies for the manufacture of transducers is also discussed.

Chapter 7 discusses aspects of offset and drift in basic electronics active components. The influence of temperature on bipolar transistors, FETs and differential input stages is treated in detail. Offset and drift are treated separately.

Chapter 8 is devoted to the problems concerning guarding and shielding systems from electromagnetic noise. With the increasing trend to handling all kinds of information electronically, the environment is charged with an ever-increasing electromagnetic pollution. Moreover, electronic measurement systems are highly sensitive to electromagnetic interference and effective precautions should be taken to avoid these influences. All kinds of preventative remedy are discussed in detail. The influence on the common-mode rejection ratio of sources of interference is also discussed.

Chapter 9 presents a discussion of noise calculations. It is not necessary to have a detailed knowledge of the physics of noise for this, and it suffices to use only the specifications of the manufacturer. The theoretical discussion is left to the following chapter.

Chapter 10 treats the physics of noise in three important noise mechanisms and the noise behaviour of the active components to be met in electronics today, such as the bipolar transistor and the field effect transistor.

Finally, chapter 11 is devoted to human engineering or ergonomics. Some interesting aspects concerning human–machine relations are presented, in particular the ergonomics of the visual and auditory organs for the presentation of information.

Almost every chapter can be read independently from the others; hence a flexible presentation of subjects can be realized. The consequence of this approach is that some details are repeated in different contexts, but this can only improve understanding. Furthermore, every chapter is provided with examples and exercise material.

It will be clear that writing a book is a tremendous task and cannot be realized without the help of others. I therefore would like to thank all the persons who have given advice directly and indirectly in the preparation of this book. Special thanks are due to Jim Sherwin of National Semiconductor Corp. for his permission to use material about noise calculations. I am also indebted to A W Wijkman, P J van den Akker and K E Kuijk from Philips Corp. for their permission to use relevant material concerning shielding and guarding. I also want to thank Paul A M Maas of Delft University Press and Paul P L Regtien and Professor Simon Middelhoek, both of Delft University of Technology, for providing material about transducers and related topics. Special thanks are due to Dr Mark Browne of the University of Manchester Institute of Science and Technology for his careful reading of the manuscript and his valuable advice.

PREFACE TO THE SECOND EDITION

Since the first edition of *Electronic Measurement Systems* was published in 1988 many technological changes have influenced the design of electronic measurement systems. Hardware and software developments have shifted design frontiers. Further miniaturization and advanced features of micromachining have made possible the design and realization of three-dimensional structures and devices performing tasks unpredicted at that time.

In chapter 2 the section about signal structures has been extended and most-used I/O bus systems, DAQ boards and their architecture have been discussed in detail.

In chapter 3 the user-programmable devices (UPLDs) have been introduced to give a flavour of their power.

In chapter 4 the use of microprocessor principles in instrumentation has been introduced.

In chapter 5 on reliability we have outlined some new approaches. Because of the tremendous increase in transistor density per chip area completely new procedures must be followed where built-in testability has become a major feature.

In chapter 6 can be found the most dramatic changes over the last decade. To make it more readable for the student, each transducer energy domain starts now with a brief introduction to the related physics in order to understand what the principle of operation is. Many transducers are now becoming smarter and smarter with built-in intelligence. Here instrumentation becomes microsystem technology.

In chapter 7 on offset and drift attention has been devoted to drift in sensors, because this has had a major impact on their proper functioning. A new method for drift elimination, the ADM method, has been introduced.

In chapter 8 on shielding and guarding we have given space to the new European electromagnetic compatibility legislation which came into effect in January 1996. Also the ISO 9000 system is introduced in brief. For students and manufacturers it is important to know of the existence of these regulations.

Chapter 9 has been extended with some other noise calculation techniques to fulfil the need of some users.

Noise in sensors is also added and discussed in chapter 10.

Chapter 11 on interfacing to sensors has been introduced, where in particular attention is devoted to bridge circuits for modulating transducers. Also the feature for autozeroing transducers is discussed.

All remarks from referees, colleagues and students have been taken into account where possible to improve and clarify missing points. These are appreciated greatly and I am indebted to all of them. Acknowledgment is made to Professor Johan Huijsing from Delft University of Technology for providing material. Special thanks are due to Dr Ramon Pallas Aremy from the University of Barcelona for sending me his valuable comments and scrutinizing the first edition. Also thanks to Professor Yves Danto of the Université de Bordeaux I, IXL, who has given valuable comments on the new approaches in reliability engineering for ICs. Special thanks to my sons Michael van Putten MD, MSc and Maurice van Putten MSc, PhD, who invented the ADM principle. I would also like to thank the staff of the commissioning and production departments of Institute of Physics Publishing, in particular Peter Binfield and Katie Pennicott.

This textbook has proven to be valuable for students from various backgrounds, including mechanical, electrical and electronics engineering, mechatronics and physics, but students studying the physics of the building environment may also benefit from this textbook. Engineers and technicians may also find relevant material and benefit from useful information.

The author would be grateful to receive any comments for improving this textbook. The references mentioned are not exhaustive, and anyone requiring further detail should contact the author. It is mainly relevant textbooks that are cited in the Bibliographies; appropriate material may be found in the following journals, or by searching library databases.

IEEE Circuits and Devices
IEEE Solid State Circuits
IEEE Transactions on Instrumentation and Measurement
Sensors and Actuators
IEEE Proceedings (with excellent review articles)
Journal of Micromechanics and Microengineering
Photonics Spectra
Measurement Science and Technology

LIST OF SYMBOLS

Symbol	Description	Units
a	acceleration	$\mathrm{m\,s}^{-2}$
A	gain factor	
$A(t)$	availability (momentarily)	
A	surface area	m^2
B	bandwidth	Hz
B	magnetic induction	$\mathrm{Wb\,m}^{-2}=\mathrm{V\,s\,m}^{-2}$
c	speed of light	$\mathrm{m\,s}^{-1}$
C	capacity	F
$[C]_i$	ion density	m^{-3}
D	specific surface charge	$\mathrm{C\,m}^2$
$D_{\mathrm{n,p}}$	diffusion coefficient	$\mathrm{m}^2\,\mathrm{s}^{-1}$
$\overline{e}_\mathrm{n}$	noise voltage	$\mathrm{nV\,Hz}^{-1/2}$
$\overline{e}_\mathrm{N}$	total noise voltage	$\mu\mathrm{V}$
E	modulus of elasticity	Pa
E	electric field	$\mathrm{N\,C}^{-1}=\mathrm{V\,m}^{-1}$
$E(x)$	expectation	
E_F	Fermi energy	eV
E_g	bandgap energy	eV
F	discrimination factor	dB
F	Faraday constant	$\mathrm{C\,mol}^{-1}$
f	frequency	Hz
g	radix or base	
g_m	transconductance	$\mathrm{mA\,V}^{-1}$
h	Planck constant	$\mathrm{W\,s}^2$
H	transfer function	
H	magnetic field strength	$\mathrm{A\,m}^{-1}$
H	rejection factor	dB
$\overline{i}_\mathrm{n}$	noise current	$\mathrm{pA\,Hz}^{-1/2}$
I	intensity of radiation	cd
I	current	A
J	current density	$\mathrm{A\,m}^{-2}$
J_ab	charge density	$\mathrm{C\,m}^{-2}$
k	Boltzmann constant	$\mathrm{J\,K}^{-1}$
Δk $(=k_2-k_1)$	change in wavenumber	
K	gauge factor	
l	length	m
L	inductance	H

Symbol	Description	Units
L_e	radiance	$W\ m^{-3}\ sr^{-1}$
m	specific mass	$kg\ m^{-3}$
M	modulus	t.b.s.†
M	measurand	t.b.s.
M	magnetic moment per unit of volume	$A\ m^{-1} = A\ m^2\ m^{-3}$
N	real number	
N_x	carrier concentration	m^{-3}
$N_{c,v}$	state density of conduction or valence energy band	
n	index of refraction	
n	number of charge carriers per volume	m^{-3}
p	concentration of holes per volume	m^{-3}
p	magnetic dipole moment of atom	$A\ m^2$
P	probability	
P	pressure	$N\ m^{-2} = Pa$
P	power	W
q	specific charge of electron	C
Q	radiant energy	J
Q	quality	
Q	charge	C
$Q_{n,p}$	phonon drag	
R	reliability	
R	resistance	Ω
R_H	Hall coefficient	$m^3\ C^{-1}$
R	universal gas constant	$J\ mol^{-1}\ K^{-1}$
S	entropy per unit volume	$N\ m\ K^{-1}\ m^{-3}$
s	position	m
S	sensitivity	any y/x
S	cross-section	m^2
t	time	s
T	period	s
T	absolute temperature	K
v	electron velocity	$m\ s^{-1}$
V	voltage (difference)	V
V_H	Hall voltage	V
V_{GAP}	bandgap voltage	V
V_{GS}	gate–source voltage	V
V_p	pinch-off voltage	V
W	energy	$J\ s^{-1} = W$
W_B	base thickness	m
X	reactance	Ω
z_i	valence of ion	
Z	impedance	Ω
α_S	Seebeck coefficient	$\mu V\ k^{-1}$

Symbol	Description	Units
β	gain factor	
γ	Thomson coefficient	$V\,K^{-1}$
ε	relative change	m
ε	relative permittivity	
ε_0	permittivity of vacuum	$F\,m^{-1}$
λ	failure rate	h^{-1}
λ	wavelength	m
$\mu_{n,p}$	mobility of charge carriers	$V^{-1}\,s^{-1}\,m^2$
μ	mean	t.b.s.
μ	mobility	$m^2\,V^{-1}\,s^{-1}$
μ_0	permeability of vacuum	$V\,s\,A^{-1}\,m^1 = H\,m^{-1}$
μ_r	relative permeability	
ν	frequency	Hz
ν	Poisson constant	
Π_{ab}	Peltier coefficient	$J\,C^{-1}$
ρ	specific mass	$kg\,m^{-3}$
ρ	specific resistivity	$\Omega\,m$
σ	Stefan–Boltzmann constant	$W\,K^{-4}\,m^{-2}$
σ	standard deviation	
σ	specific conductivity	t.b.s., e.g. $(\Omega\,m)^{-1}$
σ	specific force	$N\,m^{-2}$
τ	relaxation time	s
φ	phase shift	rad
Φ	flux of radiation	lm
ω	angular frequency ($2\pi f = 2\pi/T$)	$rad\,s^{-1}$

† t.b.s. to be specified, dependent on application.

ACKNOWLEDGMENTS

The author is grateful to the following for granting permission to reproduce material included in this book.

National Instruments for parts of sections 2.5 and 2.6
Eindhoven University of Technology for figures 3.24, 3.35, 3.36, 3.37, 3.38, 3.39, 3.40 and 3.41
Fluke Nederland for figure 12.2
Institute of Optical Research for figure 6.26
Chalmers University of Technology for figures 6.31 and 6.42
Department of Signals, Sensors and Systems for figure 6.43
Fraunhofer G-ISIT for figure 6.15
TU Denmark for figure 6.17
Micro Parts, Germany, for figure 6.16
IEEE for figure 6.29
Berkeley Sensors & Actuator Center for figure 6.25
John Wiley & Sons Inc. for figure 6.61
Academic Press for figures 6.33, 6.34, 6.60 and table 6.22
Intel Corporation for figures 4.21, 4.22 and 4.27
Institute for Technology, Sweden, for figure 6.41
Sweden Report for figure 6.32 and table 6.18
Elektrotechnische Materialen for figures 6.57 and 6.58
Artech House Books for figure 6.74
Peter Peregrinus Ltd for table 8.12
Measurement Systems: Applications and Design for figure 4.20
Twente University for figure 6.46
Delft University Press for figures 7.11(a) and (b)
IMEC for figures 6.37(b) and 6.55

One

INTRODUCTION TO ELECTRONIC MEASUREMENT SYSTEMS

Ever since human beings began to think, we have exchanged information and have given measures to quantities in an attempt to understand our surroundings. We appear to have a basic curiosity about and a need to investigate our environment. We recognize this with our senses, but our natural capacities are limited, so we have developed tools to help us to fulfil our measurement needs. In a continual process towards perfection, we are still refining our tools to improve our understanding of all kinds of mechanism. Often models are shaped which describe more or less accurately the real world. This requires collection of information about the environment, a system or a process. Knowledge is related to information and, with collected information concerning a system or a process, we can increase our knowledge. This is a continuous story of interaction: model making and performing measurements and vice versa. Besides these aspects our society has become an electronic society and, generally speaking, the complexity of our society forces us to measure and to control energy and information in a large variety of applications.

1.1 INFORMATION

Information is anything that increases our image of knowledge of a system or process (MacKay), for instance, a type of arrangement or regularity which can be recognized in the environment. The letters on a page show a certain regularity and the message involved is recognized by the sequence in which the letters have been printed. For instance, information content is implemented in the order in which the letters appear on paper, and a limited quantity of information is conveyed. We can also say information is anything which reduces uncertainty of the source sending this information. We call this type *semantic* information and it is language related. If the same letters appear completely at random, all semantic information is lost and interpretation is no longer possible. The amount of semantic information has become zero. It is also said the left *redundancy* is zero.

*   *   *

Example 1.1

To illustrate the concept of information redundancy in a language consider the following sentence: 'N t mrng t bkr bks brwn brd'. Although characters are missing you still can read this sentence; in other words in most sentences or words redundancy is present.

*   *   *

However, no quantitative measure for semantic information is known so far. This example explains why it is said that information always involves a certain amount of regularity. Put in a reverse way, every regularity contains a limited quantity of information. Here order and disorder or information and chaos are each other's opposites.

A second way to define information is based on a *technical* relationship for information. It is based on the number of independent degrees of freedom in a source of information. This is called *structural* information and it is expressed in *logons*. In its turn, each logon can be expressed in *metric* or *selective* information. Usually, metric information is connected to energy and it is expressed in a dimensionless number called *metrons*. Selective information is related to the content of a message consisting of m possible symbols each with a probability of occurrence.

Note that uncertainty is an event that can occur, and information is the event that has occurred. For an information source the following diagram reviews the different concepts used in information theory.

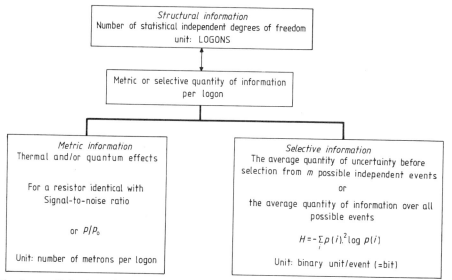

The base of the logarithmic unit is two. An example can illustrate the concepts.

*   *   *

Example 1.2

A source produces symbols from an alphabet with 26 different characters with the following probabilities: $p = 1/8$ for 1 character; $p = 1/16$ for 8 characters; $p = 1/32$

for 9 characters; $p=1/64$ for 4 characters; $p=1/128$ for 4 characters. Then the selective information is

$$H = -\frac{1}{8}\log\frac{1}{8} - \frac{8}{16}\log\frac{1}{16} - \frac{9}{32}\log\frac{1}{32} - \frac{4}{64}\log\frac{1}{64} - \frac{4}{128}\log\frac{1}{128} = 4.375 \text{ bit/char.}$$

The maximum amount of selective information is found for a symmetric distribution, hence $H_{max} = \log 26 = 4.700$ bit/char. Then the redundancy, Red. $= (H_{max} - H)/H_{max} = (4.7 - 4.375)/4.7 = 0.0691$. Note that when $H = H_{max}$ then Red. $= 0$!

*　　*　　*

When a certain event occurs a limited amount of information is received. The interrelationship between information and chaos was first shown by Shannon, who has connected the idea of information to entropy. In accordance with the second law of thermodynamics in a closed system the entropy increases towards a maximum, which corresponds to a minimum amount of energy in that system. Hence, entropy is defined as a measure for chaos, or disorder. Information can be quantified accordingly, based on the uncertainty of possible events and their probability of occurrence. On the basis of this concept a smallest amount of selective information is defined. We will not go into further detail here, but reference is made to the relevant literature. In this respect another important definition to distinguish is the concept of *knowledge*. Knowledge is anything that reduces uncertainty of a system or process. It is the same concept as information. However, we can distinguish knowledge in a different respect. Firstly, we can distinguish *perceptual knowledge* as parametric knowledge, which is based on by whom, where and when the information is interpreted. Secondly, we can distinguish knowledge revealed from natural laws and not dependent on human interpretation, for instance Newton's law on gravitation or Ohm's law. If an experiment is repeated under the same conditions the same results are always found. Finally, we can distinguish knowledge known by *intuition*. It is the way artists know how to make something, or how sometimes an invention is born. It is just there and the explanation 'why' follows (sometimes) later on. A nice example is the discovery of superconductivity in 1911 by Kamerling Onnes explained in 1957 by Bardeen, Cooper and Schiefer (the BCS theory).

Another important point to consider is that information can be bound to six different energy forms. In particular, we have electrical, thermal, radiant, magnetic, chemical and mechanical energy. In general, energy is defined as anything that can perform labour. We will state here that information is always bound to a certain type of energy carrier or mass. For instance, electrical information is connected to electrons, magnetic information is connected to magnetic dipoles. Information can also be printed on paper. Radiant information is connected to photons, mechanical information is connected to mass, and so on. In the electrical energy domain the electrons are the smallest information carriers. How information is connected to electrons is not well understood.

Today everyone is familiar with the fact that information can be measured, handled, stored, retrieved and represented, just depending on its application and requirements demanded of it. Also, information can be converted from one energy form into another by so-called *transducers*. For measurement purposes this feature is of utmost importance as we will see.

Our world is a very complex one with a tremendous amount of different types of information available, coming from all kinds of process and physical mechanism, often so complex that it is not possible to understand all collected information directly. A good example of such a process is the climate, which is very complex indeed. If statements have to be made about the weather over a particular area and at a specific time, a lot of data are collected concerning temperature, pressure, humidity and wind velocities and directions at different heights. Finally, we make a weather forecast, which always involves a certain amount of uncertainty. This uncertainty is a basic characteristic for every statement which is based on data collected by measurement techniques. The reason for this may be that the process is too complicated, the way the data are collected is imperfect or the amount of data collected is insufficient. In the above example this uncertainty is quite obvious, but errors are inherent in every measurement technique. The result is that we create an image of a part of the world with a limited amount of information. However, despite this implicit imperfection measurements remain the heart of science and technology and without them no progress can be made.

The first thing to do when performing measurements is to characterize the type of process that is necessary to investigate and to determine the physical parameter(s) which are to be measured. Also, any possible interference disturbing the measurement must be isolated from the measurements. This means that criteria must be formulated for every measurement concerning type, response time, accuracy, etc. After the process has been characterized, decisions can be made concerning which type of measurement will be applied and then collection of data can begin. After the collection of the data we have to analyse, synthesize and evaluate the measurements. Finally, an interpretation and check of the results must follow. Collecting data corresponds to receiving information about the system or process. As more information is received, uncertainty concerning the system is decreased.

In this respect again two different kinds of information have to be distinguished. Firstly, in the process under consideration, the type of information must be recognized. This concerns the characteristics of the process. It is a *qualitative* description of the information involved and describes what sort of energy is present. This is again called the structural information of the process. Secondly, the magnitude or intensity of the relevant characteristics must be discovered, for instance scaling factors concerning how much and how fast, and must be fixed. This is called *metric* information. The process by which this information is collected is a *quantitative* process. It will be clear that a qualitative process must precede the quantitative process. Ultimately, the measurement process can be interpreted as making assignment statements in accordance with predefined procedures by technical means. The result is a collection of numbers assigned to the parameters under consideration by a well chosen relationship.

1.2 TYPES OF MEASUREMENT

In practical circumstances measurements can be performed in five different ways with five different levels of accuracy and degrees of freedom.

1 Given two unknown quantities of the same kind, if only a statement of whether they are equal is required, a *nominal* measurement is involved. Examples are a colour impression of two different objects or the acidity of two liquids.

2 If additional information is collected about relative sizes, an *ordinal* type of measurement is performed. For instance, in a sample of boys and girls, the height and weight can be fixed and classified according to this kind of ordering.

3 When more specific information is required such as the magnitude of the intervals between the quantities of heights and weights in the above example, without making any statement about a kind of zero reference, a type of *interval* measurement is performed. There is an increase in information content received. Interval measurements can take the form of a mass measurement in kilograms or milligrams, a temperature measurement in degrees Celsius or a time measurement in hours or seconds. There is no zero or floating reference point.

4 We can choose a certain point as a reference of zero level and then perform *ratio* measurements by dividing every measurement by an arbitrarily chosen reference value and determining their relative magnitudes. For instance, all measurements are divided by the maximum measured value and then a so-called normalized measurement is performed; the largest value will be 1 or 100%; all the others will be less than 1. This type of measurement is very often applied in physical experiments and again provides more specific information.

5 Finally, statements can be made about the magnitude of the reference compared with a standard reference. When a standard system is available, every physical parameter can be measured compared with its respective standard reference quantity with no degree of freedom and no possible ambiguity. These types of measurements are called *cardinal* measurements. The SI system is just such a well accepted reference system, although in aircraft industries other reference systems are still in use, for instance mph, knots, gallons, pounds and feet.

It is interesting to note that in the real world it is often not possible to perform cardinal measurements. For instance, in biological and economic systems the processes are so complex that the available tools fail to provide accurate predictions and statements. Here, time and the number of interacting parameters are the most disturbing and complicating factors. Last but not least often every measurement disturbs the process under consideration. However, note the word 'often' here. Does a completely optically executed measurement technique disturb the process under consideration?

1.3 WHY ELECTRONIC MEASUREMENT SYSTEMS?

So far in this introduction, basic concepts have been discussed which you can find in most textbooks about measurement techniques and systems, but the next question to answer is, why is it such a great advantage to handle information by electronic means using electric charges as information carriers? Information can also be handled by pneumatics, hydraulics or mechanical means based on exactly the same principles as in electronics. Well known are pneumatic amplifiers and switches such as fluidics, sometimes still in use where the highest safety precautions have to be taken into account. Here are the main reasons why it is preferable to handle information by assigning it to electrical carriers.

1 A very large dynamic power range can be covered by electronic means. The available power range is 10^{-9} to 10^{+9} W, which means a dynamic range of 10^{+18}.
2 For electric charges a very high speed and acceleration are available owing to the high ratio of charge to mass of the smallest possible electrical charge q, the electron:

$$q/m = 1.6 \times 10^{-19}\,\text{C}/9.1 \times 10^{-31}\,\text{kg}$$

$$= 0.18 \times 10^{+12}\,\text{C kg}^{-1}.$$

The acceleration equals

$$a = (q/m)E\ (\text{m s}^{-2})$$

where q again is the specific electrical charge and m the specific mass and E is the electric field strength to which the charge is subjected. If, for instance the field strength is $1\ \text{V m}^{-1}$ we find

$$a = 18 \times 10^{+10}\,\text{m s}^{-2}$$

which is an exceedingly large acceleration. Note that in semiconductors we meet very thin layers ($<10\ \mu\text{m}$); the field strength can easily exceed $100\,000\ \text{V m}^{-1}$.
3 A very large time domain can be controlled from picoseconds (10^{-12} s) to hours which means a time range of 10^{+15}. No other energy carrier so far known can handle information in such an easy way and with such high speeds and accelerations with such huge dynamic ranges.
4 Information can be transported very efficiently by means of copper or aluminium cables, and via satellite or radio channels. When fibre optics are used for trans-port-information purposes, the highest specific transport-information density over long distances is available. However, here we enter the optical domain, which will be discussed later.
5 Information can be amplified by electronic means with very large gain factors from very low levels of signal amplitude (nanovolts) to hundreds of volts, which is the process occurring most in electronic information handling. Gain factors of 10^{+8} are possible with one single device.
6 Finally, information can be modulated in other shapes, stored, retrieved and modified. All known mathematical operations can be performed either in an analogue way or in a digital way at very high speeds in very small volumes.

With these striking advantages compared with other techniques it is not surprising that we prefer to handle information electronically. The consequences are a tremendous growth in applications of electronic means at all levels and in all disciplines of our society. It is a strength and simultaneously a weakness that our society has become completely dependent on electronic means and tools. We encounter the influence of electronics in the field of medicine, throughout industry, in research and development, in banking, in insurance, in business and in our private lives.

Table 1.1 Our electronic society.

Description	Examples
Telecommunications	Land, satellite and mobile
Banking systems	Stock market, money selling points
Airco industry	Climate control, clean rooms
Education	Interactive audio and visual equipment
Meteorology	Temperature, humidity, pressure
Computer industry	Process and personal computers
Industrial	Process industry and car manufacturing
Consumer electronics	Audio, video, PCs, household appliances
Data communications	Fax, satellite
Medical equipment	Diagnostic and therapeutic equipment
Toy industry	Cars, airplanes, dolls, games
Power generation	Nuclear, fossil, hydrocarbon, solar
Military applications	Radar, navigation, weapons
Transportation	Land, sea and air

In table 1.1 we summarize a (non-exhaustive) survey, where we find electronic means and applications today. Note that in all these applications we measure and control information and energy.

Our society has really become an information society as Toffler describes in his *The Third Wave*. This dependence is the main reason that there is a chapter on reliability in the current book. There is an increasing need for statements about the reliability of electronic systems we depend on so heavily.

When considering the development of specific electronic tools or components for electronic measurement systems, the birth of the 4-bit microprocessor in 1971 cannot be overlooked. This component, now considered to be obsolete, was a simple 4-bit p-channel MOS chip, initially manufactured as a custom-designed chip for a Japanese calculator manufacturer. We now have arrived (1996) in the era of 32-bit and 64-bit microprocessors. Examples are the Intel P6 and P7, with over 3×10^6 components on a single chip. The impact will again be enormous and cannot be precisely foreseen. The 64-bit microprocessor has been introduced by several other manufacturers already, such as the Alpha 21164 processor from Digital Equipment with over nine million transistors on chip operating at 320 MHz. The economic life cycle of electronic microprocessors is still decreasing. How far will this go? It will be obvious that this processing power will influence also the science of measurement and control of energy and information. Also, software tools are still acquiring more power for testing, design and production control.

In the meantime the microprocessor has earned its place in measurement applications and technology owing to its great flexibility in collecting and processing data. One may well ask whether there are any disadvantages to mention. There are a few. The first is that microprocessors and all other silicon-made components can only function within a limited temperature range. The lower limit is $-50\,°C$ and the upper limit is about 175 °C. They are also not suitable for applications in situations demanding very high degrees of intrinsic safety, for example in explosive environments. Intrinsic safety means that installed electric power may not exceed 200 mW.

Electronic systems will also fail when very large forces are required, for example in offshore industry.

An electronic system can be best described as types of information channel consisting of three parts. The first is the input part, where the information is picked up. This information is very often a non-electrical physical quantity to be converted into an electrical quantity by a type of transducer. Secondly, there is a processing part where the information is processed and modified into the desired shape and form. Thirdly, there is an output part where the information leaves the channel in order to be perceived by the observer or other systems. Measured data should be presented in a form easy to recognize, which is primarily visual. Many physical principles can be employed for display purposes, the choice being dependent on different criteria such as price, power consumption, response time and ergonomic aspects; the latter is of particular importance in an aircraft where a lot of data must be recognized and interpreted sometimes in a split second. In the aircraft industry, cockpits have undergone a complete change of display, saving a complete crew member.

1.4 FUTURE TRENDS

It is useful to discuss some future technological trends resulting from the great technological push of electronic tools and systems. A lot of attention is paid to speech processing and pattern recognition. This has already resulted in a complete automatic talking dashboard in a motor-car. In the future it may be expected that the way typewriting is executed will be replaced by speech commands being given to a type of processor able to recognize the human voice. On a macroscopic scale we can expect large changes in the field of data communication and telecommunication, solid state technology, consumer electronics, small and large computers, power and energy, environmental control, software engineering and its applications, transportation, aerospace and military electronics, medical electronics, industrial electronics and test and measurement systems. All these developments are made possible by the underlying technological advances made in the IC semiconductor industry. We will mention some important developments.

There is a further decrease in line width in IC technology by means of deep UV, electron or x-ray photolithographic techniques. These techniques make it easily possible to reduce the line width below 0.5 μm. Another major development is the progress made in wafer and die size; wafer sizes of 8 in (200 mm) and die sizes of over 900 mm^2 (NEC) are being used. *Microsystem technology* (MST) is the discipline of making small functional systems. Application of micromachining, micromechanics and electronics can provide three-dimensional structures on a single piece of silicon. Acceleration sensors are manufactured for the automotive industry. They contain a microactuator and they have a built-in self-test capability. Moving parts such as micromotors with diameters of 100 μm and accelerometers are manufactured in silicon with the help of micromachining. *Mechatronics* is also a buzz word: it is the combination of mechanics and electronics and may range from micron-scale to large systems. We will come to these aspects again in chapter 6.

Another completely new approach is the making of transistors in polymer technology without using silicon. These transistors have larger structures, but they can be bent without losing their electrical characteristics and properties. It is said to become a technology on the molecular level.

The concept of *smart sensors* must be mentioned here too. They are defined as sensors with built-in intelligence and form one complete functional system with sensor, electronic processing circuitry and output. For instance, pressure sensors are manufactured now with all needed electronics integrated on a single piece of silicon.

Memory size and speed will further increase especially in NMOS and CMOS technology. Up to 1 Gbit dynamic RAMs are already being manufactured. The compact disc not only is useful for recording music but is used for storage applications as well and is able to provide a storage capacity of >500 Mbytes on a single disc of 4 in diameter. For communication purposes fibre optics will replace copper cables in most cases because of cost and bandwidth capacity advantages [1].

Rapid progress is being made in gallium–arsenic technology and it is already being asked whether GaAs will beat silicon. Cost aspects play a dominant role here. For digital electronics GaAs appears to be promising, especially with respect to high-frequency behaviour and speed. In telecommunications the fastest growth is in the interface with optical fibres where digital data are processed at 10 Gbit s^{-1}. Multipliers and RAM chips are already available [2]. One other aspect must also be mentioned here. GaAs is a direct bandgap material. This means that with this material optical and electronic devices can be combined on a single chip. For instance, an optoelectronic bus system has been designed by Delft University, The Netherlands, for parallel computing purposes.

There is an increasing trend for information handling to be performed in the optical domain. Not only is transportation of information with fibre optics more efficient than with copper cables, but also numerous optical devices, such as optical amplifiers, optical switches and other optical logic components, have become available. This is called *photonics* and it is defined as information processing and handling with the help of photons. An intermediate connection between electronics and the optical domain is called *optoelectronics*. We will pay attention to that subject too.

In electronic measurement systems an increasing shift to digital data processing is taking place. This digital processing happens more and more directly after the data are collected from the outside analogue world. Every physical process is analogue in itself but these quantities are converted into an electrical digital quantity as soon as the minimum required signal level is acquired. Moreover, systems designed, developed and manufactured must be tested before shipping. This demands automated test procedures and in the design stage we talk about design for testability (DFT). For this reason measurement equipment has shifted from manually operated equipment into completely automated test systems. In this respect built-in self-test (BIST) structures must also be noticed. A circuit is able to check its own performance. Bus-based measurement systems are common practice now. These developments have also strongly influenced all kinds of industry and organizational structure.

With this in mind we can continue our story and concentrate on important aspects related to designing measurement instruments. If a measuring instrument has to be constructed it is not a matter of merely connecting the inputs and outputs of a set of subsystems. To start with, a careful and precise description of the quantity to be measured must be made. Important questions to be answered are as follows.

- What quantity(ies) has(ve) to be measured?
- What is the measuring object?
- What are the environmental conditions ?
- What accuracy is required?
- What reliability is required?
- What is the required display
- What will be software and what hardware?

We can conclude the introduction with the following statement: 'No system is better than the weakest link'. Thus a check and comparison for every subsystem must be executed on all the relevant aspects mentioned above. The same questions will hold for the user also.

PROBLEMS

1   Discuss the concepts of semantic and technical information.

2   Mention at least four kinds of energy carrier to which information can become bound.

3   Discuss the relationships between information, entropy and uncertainty.

4   What is meant by the characterization of the kind of information? What is meant by structural information? What is meant by metric information?

5   Discuss the main advantages of assigning information to electronic charge carriers.

6   Give a definition of analogue information and digital information.

7   Which five types of measurement can be distinguished?

8   Why has reliability become an increasing factor of importance in the design of measurement systems?

BIBLIOGRAPHY

[1] Technology 1995 Analysis and Forecast Issue, *IEEE Spectrum* January 1995
[2] Deyhimy I Gallium arsenate joins the giants *IEEE Spectrum*, pp 33–40
[3] Ristic L 1994 *Sensor Technology and Devices* (Boston, MA: Artech) ch 4
[4] Bryzek J and Peterson K 1994 Micromachines on the march *IEEE Spectrum* May 1994
[5] Mackay D M 1969 *Information, Mechanisms and Meaning* (London: MIT)
[6] Toffler A 1980 *The Third Wave* (London: Collins)
[7] Goldman S 1955 *Information Theory* 3rd edn (New York: Prentice Hall)
[8] Shannon C E and Weaver W 1971 *The Mathematical Theory of Communication* 12th edn

Two

TYPES OF MEASURING SYSTEM

2.1 INTRODUCTION

In a given process, in a certain arrangement, a limited amount of information is available, generally as a function of time and place. Examples are the temperature division in a radiator, the flow velocity in a tube, the radiant intensity of a surface, mechanical stress in structures and so on. Information retrieved from a process is obtained by applying a measurement. It can be said that measuring is making a kind of image on a well defined scale of symbols. More precisely, measuring is creating a representation of a physical quantity, which can be described as

$$y(t, s) = f[x(t, s)] \qquad (2.1)$$

where t and s denote time and place respectively and f represents the functional behaviour of the applied instrument.

The physical parameter of which information has to be retrieved is always bound to a certain energy carrier. Of the different types of energy, the electrical form is preferred because of its flexible way of easy handling. In many cases the quantity to be measured appears to be of a character other than an electrical one. Therefore an energy conversion is required. This conversion is realized by the appropriate transducer or sensor. At the output of the measuring chain again a type of transducer can be found to make the measurand perceivable for humans or suitable for registration. The centre of the measuring chain between input and output is characterized by an electrical process.

Measurement systems have become a complex mixture of analogue and digital electronics. To be able to discuss the several aspects and impacts of electronic measurement systems we have to introduce many (theoretical) tools. In this chapter we will present a discussion of the classification of measuring structures. In the first instance, measuring instruments can be divided into four classes with respect to their functional structure, the type of signal structure information, their spatial structure or architecture and automated structures. In particular we distinguish the following.

1 *The functional structure*
The functional structure describes the way the measurement is performed and depends on the number of unknown related physical quantities to be measured and whether there is an *active* or *passive* quantity involved.

2   *The signal structure*

Measurement systems can be distinguished related to the type of signal to be measured or processed. This ranges from signals with a unique character, to periodic, sampled and stochastic signals.

3   *Spatial structures*

In this case the spatial layout of the input quantities determines the structure and architecture. Data-acquisition systems, for instance, can lead to very complex spatial structured systems.

4   *Automated structures*

Here the digital structure of the measurement system is emphasized. This arises from the present trend from performing manual measurements, via semiautomatic systems to the use of fully automatic and computer-controlled systems. Complete software-configured measurement systems can be put together with use of a computer and some interfacing cards.

2.2   FUNCTIONAL STRUCTURES

There are two basic types of functional structure. This distinction depends on the measurand's ability to carry and transport power or energy. The measurand can be an active or a passive quantity. This characteristic determines the way the measuring system is composed and prescribes the functional structure of the system.

2.2.1   Measuring structures for an active physical quantity

A physical quantity is defined as *active* or *intensive* if it can carry and transport power. Examples of active quantities are force, electrical field, pressure, a light intensity and so on. As will be seen, for such signal sources, no auxiliary energy is required.

A set-up for measuring an active physical quantity is composed of a limited well defined number of functional blocks and can be represented as shown in figure 2.1. If a non-electrical physical quantity has to be measured, then in the first functional block an energy conversion to the electrical domain must be realized. This first functional block is often called a sensor or transducer. Transducers will be discussed in chapter 6. For the moment it suffices to say that the sensor is able to produce an electrical signal which is a good representation for the physical quantity to be measured. The second functional block is the measuring system assigned to the process, in which the electrical signal is modified into the desired shape. This electrical modification process can accomplish many different functions. A collection of

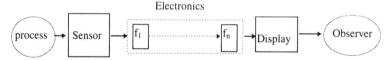

Figure 2.1   General arrangement for a measuring set-up.

Table 2.1   Review of some electronic functions.

Functions
Amplification (of the small source signal)
A/D and D/A conversion
Modulation and demodulation
To determine a maximum or minimum (peak detection)
Calculation of the average or RMS value
Derivative or integral of a signal
Sampling
Filtering
Measuring zero crossing, or timing intervals
Summing, multiplying
Counting

possible functions, sometimes called signal-conditioning functions, is summarized in table 2.1. We will discuss several of these functions in chapter 3. In general, the electrical process can be written as

$$f_E = f_1 f_2 \ldots f_n \tag{2.2}$$

where $f_1 \ldots f_n$ describe the different electrical functions to be distinguished in the second functional block. The third functional block is composed of a type of display able to represent the measurand in a suitable form which can be perceived by the human senses, or in a shape suitable for storage. Generally, the function of a system measuring an active physical quantity can be written as

$$M = f_{OT}\{f_E[f_{IT}(p)]\} \tag{2.3}$$

where M is the desired measurand, $f_{IT}(p)$ represents the input transducer's function of process p, f_E represents the electrical process and f_{OT} represents the output transfer function.

When more than one unknown related physical quantity must be measured the situation is more complex. We can use the circuit as shown in figure 2.2. Assume a pressure and a temperature must be measured simultaneously, where it will be clear that the pressure is related to temperature. Then the measurement set-up should provide two signals and hence two equations must be solved to obtain the required information.

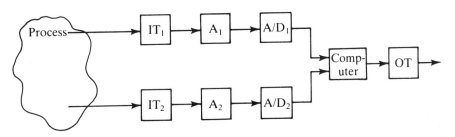

Figure 2.2   Measurement set-up for two unknown signals. The computer will calculate both quantities from two digital signals.

   In figure 2.2 two parallel measuring chains are shown. At the end of the chain a computer will solve both equations, providing the requested information. In this figure IT_1 represents a pressure transducer delivering a temperature-dependent magnitude. Transducer IT_2 represents a second temperature-dependent pressure transducer, delivering another and different dependence between pressure and temperature. So, two independent equations can be formed. Often amplifiers are needed to amplify the very weak transducer signals of sometimes several microvolts. They are denoted by A_1 and A_2. The analogue–digital conversion is executed by A/D_1 and A/D_2. Finally, the output transducer OT will display the magnitude of pressure and temperature calculated by the computer.

2.2.2   Measuring structures for a passive physical quantity

A *passive* or *extensive* physical quantity is not able to transfer energy to a sensor directly. The relevant information is implied in the structure of matter itself and can only be retrieved by measuring the response gained when the object is subjected to a suitable stimulus. A well known application is the forced seismic response to an explosion. Other examples of these are mass, the elasticity of a spring, the strain in a bridge girder, the colour pattern of an image, an impedance and so on. If such a physical quantity has to be measured, an auxiliary energy source must be attached on that object in order to force the information through the measuring system. This configuration is illustrated in figure 2.3, which shows a symmetric lay-out.

   A signal generator drives an actuator, which stimulates the object to generate a response from which the required physical quantity can be retrieved. A generator producing mechanical vibrations in a steel rafter is a good example. In this measuring set-up both the stimulus and the response are measured and the ratio of the two values produces the result.

   This configuration has several advantages. If the circuit shows a good symmetry and both the stimulus and the response have the same dimensions the ratio delivered by the divider will cancel several errors. The influence of the ambient or medium temperature may be well cancelled if the symmetry is guaranteed. The measurand can be expressed in the following function:

$$M = f_{OT} \frac{\{f_{E1}[f_{IT1}(P)]\}_{\text{response}}}{\{f_{E2}[f_{IT2}(P)]\}_{\text{stimulus}}} \tag{2.4}$$

where the different functional blocks are M measurand, f_{IT1} transduction by sensor 1, f_{IT2} input transduction by sensor 2, f_{E1} electronic modifier including gain in response branch 1, f_{E2} electronic modifier including gain in stimulus branch 2 and f_{OT} output transducer.

   A very simple example can illustrate the concept although here the division is performed manually.

                                    *   *   *

Example 2.1

Suppose an unknown impedance Z_2 must be determined; then the circuit of figure 2.4 can be used. An impedance is a passive quantity and cannot carry or transport

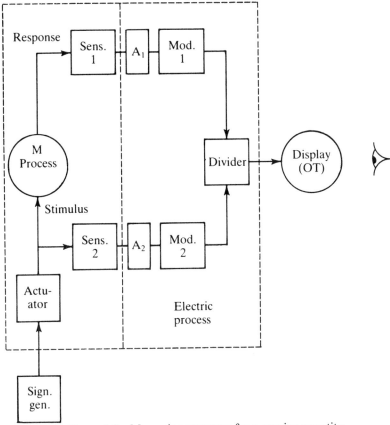

Figure 2.3 Measuring structure for a passive quantity.

power, so an external power source V_p must be applied. If a second and known impedance Z_1 is connected to the unknown impedance a ratio measurement can be executed.

The external power source will force a current I_0 to flow through both impedances generating voltages across Z_1 and Z_2 and the result can be derived from the ratio of

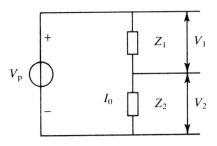

Figure 2.4 Measuring an impedance as an example of a passive quantity.

the two voltages V_1 and V_2:

$$Z_2 = Z_1 \frac{V_2}{V_1} \tag{2.5}$$

where

$$V_1 = I_0 Z_1 \qquad V_2 = I_0 Z_2.$$

Note that I_0 and V_p are not implied in the expression (2.5) and are cancelled; hence any related interference is absent. The same holds for the influence of the ambient temperature on the resistive part of the impedances. It is said that Z_1 acts as a reference impedance.

*   *   *

Example 2.2
Concerning sensors, for high-accuracy measurements another interesting application is the use of vector sensors. Vector sensors have a bi-directional input characteristic for a given physical parameter S, and have isotropic characteristics for any other possible physical input. Any Wheatstone bridge including at least two sensing elements can act as such. A measurement value is obtained as the algebraic sum of two opposite measurements in an arbitrary short time interval using the same sensor. The measurement technique is called the *alternating direction method* (ADM) and it ensures very reliable, drift-free long-term measurements. See also chapter 7.

*   *   *

2.3 SIGNAL STRUCTURES

The second possible distinction classifying electronic measurements is related to the type of signal they must process. In general, a signal can be defined as a function representing a physical quantity or process. Usually, when we want to quantify a signal we talk about amplitude, peak value, mean value and so on. Usually, we use the parameters indicated in table 2.2.

The types of signal to be processed can have quite a different character and of course can have a great impact on the structure of the measurement system itself. We start with an overview of possible signal classifications. Four different types of signal can be identified:

1 Signals with a unique character
2 Periodic signals
3 Sampled signals
4 Stochastic signals.

Each of these will be discussed in turn and examples will be given. Other classifications of signals are possible, for instance (1) AC versus DC, (2) analogue versus digital, (3) stochastic versus deterministic, (4) continuous versus discrete. However, we will discuss the distinction as is given.

Table 2.2 Review of frequently used signal values.

Description	Expression for $(0 < t < \tau)$
Peak value	$y_p = \max[|y(t)|]$
Peak-to-peak value	$y_{pp} = \max[y(t)] - \min[y(t)]$
Mean value	$y_m = \dfrac{1}{\tau} \displaystyle\int_0^\tau y(t)\,dt$
Mean absolute value	$|y_m| = \dfrac{1}{\tau} \displaystyle\int_0^\tau |y(t)|\,dt$
Root mean square (RMS) value	$y_m = \left[\dfrac{1}{\tau} \displaystyle\int_0^\tau y^2(t)\,dt\right]^{1/2}$
Mean signal power	$y_m = \dfrac{1}{\tau} \displaystyle\int_0^\tau y^2(t)\,dt$

2.3.1 Signals with a unique character

Signals with a unique character are characterized by their unique occurrence. Sometimes they are strictly dependent on the process or system under consideration. For instance, when weather conditions are observed all signal parameters show a unique character. To characterize these types of signal two main subdivisions can be made:

1 *Time analysis*
Time analysis is related to a process which shows a unique time-dependent character. Here a given source producing these types of signal is considered.

2 *Response analysis*
Here, a given system is observed which must be characterized. As will be discussed three other subdivisions can be made here.

2.3.1.1 *Time analysis*
The behaviour of the weather with respect to time shows a unique character concerning pressure, temperature and humidity. Measurements are performed analysing the behaviour of pressure, temperature and humidity with respect to time. Commonly applied characteristic signal operations are summarized in table 2.3. They will be discussed in more detail in chapter 3. Another good example of these signals is earthquake phenomena. They also show a unique character with respect to time and amplitude and until now are completely unpredictable.

2.3.1.2 *Response analysis*
To make calculations possible in electronic systems it is well established that every electronic system is considered as a linear system. In fact this is not true, but the

Table 2.3 Review of commonly applied signal
operations on unique time-dependent signals.

Zero-crossing detection with respect to time
Peak detection with respect to time
Taking the average or RMS value of the signal
Dynamic logarithmic compression
Dynamic exponential expansion
Integration or taking the derivative
Summing, multiplying, division, etc.
A/D and D/A conversion

---

assumptions are allowed if we state that all signal changes in a system are small
enough with respect to their biasing levels. For small signal changes the system
behaviour is said to be linear. By definition a system is called a linear system if it
can be described with a linear differential equation (LDE), or

$$a_0 S_i + a_1 \frac{dS_i}{dt} + a_2 \frac{d^2 S_i}{dt^2} + \ldots = b_0 S_o + b_1 \frac{dS_o}{dt} + b_2 \frac{d^2 S_o}{dt^2} + \ldots \qquad (2.6)$$

where it will be assumed that all coefficients in the differential equation are constant
and where S_i is the input signal and S_o is the output signal. The consequences are
as follows.

1   A proportional change of the input signal results in a proportional change of the
    output signal, or

$$aS_i \Rightarrow aS_o \qquad (2.7)$$

where a is as an arbitrary number.

2   The superposition theorem holds, or

$$a_1 S_i 1 + a_2 S_i 2 \Rightarrow a_1 S_o 1 + a_2 S_o 2. \qquad (2.8)$$

3   An input signal $S_i = \exp(\alpha t) \sin(\omega t + \phi)$ results in the same shape of output signal
    apart from a possible time shift ϕ.

Hence, we will talk about linear systems only and how these systems can be characterized with the help of different types of input signal.

1   *The impulse function: $\delta(t)$*
It is well established that every linear system can be characterized by its *transient* or
impulse response $h(t)$ to an impulse function $\delta(t)$. The impulse function denoted by
$\delta(t)$ is also known as the *Dirac delta function*. The impulse function $\delta(t)$ is defined
as a signal of unit area vanishing everywhere except in the origin, or

$$D_s = \int_{-\infty}^{\infty} \delta(t)\, dt = 1 \text{ with } \delta(t) = 0 \text{ for } t \neq 0. \qquad (2.9)$$

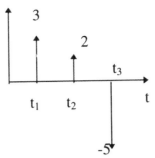

Figure 2.5   Impulse sequence.

The impulse function is illustrated in figure 2.8(a). There are two other characteristics to mention for $\delta(t)$:

1   the function $\delta(t)$ is even, and hence $\delta(t) = \delta(-t)$;
2   the function $\delta(t - t_0) = \delta(t_0 - t)$ is an impulse function centred at t_0 of unit area.

* * *

Example 2.3
Suppose we have the sequence of impulses shown in figure 2.5. Then this sequence can be regarded as the superposition of three separate pulses:

$$f(t) = 3\delta(t - t_1) + 2\delta(t - t_2) - 5\delta(t - t_3).$$

* * *

Of course the Dirac pulse function is only a theoretical mathematical function, because it cannot be realized in a physical system, but it appears to be applicable in many cases to describe systems in terms of the superposition of impulse responses. In general we say that every linear system reacts on the impulse function $\delta(t)$ in a unique characteristic way denoted by $h(t)$. This is illustrated in figure 2.6.

If the transient response $h(t)$ of a linear analogue system is known it can be shown that a response $y(t)$ to an arbitrary input $x(t)$ can be calculated by applying the convolution integral of the input signal $x(t)$ and the transient response $h(t)$. By definition *convolution* implies a binary operation between the functions $x(t)$ and $h(t)$. It finds its basis in the previously mentioned consequences for linear systems. Apart from this the system is considered as *time invariant* also, or a delay at the input results in an equal delay at the output. This is expressed in the following equation for analogue systems. The same kind of statement can be made for

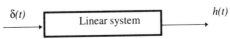

Figure 2.6   Unique response $h(t)$ of a given linear system to the impulse function $\delta(t)$.

digital systems.

$$y(t) = \int_{-\infty}^{\infty} h(\tau)x(t-\tau)\,d\tau. \tag{2.10}$$

Equation (2.10) is also written $y(t) = h(t) * x(t)$. We can say that the zero-state response of any linear system equals the convolution of the input $x(t)$ with the impulse response $h(t)$ of the system. Hence, to determine $y(t)$ it suffices to apply the mathematical operation (2.10), known as convolution, with the input signal $x(t)$.

*   *   *

Example 2.4
Assume that the impulse function $h(t)$ of a system is known as

$$h(t) = \exp(-at)\varepsilon(t) \tag{2.11}$$

where $\varepsilon(t)$ is defined as the step function (see further) starting at time $t=0$. Then the response $y(t)$ to an arbitrary function $x(t)$ may be written as

$$y(t) = \int_{0}^{t} x(\tau)h(t-\tau)\,d\tau. \tag{2.12}$$

If $x(t) = 0$ for $t < 0$ then the limits are as indicated. Substitution of (2.11) in (2.12) with the same boundary conditions, since $\varepsilon(t)$, gives

$$y(t) = \int_{0}^{t} x(\tau)\exp[t-\tau)]\,d\tau. \tag{2.13}$$

*   *   *

So, the boundary conditions of the integral change in accordance with $\varepsilon(t)$. To calculate convolution it is often more convenient to apply Laplace transforms. It can be shown that a convolution of two functions in the time domain corresponds to the product of their Laplace transforms, or

$$h(t) * x(t) \Leftrightarrow H(s)X(s).$$

Remember that the Laplace transform of a function is defined by the following equation (more precisely it is called the unilateral Laplace transform):

$$F(s) = \int_{0}^{\infty} f(t)\exp(-st)\,dt. \tag{2.14}$$

The output signal can be interpreted as the sum of the transient responses to a continuously input series of the amplitude of the absolute value of the input signal $|x(t)|$. Another interpretation says that the *zero-state response* $y(t)$ of any linear system equals the convolution of the input $x(t)$ with the impulse response $h(t)$ of the system. A zero-state response means response to an external source when the internal system energy is zero. Hence, to determine $y(t)$ it suffices to find $h(t)$ and to convolve it with the input $x(t)$.

Note that measuring transient responses has a great disadvantage because a large amount of energy must be supplied to the system in a very short time. Infringement of the allowed amplitudes can occur easily, which often results in jamming of the system.

2 The step function: $\varepsilon(t)$

A more easily applicable stimulus is the *step stimulus*, $\varepsilon(t)$, defined as

$$\varepsilon(t) \overset{\text{def}}{=} \begin{cases} 0, & t < 0 \\ 1, & t \geq 0. \end{cases} \tag{2.15}$$

From the system a step response is obtained, which offers information about the speed at which the system achieves a stable state. Note that the step function $\varepsilon(t)$ is the integral of the impulse function, or

$$\varepsilon(t) = \int_{-\infty}^{t} \delta(t)\, dt. \tag{2.16}$$

The unit step function is depicted in figure 2.8(b). Alternatively the derivative of the step function $\varepsilon(t)$ gives the impulse function $\delta(t)$, or

$$\delta(t) = d\varepsilon(t)/dt.$$

*   *   *

Example 2.5
The following staircase function, $f(t)$, is given. This function can be written as $f(t) = 3\varepsilon(t - t_1) + 2\varepsilon(t - t_2) - 5\varepsilon(t - t_3)$ and is depicted in figure 2.7. Then the derivative of this function is $f'(t)$ and is the superposition of three delta functions. This function has already been illustrated in figure 2.5. It can be written as

$$f'(t) = 3\delta(t - t_1) + 2\delta(t - t_2) - 5\delta(t - t_3).$$

*   *   *

The same discussion is valid for signals with a discrete time character. Discrete time signals are obtained when sampling is involved. This will be discussed later in more detail.

3 The ramp function: $t\varepsilon(t)$

A third and often applied function is the *ramp function* (figure 2.8(c)), by which means a system or process can be characterized. A ramp function is

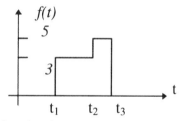

Figure 2.7   Staircase function $f(t)$ as the superposition of three step functions.

defined as

$$x(t) \stackrel{\text{def}}{\Rightarrow} t\varepsilon(t). \tag{2.17}$$

The response to a ramp function provides information about the linearity or non-linearity of a system. The different types of test function are depicted in figure 2.8.

2.3.2 Periodic signals

Periodic signals are one of the most frequently occurring and applied signals in electronic systems. They are characterized by their periodic and often harmonic behaviour and frequently occur in natural environments. For instance, the strings of a violin or mechanical rafters can produce periodic harmonic vibrations. Periodic signals can be described with amplitude, frequency and phase. It can be shown that every periodic function with period T can be expanded into an infinite series of *harmonics*. This infinite sum of terms is called a *Fourier* series and is expressed as

$$f(t) = a_0 + \sum_{n=1}^{\infty} [a_n \cos(n\omega_0 t) + b_n \sin(n\omega_0 t)]. \tag{2.18a}$$

It is also possible to write $f(t)$ in short form as

$$f(t) = a_0 + \sum_{n=1}^{\infty} c_n \cos(n\omega_0 t + \varphi_n). \tag{2.18b}$$

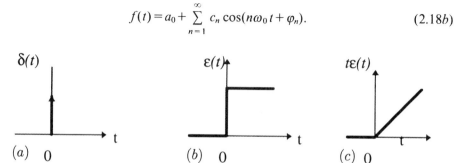

Figure 2.8   (a) The impulse function $\delta(t)$; (b) the unit step function $\varepsilon(t)$; (c) the ramp function $t\varepsilon(t)$.

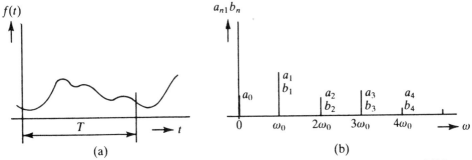

Figure 2.9   (a) Periodic signal $f(t)$ with period T. (b) Amplitude spectrum of $f(t)$.

In these equations the constants are defined by

$$a_0 = \frac{1}{T} \int_{-T/2}^{T/2} f(t)\, dt \tag{2.19}$$

$$a_n = \frac{2}{T} \int_{-T/2}^{T/2} f(t) \cos(n\omega_0 t)\, dt \tag{2.20}$$

$$b_n = \frac{2}{T} \int_{-T/2}^{T/2} f(t) \sin(n\omega_0 t)\, dt \tag{2.21}$$

and for the c_n in equation (2.19b) we find

$$a_n = c_n \cos \varphi_n$$

$$b_n = c_n \sin \varphi_n$$

$$c_n = (a_n^2 + b_n^2)^{1/2}$$

$$\varphi_n = \tan^{-1}\left(\frac{b_n}{a_n}\right).$$

The first harmonic is obtained by substituting $n=1$ in (2.18) and ω_0 is called the fundamental frequency and equals $2\pi/T$. Observe that a_0 is just the mean of the function $f(t)$. For the purposes of analysis it suffices to determine the period T and the coefficients a_n and b_n of the Fourier series expansion. This is illustrated in figure 2.9 in which the coefficients represent the different amplitudes of the different frequency components. The coefficients of the amplitudes of the different frequencies can be measured either by analogue integration or by digital computation.

*   *   *

Example 2.6

We give a simple example of the calculation of these Fourier coefficients. Given is the square-shaped signal depicted in figure 2.10(a). The coefficients are calculated with the help of equations (2.19), (2.20) and (2.21). The coefficients are $a_0 = 0$, $a_n = 0$ and $b_n = 2E/n\pi\ (1 - \cos n\pi)$.

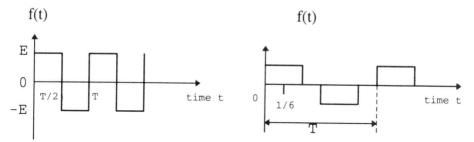

Figure 2.10 (a) Example of a pulse-shaped signal. (b) A pulse-shaped signal with interrupts.

Note that the $a_0 = 0$ because it represents the average value of the square-shaped pulse. Hence the Fourier series becomes

$$F(t) = \frac{4E}{\pi}\left(\sin \omega_0 t + \frac{1}{3}\sin 2\omega_0 t + \frac{1}{5}\sin \omega_0 t + \ldots\right).$$

A more complicated example is the following pulse-shaped signal in which interrupts are present. This is illustrated in figure 2.10(b). The coefficients are, using the same equations

$$a_0 = 0$$

$$a_n = \frac{4E}{\pi n}\sin \frac{\pi n}{2}\cos \frac{\pi n}{6}$$

$$a_n = 0 \qquad \text{for } n \text{ is even and for } n = 3, 9, 15, \ldots$$

$$\frac{2E}{\pi n}\sqrt{3} \qquad \text{for } n = 1, 7, 13, \ldots$$

$$-\frac{2E}{\pi n}\sqrt{3} \qquad \text{for } n = 5, 11, 17, \ldots.$$

For b_n we find $b_n = 0$; then the Fourier series becomes

$$F(t) = \frac{2\sqrt{3}}{\pi}E\left(\cos \omega_0 t - \frac{1}{5}\cos 5\omega_0 t + \frac{1}{7}\cos 7\omega_0 t\right.$$

$$\left. -\frac{1}{11}\cos 11\omega_0 t + \frac{1}{13}\cos 13\omega_0 t - \ldots\right).$$

*   *   *

There is another reason for applying (co)sine functions to a system. (Co)sine functions are preferred for measurement of transfer functions of a system because, inherent to their character of continuity, there is less risk of jamming. Furthermore, sine-shaped test signals can reveal the transfer function of a system. Ultimately, we can find the complex frequency response $H(j\omega)$ of a system if sine-shaped input signals are chosen as test (input) signals. From the frequency response the Bode plots can be retrieved, which describe the behaviour of a system. See also chapter 3.

It can be shown that the complex frequency response $H(j\omega)$ equals the *Fourier-transformed* impulse response $h(t)$ of the system, or

$$H(j\omega) = \int_{-\infty}^{\infty} h(t) \exp(-j\omega t)\, dt \tag{2.22}$$

and, conversely,

$$h(t) = \frac{1}{2\pi} \int_{-\infty}^{\infty} H(j\omega) \exp(j\omega t)\, d\omega \tag{2.23}$$

where

$$\begin{aligned}\exp(-j\omega t) &= \cos(\omega t) - j\sin(\omega t)\\ \exp(j\omega t) &= \cos(\omega t) + j\sin(\omega t).\end{aligned} \tag{2.24}$$

If the input signal $x(t)$ equals $\exp(j\omega t)$ then it can be shown that the output signal

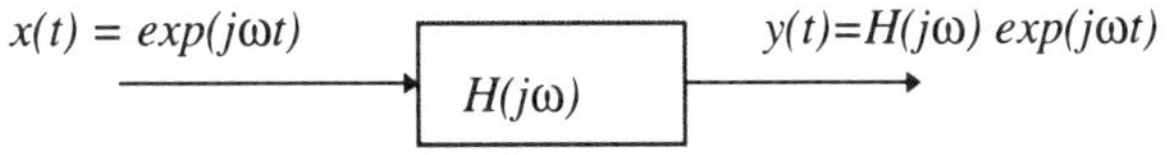

$$x(t) = exp(j\omega t) \longrightarrow \boxed{H(j\omega)} \longrightarrow y(t) = H(j\omega)\, exp(j\omega t)$$

Figure 2.11 The complex frequency response $H(j\omega)$ represents the transfer function of a linear system.

$y(t)$ is given by equation (2.25). This is illustrated in figure 2.11, where $X(j\omega)$ and $Y(j\omega)$ represent the Fourier-transformed input and output signals respectively, or

$$Y(j\omega) = H(j\omega)X(j\omega). \tag{2.25}$$

By definition the real parts of $X(j\omega)$ and $Y(j\omega)$ are, respectively,

$$Re[X(j\omega)] = \cos(\omega t)$$

$$Re[Y(j\omega)] = Re[H(j\omega)\cos(\omega t)] - Im(H(j\omega))]\sin(\omega t).$$

Multiplying the real part of the response, $Re[Y(j\omega)]$, by $\cos(\omega t)$ and taking the time average, we find the real part of $H(j\omega)$. Conversely, if we multiply the real part of the response, $Re[Y(j\omega)]$, by $\sin(\omega t)$ and execute a time average the imaginary part of $H(j\omega)$ is found. This process is implemented in so-called *synchronized detectors*. Using synchronous detection has great advantages with respect to offset, drift, a bad signal-to-noise ratio and/or low-frequency noise, in particular when small voltages must be measured. The principle is based on the multiplication of the AM input signal with its original carrier frequency. Basically, it is a demodulation technique. (See also example 2.7.)

Where (co)sine signals are involved, it is often convenient to depict a transfer function of the system in a polar plot or in Bode plots. This will be discussed in chapter 3. A (co)sine-shaped signal will not be distorted when it passes a linear system; only amplitude and phase changes may occur. This is a unique characteristic for sine-shaped input functions.

There is a third reason why applying a sine-shaped signal is preferred. They can be manipulated easily for modulation purposes in order to *transport* measurement signals. Modulation can be executed in one of the parameters of the sine function, i.e. the frequency, the phase or the amplitude. When using pulses as a carrier, pulse width and pulse height modulation can also be used. When modulation is applied, a multiplication is usually involved. For instance, multiplying two sine functions in the time domain results in an amplitude which is proportional to the amplitudes of both signals. The generated frequencies are the sum and the difference of both source frequencies.

$$* \qquad * \qquad *$$

Example 2.7
Suppose the input signal is represented by $x_i(t) = \hat{E}_i \cos(\omega_i t)$ and the carrier signal is represented by $x_c(t) = \hat{E}_c \cos(\omega_c t)$; then the AM output signal $x_o(t)$ can be described as

$$x_o(t) = \hat{E}_i \cos(\omega_i t) \hat{E}_c \cos(\omega_c t) = \tfrac{1}{2} \hat{E}_i \hat{E}_c \left[\cos(\omega_i + \omega_c)t + \cos(\omega_i - \omega_c)t \right].$$

We see that the generated output signal $x_o(t)$ contains the sum and the difference of both frequencies.

$$* \qquad * \qquad *$$

In frequency modulation the zero crossings are the only parameter of importance. Very accurate conversion of amplitude into the time domain can be realized by frequency modulation.

2.3.3 Sampled signals

Sampled signals are characterized by the values of the primary signal at certain equidistant times and places, allowing for instance the transmission of more than one signal through one transmission channel only. This is called time sharing or time multiplexing of signals and can be realized with different types of modulation. Pulse amplitude modulation (PAM), pulse width modulation (PWM), pulse position modulation (PPM) and analogue-to-digital (A/D) conversion are different possibilities. This is illustrated in figure 2.12 where $x_o(t)$ is the output signal. With the increasing importance of processing signals in a digital form, the A/D converter and the D/A converter have both become extremely important components which can be found in a large variety of instruments and systems. We will discuss the A/D and D/A

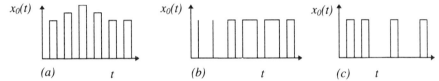

Figure 2.12 Illustration of an input signal with (a) pulse amplitude modulation, (b) pulse width modulation and (c) pulse position modulation.

conversion in more detail in Section 3.4, but we first have to explain the principle of *sampling*.

The signals involved in conversion by an A/D converter are:

1   the continuous-time signal $x(t)$;
2   the samples or discrete-time analogue signal $x(nT)$ and
3   the output signal $x(n)$, which can be characterized as a discrete-time digital signal.

Ultimately the output of the A/D converter will represent a rounded-off sequence of numbers. The values are multiples of the smallest available digit. Thus in an A/D converter a sampling process and a quantization process are executed. This conversion process is illustrated in figure 2.13.

The quantization process implies a kind of *binary coding*. Binary coding is the assignment of a binary number to samples of the discrete-time analogue signal $x(nT)$. Samples of the analogue signal $x(t)$ are obtained by sampling the function $x(t)$. Generally, in each analogue signal $x(t)$ we have available time and amplitude to assign numbers to. Hence, in sampling this signal, we can make four different possible combinations:

1   continuous in time and amplitude $(t_c a_c)$;
2   discrete in time and continuous in amplitude $(t_d a_c)$;
3   continuous in time and discrete in amplitude $(t_c a_d)$;
4   discrete in time and discrete in amplitude $(t_d a_d)$.

Sampling can also be interpreted as a kind of modulation and is realized by applying the basic sample-and-hold circuit depicted in figure 2.14. S_1 represents an electronic switch, controlled by the digital S/H signal. During the time that the switch S_1 is closed, a buffered signal from the storage capacitor C_H containing the input signal $x(t)$ is obtained. The output signal follows the input signal. When the switch is

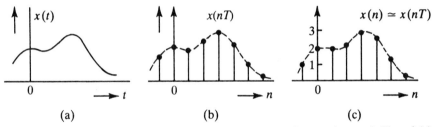

Figure 2.13   (a) Continuous-time signal $x(t)$, (b) discrete-time *analogue* $x(nT)$ and (c) discrete-time *digital* signal $x(n)$.

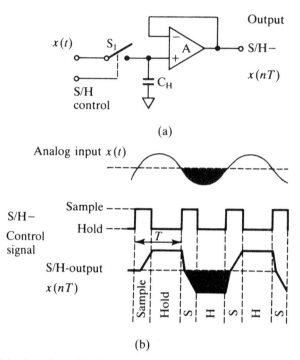

Figure 2.14 Sample-and-hold circuit for digitizing an analogue signal $x(t)$.

opened the input signal is stored in C_H and the output signal will be held, until the next sample period starts. This illustrates the ideal situation; in practice, possible errors are: transfer errors in sample-to-hold state and vice versa, time-varying sampling intervals, time-varying hold intervals. Furthermore, an ideal amplifier and capacitor and no external interference are assumed. Now assume that $x_1(t)$ is a periodic signal which is modulated (i.e. sampled) with a periodic Dirac impulse $x_2(t) = \Sigma\delta(t - nT_0)$, then this will result in an output signal $y(t)$, which is the product of $x_1(t)$ and $x_2(t)$, or

$$y(t) = x_1(t)[\Sigma\delta(t - nT_0)] \qquad \text{for } n = 0, 1, 2, 3, \ldots. \tag{2.26}$$

Basically, a modulation mechanism is involved, where modulation is realized using an (electronic) switch generating a kind of square-wave carrier frequency.

This result can be transformed in the (complex) frequency domain which is often more convenient. In particular for *non-periodic* signals these types of signal will deliver a continuous spectrum. In general, the transformed function of an arbitrary time function $f(t)$ is found using (2.27), or by definition

$$F(j\omega) = \int\limits_{-\infty}^{\infty} f(t)\, e^{-j\omega t}\, dt \tag{2.27}$$

and back again from the frequency domain into the time domain we have

$$f(t) = \frac{1}{2\pi} \int_{-\infty}^{\infty} F(j\omega) \, e^{j\omega t} \, d\omega. \tag{2.28}$$

$F(j\omega)$ can be interpreted as the spectrum of $f(t)$ and conversely $f(t)$ is the spectral decomposition of $F(j\omega)$, or the superposition of exponentials.

* * *

Example 2.8
For a rectangular pulse with amplitude $\hat{E}$ between $t_1 = -a$ and $t_2 = +a$ we find for the complex Fourier transform (figure 2.15)

$$F(j\omega) = \int_{-a}^{a} \hat{E} \, e^{-j\omega t} \, dt = 2\hat{E} \, \frac{\sin a\omega}{\omega}.$$

* * *

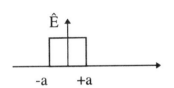
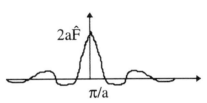

Figure 2.15 The complex Fourier transform of a rectangular pulse.

Applying the Fourier series expansion for $x_1(t)$ we obtain $x_1(j\omega)$. Reasoning in the same way, but now applying the *Fourier integral transformation* to the periodic Dirac pulse, we find

$$X_2(j\omega) = \frac{2\pi}{T_0} \sum_{-\infty}^{\infty} \delta(\omega + n\omega_0) \tag{2.29}$$

where $\omega_0 = 2\pi/T_0$.

The output signal in the frequency domain $Y(j\omega)$ can be obtained by applying the frequency *convolution* theorem to (2.26). This yields

$$Y(j\omega) = \frac{1}{2\pi} X_1(j\omega)^* X_2(j\omega). \tag{2.30}$$

Convolution is applied here in the frequency domain with the same basic concept as we have seen before in the time domain. A graphical representation of this process is depicted in figure 2.16, where it can be seen that the frequency spectrum of $X_1(j\omega)$ is symmetrical with respect to the middle of the sample frequency $1/T_0 = f_0$. All the periodic frequency spectra of $X_1(j\omega)$ are equal and contain the same information. By definition, every periodic frequency spectrum of $X_1(j\omega)$ is called an *alias*. It can

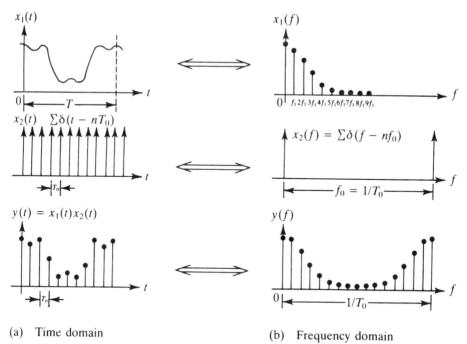

(a)   Time domain   (b)   Frequency domain

Figure 2.16   Illustration of the sampling theorem. The time domain representations are shown in the left figures, the frequency domain is shown right. Note that $f = \omega/2\pi$.

be deduced from the same figure that, if the sample frequency f_0 is chosen too low, an aliasing error will occur. The periodic spectrum of $X_1(j\omega)$ will show overlaps in the middle of the frequency spectrum. This is called an *aliasing error* and must be avoided to prevent loss of information. The sampling theorem says that, if the sampling frequency is chosen high enough, the frequency spectrum will show no overlap and no aliasing errors will occur. This can be realized by confining the frequency band B of $x_1(t)$ to $f_0/2$ with the aid of a low-pass filter. To prevent loss of information, in practical circumstances, the sampling frequency will always be chosen to be more than twice (up to 20 times) the highest-occurring signal frequency in $x_1(t)$, or $f_0 > 2B$. This theorem was first formulated by Shannon and Nyquist and this frequency is known as the *Nyquist frequency*.

2.3.4   Stochastic signals

In this case the instantaneous value of a signal is described by its *specific probability density function* $p(x, t)$ with respect to place and time. It is called a stochastic signal for the reason that only a statistical description of the signal can be given. For instance noise is a representation of this kind of signal. Because of its very nature, a noise signal is only predictable in terms of amplitude probability density functions of place and time, $p(x, t)$. It will be obvious that the sum (i.e. integral) of the density function over all possible events must equal unity.

The signals or noise generators are investigated by measuring the amplitude distribution $p(x, t)$ and the power spectrum $S(j\omega)$. The signal is fully described by these parameters. A detailed and comprehensive discussion concerning noise calculations in electronic systems and components is presented in chapter 9. Here we shall only discuss two applications of noise.

Noise can be used as a *test* signal for response measurements of linear systems. If so-called *white noise* is used as the test signal, $x(t)$, it can be shown that the *correlation function* R_{xy} of $x(t)$ and the response $y(t)$ of that system equal the previously defined impulse function $h(t)$, or

$$h(\tau) = R_{xy}(\tau) = \lim_{T \to \infty} \frac{1}{2T} \int_{-T}^{T} x(t)y(t+\tau)\,dt. \qquad (2.31)$$

White noise can be defined as noise with a constant power per unit bandwidth. (See also chapter 9.) *Correlation* can be understood as a measure of the dependence between two signals or physical quantities. For instance, there exists a certain dependence or correlation between length and weight of human beings. The correlation factor is zero if the signals are independent and unity when the signals are completely dependent. An advantage of measuring the impulse response caused by noise is the natural phenomenon of noise. Adding noise to a system will not disturb the system's process under consideration. A block diagram of an analogue correlator is shown

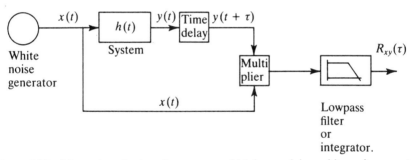

Figure 2.17 Measuring the impulse response $h(t)$ by applying white noise as a test signal.

in figure 2.17. White noise can also be used as a test signal for testing digital systems, generating a unique response image when the system is functioning correctly.

There is another interesting application in which noise is involved. When a periodic signal with unknown form is buried in additional noise it is possible to detect the periodic component by applying the *autocorrelation function* $R_{xx}(\tau)$, which is defined as

$$R_{xx}(\tau) = \lim_{T \to \infty} \frac{1}{2T} \int_{-T}^{T} x(t)x(t+\tau)\,d\tau. \qquad (2.32)$$

At large values of the delay time τ, the autocorrelation function of the noise components is zero, but that of the periodic signal will be non-zero. This application is

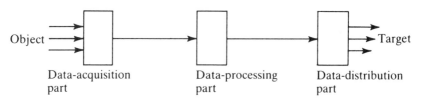

Figure 2.18   The measuring system considered as an information transport channel.

used for instance in long-distance data channels and in circuits in which very weak signal sources are involved.

2.4   SPATIAL STRUCTURES

In this approach to electronic measurement systems we can interpret a system as a transport channel transporting information. Three functions can be distinguished here, (1) a *data-acquisition* part, (2) a *data-processing* part and (3) a *data-distribution* part. This is illustrated in figure 2.18.

In the data-acquisition part, information is acquired from the measuring object and converted into data. Usually, the measuring object produces an analogue signal. Often more than one quantity must be measured, or at different places a multiple of the same physical quantity must be monitored, for instance the temperature, strain or pressure in an oil refinery plant. Another good example is the measurement of the convective flow profile on a wall surface.

The data-processing part contains all the required parts to bring the signal into the required form and condition. This part often starts with filtering and amplifying. The possible operations are already summarized in tables 2.1 and 2.2.

The third part performs data distribution towards the target objects, where the information is required. Often a multiple output is involved.

The first and the last parts can be expanded as illustrated in figure 2.19. The data-acquisition part can be subdivided into a sensor part, a signal conditioner part and an A/D converter. The sensor part produces a signal which is a continuous representation of the measurand and converts it into an analogue electrical signal. The signal conditioner conditions the signal and makes it suitable for conversion

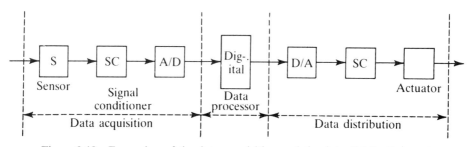

Figure 2.19   Expansion of the data-acquisition and the data-distribution parts.

into a digital signal by the A/D converter. The operations involved can be found in tables 2.1 and 2.2.

In many cases digital processing of the signal is preferred. The consequence of this is that, after the data are processed, we have to convert the digital signal again into an analogue signal. Finally, the analogue signal is delivered to an actuator or output transducer. In the output transducer the information is often converted into a non-electrical form, able to activate a switch, a valve, an indicator, a type of display, etc. In the following section a multiple-input and output config____ is discussed. We will discuss the following structures:

1 multiple-input/output configuration
2 centralized data-acquisition system
3 decentralized data-acquisition system with digital multiplexer
4 distributed multiplexer data-acquisition system
5 telemetry system.

2.4.1 A multiple-input/output configuration

In figure 2.20 a multiple-input, multiple-output system is shown. At the input an analogue *multiplexer* connects the different input signals to the digital processor in

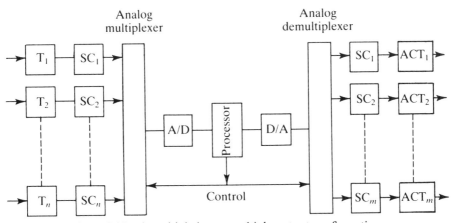

Figure 2.20 A multiple-input, multiple-output configuration.

a time-programmed serial sequence. A multiplexer can be defined as an electronically controlled switch with multiple inputs and one output. The switch is controlled in time and therefore the operation can be called *time multiplexing*. At the output an analogue *demultiplexer* is installed to provide multiple output signals. Also the demultiplexer is a type of electronic switch, time programmed in a serial sequence of switching operations. In this configuration only one A/D and D/A converter is installed, which is cost effective of course, but a disadvantage might be that little time is available for conversion. Conversion has to take place during some clock cycles of the processor. This can be avoided by giving every input and output channel its own A/D and D/A converters.

Note that the numbers of input and output channels might be different. We will discuss several other configurations of the above-outlined concepts. We start with a centralized data-acquisition system by presenting an example in which many aspects of such a system are characterized.

2.4.2 centralized data-acquisition system

*   *   *

Example 2.9

Assume that we want to measure the mechanical strain in a wing structure of an aircraft with the aid of strain gauges. A strain gauge is a resistive film element consisting of thin wire. It changes its resistance when a mechanical load is applied to the gauge.

The strain in the wing can be measured at the surface of the structure and will be a function of load and position. Before the strain gauges can be attached a qualitative measurement is executed. This is to discover the best locations to attach the strain gauges. This qualitative measurement can be performed by painting the structure with a specific type of lacquer whose reflection coefficient is highly sensitive to strain variations. A measurement set-up with which this effect can be revealed is depicted

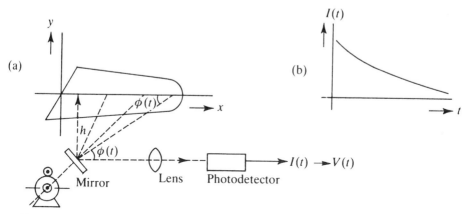

Figure 2.21 (a) Measurement and conversion of a space function into a time-dependent current; (b) its graphical representation.

in figure 2.21. It consists of a photodetector, a rotating mirror and a lens system. If the mirror rotates with a speed of n revolutions per minute, the system will convert the space-dependent strain function into a time-dependent current function $I(t)$. The photocurrent will be a function of the incident light falling on the photodetector reflected from the surface. This is expressed in the following relationship between the photocurrent and strain:

$$I(t) = f[S(x)] \tag{2.33}$$

where $I(t)$ is the current function versus time and $S(x)$ the strain function versus position. From the figure it can be seen that the x coordinate function is given by

$$x = h \tan\{\phi(t)\}^{-1} \qquad (2.32)$$

where x is the distance with respect to a certain reference point, h is the perpendicular distance from mirror to object, $\phi(t)$ is the angle between mirror and axis of the photodetector and $\phi(t)$ is given by

$$\phi(t) = \omega t + \phi(0) \qquad (2.33)$$

where $\omega = 2\pi n/60$, n is the number of revolutions per minute and $\phi(0)$ is the phase shift. The result is an electrical signal $V(t)$ representing the strain as a function of intensity in one dimension, here the x axis.

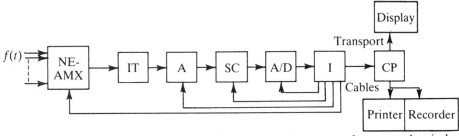

Figure 2.22 Block diagram of the qualitative measuring set-up for a non-electrical quantity.

The measuring set-up is depicted in figure 2.22. The first block represents the mechanical analogue multiplexer NE-AMX (i.e. rotating mirror) and is placed before the input transducer, IT. The photodetector is the input transducer. An amplifier stage, A, will deliver the required gain. The signal conditioner, SC, executes the appropriate electronic modification before A/D conversion can be performed. An interface, I, connects the system to the digital processor. As well as this, the *interface* controls the different mentioned functions. An *interface* can be defined as a piece of hardware between two digital systems that provides the right (digital) conversions to make communication between the two systems possible. Finally a central processor (CP) with appropriate peripherals, for instance a printer, a recorder or a display, are depicted.

Once a qualitative measurement is executed to determine where highest strains may be expected, an electrical quantitative measurement can be performed. The strain gauges are attached to the surface of the wing with a special type of glue. Sometimes hundreds of these gauges are mounted on one structural part. This is depicted in figure 2.23.

All these strain gauges can be connected with two or three wires into a Wheatstone bridge configuration. Becasue of the very weak signals for strain gauges this is a preferred configuration. Finally, the strain gauges could be connected to a central processing unit as shown in figure 2.24.

The A-MUX connects the bridge outputs sequentially to the input of the amplifier A. The result will be a time-multiplexed signal of the strain as a function of place along one axis in the structure.

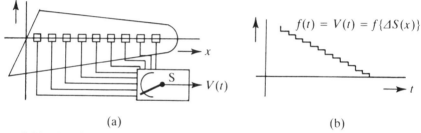

Figure 2.23 (a) The strain gauge measuring set-up for a wing structure; (b) a graphical representation of the output signal.

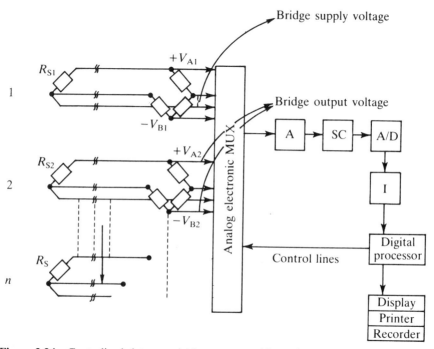

Figure 2.24 Centralized data-acquisition system with analogue electronic multiplexer (A-MUX).

*   *   *

The above-given description is an example of a complete *centralized data-acquisition* system. Such a system has several disadvantages, such as the following:

1   The transport of a weak transducer signal over long distances makes the system highly sensitive for interference, hence the result can be a bad signal-to-noise ratio.

2   Long transducer cables and the analogue multiplexer decrease in general the result's accuracy. Especially, when high demands are put forward with respect to switching speed of the MUX, there is a considerable reduction of the overall accuracy.

3   If a great number of long cables is required the system will become expensive.

Another configuration, a decentralized data-acquisition system, can improve the system specifications to a large extent. This is discussed below.

2.4.3 A decentralized data-acquisition system with a digital multiplexer

Assume each transducer is given its own signal-processing part. The signals are then converted again into a digital signal to decrease the sensitivity to interference. Between the corresponding A/D converters and multiplexer long cables are present. In this configuration no switching at the transducer's side is required. This will improve the overall accuracy because all bridges are active, avoiding any time-dependent settling effects. The multiplexer is a digital one controlled by the processor. A block diagram is shown in figure 2.25.

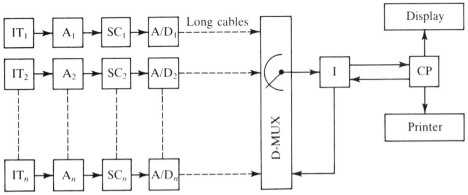

Figure 2.25 Decentralized data-acquisition system with a digital multiplexer (D-MUX).

This configuration shows some other disadvantages, for instance the following.

1   The system is very expensive because each transducer has its own processing unit. Depending on the application this is sometimes justified and can be worthwhile.
2   A large amount of cabling is involved.

A recent development shows that the first disadvantage may be overcome by the use of silicon integration technology. Integration technology makes it possible to integrate the sensor part and the signal processing part on to one chip. This will be discussed in more detail in chapter 6.

In the following section, a distributed multiplexer system configuration is discussed.

2.4.4 A distributed multiplexer data-acquisition system

In this configuration all the measuring units are switched sequentially in parallel on one long coaxial cable. Every data-acquisition part contains its own multiplexer switch operated by the processor's interface. The system can operate either with digital or with FM signals. This concept is illustrated in figure 2.26. The disadvantage of a large number of cables experienced with the configuration described in section 2.4.3 has disappeared. Only the large number of decentralized mounted electronic

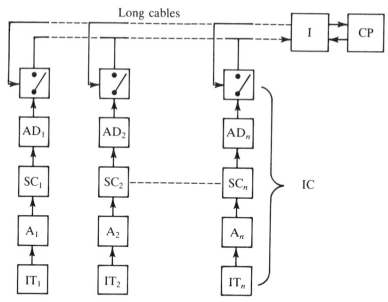

Figure 2.26 Distributed multiplexer data-acquisition system with only one coaxial interconnection cable.

components and subsystems might remain as a disadvantage, because this increases costs.

Here the same remark can be made concerning integration technology. 'Smart' sensors have been developed, and are realized already and commercially available. Note the remark here that in practice there exists a large discrepancy between what has been developed and what is commercially available. All the required electronics can be integrated onto one piece of silicon. This implies that operations such as gain, filtering, A/D conversion, digital processing and multiplexing are integrated onto one piece of silicon. This reduces cost and volume considerably and increases reliability.

There is one other important configuration worth mentioning. In some circumstances the information must be transported over extreme long distances or mobility is an absolute prerequisite. This is performed using radio channels and it is called *telemetry.*

2.4.5 Telemetry

In telemetry systems either the information has to be transported over extreme long distances or it is not possible to connect the data-processing part close to the data-acquisition part. Examples can be found in measuring complicated machinery, in a radioactive environment, in astrophysical observations and in space sciences. Also, applications of wireless data transmission are found in large buildings and in medical applications for mobile patient monitoring.

In general, there are three possible methods of transmitting the information: (1) radio channel, (2) a telephone (copper/coax) cable or (3) using glass fibre cable. It is obvious that, for satellite data transmission, a radio channel is the only possible one.

In medicine, a pressure telemetry system is a very sophisticated application. When the behaviour of the bladder pressure versus filling volume must be monitored, a telemetry 'pill' system allows the registration of the pressure without any inconvenience to the patient. The pressure sensor and the telemetry system are sealed into a miniature cylindrical ceramic container and implanted in the patient. The information is transmitted by applying an FM signal and received again by a receiver with a high signal-to-noise ratio. The pressure signal can be available in either analogue or digital form. Figure 2.27 shows a schematic block diagram.

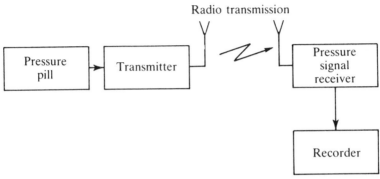

Figure 2.27 Block diagram of a telemetry pressure sensor configuration.

Very-large-scale integration (VLSI) techniques are of great value in these types of application in communication. We will consider this potential in some more detail in chapter 6.

2.5 AUTOMATED ELECTRONIC STRUCTURES

In general, automated electronic structures can be defined as a combination of a number of electronic instruments controlled by a central unit. Not only can this central unit control the operation of the individual instruments, but also it can determine the sequence of the measurements more or less automatically. Possible additional operations are sorting, recording (i.e. data logging) and calculating. In general we always have to deal with a three-stage system of activities: (1) acquisition of data, (2) analysis and (3) presentation. Automated measuring systems can be divided into different groups.

1 *Programmer-operated systems*

Programmer-operated systems are preset test and measuring set-ups with partly manual control or operation. The various test programs are preset manually, for instance by means of electronic switches, and then initiated by a push button or by automatic sequencing. All programmable logic controllers (PLCs) can be assigned to these systems.

2 Controller-operated systems

These test and measurement systems have simple processing facilities, for instance sorting and recording. No calculation facilities are present. The various test programs are stored on a magnetic tape, close to the central controller, or stored in a kind of programmable logic device (PLD). Sophisticated measurement instruments have become available which are more or less easy to program for a specific task. A complete new design philosophy has entered instrument manufacture, such as that implemented in the logic analysers of HP (HP54620A) which can be operated as simple oscilloscopes. The bottom line of this concept is the pay-per-use philosophy that asks the user to pay only for the capability that he or she actually uses, but simultaneously being able to use a top-end tester.

Another example is the Fluke 867 digital multimeter which offers facilities for in-circuit tests and can plot trends and detect logic activity. Analogue, digital and waveform signals can be displayed. The model offers also interfacing possibilities with a computer.

3 Computer-operated systems

These fully automated measuring systems have all kinds of processing facility on board such as sorting, recording and computing. The various measuring programs are stored in the computer memory. An overwhelming number of instrument makers offer all kinds of plug-in cards which can be connected to a PC. We have to mention here another interesting development. We refer to the concept of making virtual instruments (VIs), of which National Instruments is one of the best known representatives.

4 Virtual instruments

Virtual Instruments are completely software-designed instruments using a type of software driver which represents a specific type of instrument. They have numerous advantages over traditional instruments. A comparison is given in table 2.4. Basically, we can distinguish two different VI system options. Firstly, a graphical programming language option exists, which uses symbols to (re)configure a measurement system. At National Instruments (NI), this option is known as LabVIEW. Secondly, there

Table 2.4 Comparison of fundamentals between traditional and virtual instruments.

Traditional instruments	Virtual instruments (VIs)
Manufacturer defined	User defined
Function specific, stand alone, limited connectivity	Application-oriented system with connectivity to networks, peripherals and applications
Hardware specific	Software specific, easy reconfigurability
Expensive	Low cost, reusable
Limited and closed functionality	Open, flexible functionality, leveraging off standard computer technology
Life cycle 5–10 years	Fast turn-around technology, life cycle 1–2 years
Minimal economics of scale	Maximum economics of scale
High development and maintenance costs	Software minimizes development and maintenance costs

is the option of an interactive C programming environment delivering compiled performance with the development ease of Basic. This NI option is known as Lab-Windows/CVI. However, other manufacturers also offer such measurement systems. We will come back to microprocessor-based measurement systems in more detail in chapter 4, section 4.5. For further reading also reference is made to the literature, e.g. [4].

2.6 I/O INTERFACE

As well as a choice between these two options we need always a type of I/O interface. An interface can be defined as a piece of hardware/software which acts as the required connection between the computer and the outside world. Numerous I/O interfaces or bus systems exist. However, the problem with a bus system is always the acceptability of conformity by the different manufacturers and the users in the market. When not enough support is given a bus system is not viable. Four different types of I/O are mentioned here, of which three are world-wide accepted. Concerning the *plug-in* cards, hundreds of different types are available all with their specific features. The following four types of I/O are the best known:

1 IEEE 488 (GPIB, IEC 625, HP-IB)
2 VXI/VME bus
3 serial and industrial networks, RS-232 (V24)
4 plug-in data-acquisition (DAQ) boards.

To prevent confusion, some often-used alternative names are also mentioned in parentheses. In the sections below the main features of each of these I/O options are discussed in some more detail.

2.6.1 IEEE 488
Initially, Hewlett–Packard designed this interface bus (table 2.5) as HP-IB in 1965. It was later accepted as IEEE standard 488-1975 and after that as ANSI–IEEE standard 488.2-1987. In 1990 Standard Commands for Programmable Instruments (SCPI) took the command set that is used in any SPCI instrument. This I/O interface is also widely known as a general-purpose interface bus (GPIB).

Table 2.5 Main features of I/O interface IEEE 488.

Characteristics
Worldwide standard since 1975
Designed for remote control of programmable instruments
Connection with flexible cable: max. 2 m distance between instruments, maximum bus length up to 20 m
Controls up to 14 instruments
8-bit parallel protocol with ASCII command sets
Transfer rates greater than 1 Mbytes s^{-1}
Also known as GPIB and HP-IB; old name is IEC 625

Table 2.6   List of abbreviations for the GPIB.

DAV	Data valid
NRFD	Not ready for data
NDAC	Not data accepted
IFC	Interface clear
ATN	Attention
REN	Remote enable
SRQ	Service request
EOI	End or identify

The bus system involves a connector in a 24-pin configuration with a total of eight lines for signal ground, eight data lines, three handshaking lines (DAV, NRFD, NDAC) and five interface management lines (IFC, ATN, REN, SRQ, EOI). The meanings of the abbreviations are indicated in table 2.6. For this GPIB a fixed-signal 24-pin connector configuration also exists. This is indicated in figure 2.28. All pins are assigned with their specific names.

2.6.2   The VXI/VME bus

VXI denotes virtual extension instruments and is basically an extension of the world-wide VME bus standard. VME denotes virtual machine environment. The VXI bus uses a mainframe chassis with a maximum of 13 slots to hold modular instruments on plug-in boards. The VXI backplane includes the 32-bit VME computer bus and combines the IEEE 488 protocols.

DIO 1	1	13	DIO 5
DIO 2	2	14	DIO 6
DIO 3	3	15	DIO 7
DIO 4	4	16	DIO 8
EOI	5	17	REN
DAV	6	18	GND Twisted pair with DAV
NRFD	7	19	GND Twisted pair with NRFD
NDAC	8	20	GND Twisted pair with NDAC
IFC	9	21	GND Twisted pair with IFC
SRQ	10	22	GND Twisted pair with SRQ
ATN	11	23	GND Twisted pair with ATN
SHIELD	12	24	Signal GROUND

Figure 2.28   The GPIB signal 24-pin connector configuration.

Table 2.7 Main characteristics of serial interfaces.

For serial instruments use of simple ASCII protocols is possible
For PLCs and other industrial devices more sophisticated protocols are available

Distances longer than 1.2 km (4000 ft) possible, e.g. RS 485
Range of data rates 100 bit s^{-1} to more than 1 Mbit s^{-1}
Other common serial interfaces are RS 485 and RS 4822
Multidrop capabilities for industrial networks
RS 232 standard available in most computers

2.6.3 Serial and industrial networks, RS-232 (V24)

In this bus system the information transfer is executed in a serial form; hence longer distances can be covered than with a parallel bus as with the VXI and GPIB. However, transfer speeds are lower. The standard serial port on a computer is an example of a serial bus using simple ASCII strings as data.

For industrial applications, connecting more sophisticated devices, such as programmable logic controllers (PLCs) in automated test equipment, more reliable protocols are required than simple ASCII protocols. The main characteristics of the serial RS 232 bus system are summarized in table 2.7.

In figure 2.29 the RS 232 pin configuration is given for a female connector. In all PCs one or more serial communication ports are found.

	13	
25	12	
24	11	
23	10	
22	9	
21	8	data channel received carrier detector
20	7	signal ground
19	6	data set ready
18	5	ready for sending
17	4	request to send
16	3	received data
15	2	transmitted data
14	1	protective ground

connect data set to line ⟶ (pointing to row 20/7)

Figure 2.29 Female pin layout for RS 232 connection. Note also that for RS 232 the signal levels are fixed. In V24 only the signal names are specified.

2.6.4 Plug-in data-acquisition (DAQ) boards

Plug-in data-acquisition (DAQ) boards are available for many computers in a great variety. These boards combine analogue, digital and timing inputs. They also have programmable channel sampling inputs and conversion modes. Usually, a separate gain for each channel is available and flexible analogue and digital triggering features are built in. Some important characteristics which can be found in most DAQs are summarized in table 2.8.

Table 2.8 Some common characteristics of DAQ boards.

Boards directly to be installed in PCs
Transfer data directly to and from computer memory
8- to 20-bit resolution available
Sampling rates up to 1 MHz
Analogue-to-digital conversion (A/D)
Digital-to-analogue conversion (D/A)
Digital input/output (DIO)
Counter/timer operations available

2.6.4.1 *DAQ board specifications*

We will go through all the important definitions of DAQ boards just to be able to read manufacturing specifications and to be able to take the right decision for a certain application. Every DAQ board function starts with a kind of signal conditioning to prepare the data for further processing.

DAQ boards can accept any kind of electrical input signal, originating from a sensor, usually producing a weak electrical signal, from some microvolts up to several millivolts. Firstly, this weak signal must be made less sensitive for interference and noise, and hence amplified.

For safety reasons the transducer signal is *electrically isolated* from the computer. This also prevents large transients from destroying the computer's input. Isolation can be performed with optocouplers or via a transformer coupling. Isolation for each channel also avoids the making of ground loops caused by ground potential differences. See also chapter 8 for more details.

Filtering is used to remove unwanted high-frequency signals. For DC signals a 4 Hz low-pass filter may be used. For AC signals a low-pass *anti-aliasing* filter is used. In this case all signals higher than the bandwidth of the DAQ board are removed to avoid the production of erroneous signals within the bandwidth of the board.

Excitation signals are used to activate self-modulating transducers such as strain gauges, RTDs and thermistors. These devices need a voltage or current source to become operational. (See also chapter 6.) In some available software packages for DAQ boards *linearization routines* for thermistors, temperature-dependent resistors (TDRs) and strain gauges are available.

The *sampling rate* determines how often conversion can take place. Higher sampling rates give more data points per unit of time and hence produce improved accuracy in representing the original signal. Think of the Nyquist theorem which says that the sampling rate must be at least twice the highest frequency involved in the conversion. For instance, a frequency of 18 kHz requires at least a 36 kHz sampling rate. In practice the sampling rate is chosen to be much higher. DAQ boards can have more than one channel of input and output and usually can be used in single mode or differential mode. Differential-mode channels each have their own ground reference. In this case interference will act as a common mode signal at both inputs and hence any interference is more strongly repressed. This issue will be discussed in more detail in chapter 8.

Multiplexing is a measuring technique for measuring more than one signal with a single ADC and with the same electronics. The ADC switches one after the other between the different channels. It will be obvious that the *effective sampling rate* is reduced inversely proportional to the number of channels to be sampled. For instance, for a DAQ board with ten channels a sampling rate of 100 kS s^{-1} is reduced to an effective sampling rate of 10 kS s^{-1} for each individual channel.

Resolution is the number of bits that the ADC uses to represent the signal. We will also see another definition for resolution in chapter 4.

$$*\qquad*\qquad*$$

Example 2.10
When a signal of 10 V is represented with a 4-bit ADC, the resolution is 10 V$/2^4 =$ 625 mV, which gives a poor representation of the signal. Improving this to 16 bits, the resolution becomes 10 V$/2^{16} = 10/65\,536 = 152.5\,\mu$V, which gives an extremely accurate presentation of the original signal.

$$*\qquad*\qquad*$$

The *range* of a DAQ refers to the minimum and maximum voltage levels the ADC can quantize. Usually, several ranges can be set on a DAQ board to find the best accuracy and resolution for the measurand. Ranges are for instance ±1 V, ±5 V or ±10 V.

The smallest detectable voltage change is determined by the range, the resolution and the gain. This voltage is called the *code width* and represents 1 LSB of the digital value.

$$*\qquad*\qquad*$$

Example 2.11
A multifunctional 16-bit DAQ board has a selectable range of 0–10 or ±10 V. The selectable gain is 1, 2, 5, 10 or 100. When the voltage range is 0–10 V, then the ideal code width (CW) is given by

$$CW = \frac{\text{range}}{\text{gain} \times 2^{\text{resolution}}} = \frac{10\ \text{V}}{100 \times 2^{16}} = 1.5\,\mu\text{V}.$$

This figure determines the theoretical resolution.

$$*\qquad*\qquad*$$

Analogue outputs provide stimuli for the DAQ system. In this case settling time, slew rate and resolution are important specifications. *Settling time* is the time needed for the output to settle to a specified accuracy, usually specified for a full-scale change in voltage. *Slew rate* is the maximum rate of change that the DAQ can produce on the output signal. It is often expressed in volts per unit of time, e.g. $10 \text{ V } \mu s^{-1}$. (See also chapter 4.) For the DAQ to generate high-frequency signals a high slew rate and a short settling time are required. The *output resolution* is defined in a similar way as the input resolution. Hence high-resolution voltage outputs can be obtained.

Digital I/O (DIO) interfaces are used to control processes and to generate patterns for testing and communication purposes. Here the number of available digital lines, the rate at which the input and output can change and the driver capability in voltage and current are important. Often the digital output of a DAQ is 0–5 V DC and several milliamperes only but, with the right interface, high-voltage devices such as motors and valves can be controlled easily.

Timing I/O or counter/timer circuitry is also a useful feature on a DAQ board. This can be used for counting digital events and digital pulse timing and can generate square waves and pulses. Applications are found in positioning from rotary and linear encoders. Important specifications are here also resolution and clock frequency. A higher *resolution* means that you can count higher because more bits are available. A higher *clock frequency* means that you can count faster. For instance, an existing DAQ has an up/down counter/timer source with 24 bits and can produce a clock frequency of 20 MHz.

Without appropriate *software* each DAQ system is useless. To avoid compatibility problems between software and hardware it is best to buy software and hardware from the same manufacturer. Three possibilities exist to develop software.

1   The registers on the DAQ are programmed directly. This is a very time-consuming affair and not recommended.
2   The second possibility is to use *driver software*. In this case attention must be paid to the following possibilities:
    ● the driver must be able to acquire data at specified sampling rates
    ● to perform data acquisitions in the background while processing data in the foreground
    ● to use programmed I/O, interrupts and DMA to transfer data
    ● to stream data to and from disk
    ● performing some functions in parallel
    ● integrating more than one DAQ
    ● communicating with signal-conditioning equipment.
3   The third possibility is using *application software* which is normally based on the appropriate driver software for the DAQ used. This application-oriented software gives the possibility to add analysis and presentation capabilities to the driver software.

2.7   ARCHITECTURE OF BUS SYSTEMS

There is one other possibility to characterize bus systems. It is just the way we can look at the architecture that the bus systems are configured in. Some well known

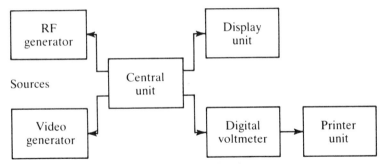

Figure 2.30 Principal circuit of a star system.

examples are mentioned here. Automated electronic measuring systems can be structured in a *star* structure or in a *bus-line* system. An example of a star system is depicted in figure 2.30.

Each instrument is individually connected to the central control unit by its own bundle of control lines. In a star system, the programming information is given as a kind of level in a certain code. The code is applied as long as the function is required. Also, a number of analogue control lines can be implemented to set continuous functions. The input and output information can be given simultaneously via separate input and output lines.

In a *bus-line* oriented system all instruments are connected in parallel to the central processing unit, CPU, via a common set of lines, called buses. In general three types of these buses can be distinguished: the *address* (*line*) *bus*, which addresses the required instrument, the *control bus*, by which lines special functions can be initiated, for instance the instructions 'input' or 'output', or read/write, and the *data bus* via which the information is transported.

Note that in smaller systems the address bus and data bus are sometimes combined into one bus. A bus-oriented system is depicted in figure 2.31. For instance, assume that this system is designed to measure automatically a part of a video system in a manufacturing process. Sources and receivers are indicated. The sources generate

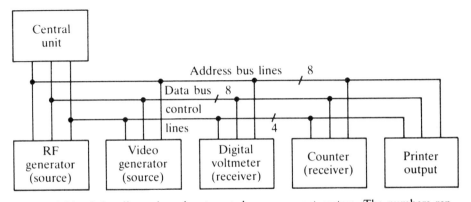

Figure 2.31 A bus-line oriented automated measurement system. The numbers represent the numbers of lines used.

the required test signals, the receivers measure the generated output signals. These measurement set-ups are known as automated test equipment (ATE) systems. For each test or measurement the appropriate instrument is instructed or 'activated' via input and output ports. This is executed by means of the appropriate input and output lines (I/O) and the address and control lines. In a bus-line system the programming information is given as sequences of binary numbers. The programming information is stored in the computer's memory. Every measurement program or routine is stored in memory in the program flow as a subroutine, which is called on by the program in the required sequence at the right time.

   In the program, precautions have to be taken to ensure that only one instrument is activated at a time. Otherwise, so-called bus conflicts, such as short circuits, may occur. Ultimately, standard protocols are implemented to prevent this happening.

   In the following chapter we will introduce electronic implementations of functions of measuring systems.

PROBLEMS

1   Indicate what is meant by an active or intensive physical quantity and give examples.

2   Discuss the measuring structure for a passive quantity.

3   A signal is given as a function of space. Which kinds of conversion have to be realized when the information is to be recorded on a magnetic tape?

4   Discuss the several reasons why it is an advantage to implement filters in a measuring chain.

5   Today in almost every measurement system an A/D converter can be found. What are the main reasons that these blocks are implemented and what is basically happening when an analogue-to-digital conversion is performed? Can you compare it with an everyday tool?

6   Determine the impulse response $h(t)$ of the Laplace-transformed system function $H = 1/(R + sL)$, where $R = 10\,\Omega$ and $L = 1$ mH.

7   The Laplace-transformed system function is given as

$$Y(s) = \frac{7s + 23}{s^2 + 7s + 10}.$$

   Determine the impulse response $h(t)$.

8   Perform the same operation for

$$Y(s) = \frac{4s + 20}{s^2 + 4s + 13}.$$

9   Find the impulse response $h(t)$, the step response $r(t)$ and the ramp response $\mathrm{rmp}(t)$ for a series connection of a resistor and a capacitor, with $R = 1$ kΩ, $C = 1\,\mu$F and $v_c(0) = 0$ V.

10   Express the Fourier series in equation (2.8) in a sum of exponentials.

11   Discuss the advantages of a bus-line-oriented automated structure. Which different hardware-implemented functions are required?

12 What are the main advantages in using a virtual instrument (VI) and what may be a drawback?

13 What kind of I/O may be preferred in an industrial environment?

14 Determine the code width of a 10-bit DAQ board with a range of 1, ±5 V and a selectable gain of 1, 2, 4, 8, 16.

BIBLIOGRAPHY

[1] Athanasios Papoulis 1980 *Circuits and Systems, A Modern Approach* (*International Editions*) (New York: Holt-Saunders)
[2] Spiegel M R 1965 *Theory and Problems of Laplace Transforms* (*Schaum's Outline Series*) (New York: McGraw-Hill)
[3] Instrumentation and Reference Catalogue 1995, National Instruments (Austin, TX)
[4] Brignell J and White N 1994 *Intelligent Sensor Systems* (Bristol: Institute of Physics Publishing)

Three

IMPLEMENTATIONS OF FUNCTIONS

3.1 INTRODUCTION

In chapter 2 we presented a review of the *types* of measuring system. This was based on the four different possible structures to be recognized in measuring systems. In the discussion of these structures, many different electronic functions were mentioned from a theoretical point of view. In this chapter, we will discuss the electronic *implementations* of functions. Most of these functions are frequently used in measuring systems. The following topics will be covered:

1 tools
2 analogue functions
3 analogue-to-digital conversion and vice versa
4 digital functions
5 miscellaneous.

3.2 TOOLS

In order to be able to present some concepts in a suitable context we start with a presentation of a collection of 'tools' by which functions can be described in a characteristic way. If these tools are already familiar this section can be skipped. The concepts are presented in an abbreviated form, without an extensive mathematical proof, and the appropriate literature should be consulted if so desired. A basic understanding of complex calculus is also assumed. However, if this understanding is lacking, a brief introduction to complex calculus can be found in the next section.

3.2.1 Complex calculus and variables
From a mathematical point of view complex numbers have been introduced to solve equations of the type

$$x^2 + 1 = 0. \tag{3.1}$$

In the collection of real numbers this equation has no solution. For this reason a special number has been introduced, $i = \sqrt{-1}$, which makes it possible to solve equations of this type.

The collection of real numbers has been extended with a so-called imaginary part. Therefore, we can introduce the concept of *complex numbers* by definition. A complex number is defined as a couple of two real numbers with the following properties:

1 $(a, b) = (c, d)$ only then if $a = c$ and $b = d$
2 $(a, b) + (c, d) = (a + c, b + d)$, which means a new complex number
3 $(a, b) \times (c, d) = (ac - bd, ad + bc)$, which means a new complex number.

Usually, an arbitrary complex number is written as

$$Z = a + jb \tag{3.2}$$

where a and b are real numbers and $j = i = \sqrt{-1}$. We also find that $j^2 = i^2 = -1$. Note that in mathematics the symbol i is used and in electronics the symbol j is used. Further, we call a the real part and b the imaginary part of the complex number Z. This is also written as $\text{Re}(Z) = a$ and $\text{Im}(Z) = b$.

$$*\qquad*\qquad*$$

Example 3.1
Assume $Z_1 = 1 + 2j$ and $Z_2 = 3 + 4j$; then the sum is

$$Z_1 + Z_2 = (1 + 3, 2 + 4) = 4 + 6j.$$

Multiplying these numbers gives

$$Z_1 \times Z_2 = (1 \times 3 - 2 \times 4, 1 \times 4 + 2 \times 3) = -5 + 10j$$

For complex numbers ordinary division is not defined, and we find for Z_1/Z_2 just a new complex number as follows:

$$\frac{Z_1}{Z_2} = \frac{1 + 2j}{3 + 4j} = \frac{1 + 2j}{3 + 4j} \times \frac{3 - 4j}{3 - 4j} = \frac{11 + 2j}{9 + 16} = \frac{11}{25} + \frac{2}{25}j.$$

$$*\qquad*\qquad*$$

We can image any complex number in a complex plane with the help of an orthogonal system of two axes. The x axis is called the real axis and the y axis is called the imaginary axis. This is illustrated in figure 3.1. Any point in the complex plane represents a complex number. From this figure we draw the following conclusions.

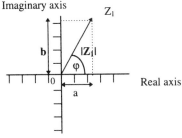

Figure 3.1 Representation of complex numbers in a complex plane.

The complex number $Z_1 = a + jb$ represents a point in the complex plane, where a is the distance on the real axis and b is the distance on the imaginary axis. The distance from to the origin to Z_1 represents the modulus of Z_1 given by

$$|Z_1| = (a^2 + b^2)^{1/2}. \tag{3.3}$$

This representation introduces a straightforward second possibility describing complex numbers. From figure 3.1 we see also that

$$a = |Z_1| \cos \varphi \qquad b = |Z_1| \sin \varphi. \tag{3.4}$$

Hence Z_1 can also be written as

$$Z_1 = a + jb = |Z_1|(\cos \varphi + j \sin \varphi) \tag{3.5}$$

where

$$|Z_1| = (a^2 + b^2)^{1/2} = \{[\text{Re}(Z_1)]^2 + [\text{Im}(Z_1)]^2\}^{1/2}.$$

It will be obvious that an arbitrary complex number $Z = a + jb$ can be understood as a complex variable. The angle φ is called the argument of Z_1 and can be denoted by

$$\arg(Z_1) = \varphi.$$

We can also write

$$\tan \varphi = \frac{b}{a} = \frac{\text{Im}(Z)}{\text{Re}(Z)} \tag{3.6}$$

and

$$\varphi = \tan^{-1} \frac{b}{a} = \tan^{-1} \frac{\text{Im}(Z)}{\text{Re}(Z)}.$$

*   *   *

Example 3.2
Assume $Z = 3 + 4j$; then

$$|Z| = (3^2 + 4^2)^{1/2} = 5$$

and

$$\varphi = \tan^{-1} \frac{4}{3} = \tan^{-1} 1.33.$$

*   *   *

There is still another way to present a complex number. This follows from the consideration that

$$e^x = 1 + \frac{x}{1!} + \frac{x^2}{2!} + \frac{x^3}{3!} + \dots \tag{3.7}$$

If we substitute $x = j\varphi$ then we find

$$e^{j\varphi} = \left(1 - \frac{\varphi^2}{2!} + \frac{\varphi^4}{4!} - \frac{\varphi^6}{6!} + \ldots\right) + j\left(\frac{\varphi}{1!} - \frac{\varphi^3}{3!} + \frac{\varphi^5}{5!} - \ldots\right). \tag{3.8}$$

This expression consists of two series; one represents the cosine and the second represents the sine function, and hence we just write

$$e^{j\varphi} = \cos\varphi + j\sin\varphi. \tag{3.9}$$

Substituting from (3.4) in (3.9) we find

$$e^{j\varphi} = \frac{a}{|Z|} + j\frac{b}{|Z|} \quad \text{or} \quad |Z|\,e^{j\varphi} = a + jb. \tag{3.10}$$

This shows that three different representations exist for a complex number or complex variable:

$$Z = |Z|\,e^{j\varphi} = |Z|(\cos\varphi + j\sin\varphi) = a + jb.$$

The representation with cos and sin is called the polar form. We will give an example.

*   *   *

Example 3.3
Assume $Z = \sqrt{3} + j$; then $|Z| = 2$ and $\tan\varphi = 1/\sqrt{3}$ which gives for $\varphi = \tan^{-1}(1/\sqrt{3}) = \pi/6$, and hence

$$Z = 2\{\cos(\pi/6) + j\sin(\pi/6)\} = 2e^{j\pi/6}.$$

*   *   *

Now assume that we have a sinusoidal voltage with amplitude $|U|$, frequency ω and phase shift φ; then this voltage can be described by

$$u(t) = |U|\,\cos(\omega t + \varphi). \tag{3.11}$$

We know already that $e^{j\varphi} = \cos\varphi + j\sin\varphi$, and so $u(t)$ can be written in the following form:

$$u(t) = \text{Re}[|U|\,e^{j(\omega t + \varphi)}]. \tag{3.12}$$

Hence we say that the sinusoidal time-dependent voltage can be written as a complex variable or, more generally speaking, we say that a complex representation of the voltage is given by the expression

$$\tilde{u} = |U|\,e^{j(\omega t + \varphi)} = |U|\,e^{j\omega}\,e^{j\omega t}. \tag{3.13}$$

This expression (3.13) can be interpreted as a left-hand-turning phasor with angular frequency ω. Hence, taking the real part of (3.13) gives again (3.11):

$$\tilde{u}(t) = \text{Re}[\tilde{u}(t)].$$

If we divide (3.13) by $e^{j\omega t}$, what remains is just an ordinary complex number

$$U = |U|\,e^{j\omega} = |U|(\cos\varphi + j\sin\varphi)$$

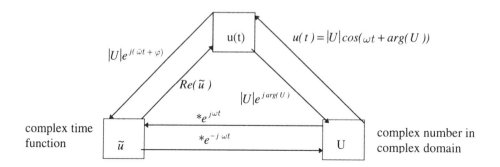

Figure 3.2 Different representations in three different domains of a time-dependent signal.

which is no longer a time-dependent function. We can put this in a little diagram as illustrated in figure 3.2. The complex number U has been assigned to the time-dependent function $u(t)$ when $t=0$. In this description no rotation is involved and a stationary state description is what remains.

Now assume that

$$\tilde{\imath}=|I|\ e^{j(\omega t+\varphi)}. \tag{3.14}$$

Then

$$\frac{d\tilde{\imath}}{dt}=|I|j\omega\ e^{j(\omega t+\varphi)}=j\omega\tilde{\imath}. \tag{3.15}$$

From (3.15) we conclude that the operator d/dt transfers into the complex domain in $j\omega$. In the same way we can see that d^2/dt^2 transfers into $-\omega^2$. It is important to note that complex calculus holds for sinusoidal functions only.

*   *   *

Example 3.4
Assume that the complex current is described with $I=8+3j$ A; then the representation in the time domain is $i(t)=\sqrt{73}\ \cos(\omega t+\tan^{-1} 3/8)$ A.

*   *   *

To make calculations more convenient the following identities are sometimes used:

$$\cos\varphi=\frac{1}{2}\ (e^{j\varphi}+e^{-j\varphi})$$
$$\sin\varphi=\frac{1}{2j}\ (e^{j\varphi}-e^{-j\varphi}). \tag{3.16}$$

In the following section we introduce a third method of describing functions.

3.2.2 Laplace transform of signals

As we have seen already, the complex representation of time-dependent signals is only appropriate for sinusoidal functions. The Laplace transform offers more flexibility and provides a tool for the description of an arbitrary signal. Additionally, it provides a tool for the solution of linear differential equations using a transformation where the differential equation is converted into an ordinary algebraic expression. The Laplace transform by definition is (see also (2.14))

$$L[f(t)] = F(p) = \int_0^\infty f(t)\,e^{-sp}\,dt. \tag{3.17}$$

Note that for the symbol s the symbol p is also used. This operation transfers a time-dependent function into the so-called p or s domain. We will use the symbol p; in general p may be real or complex. The Laplace transform exists if the integral (3.17) converges for some value of p. Some examples of Laplace transforms are given in table 3.1.

Table 3.1 Sample of Laplace-transformed time functions.

$f(t)$	$F(p)$	
1	$1/p$	$p > 0$
t	$1/p^2$	$p > 0$
$\delta(t)$	1	$p > 0$
e^{at}	$\dfrac{1}{p-a}$	$p > 0$
$\dfrac{t^{n-1} e^{a}}{(n-1)!}$	$\dfrac{1}{(p-a)^n}$	$p > 0$
$\dfrac{\sin at}{a}$	$\dfrac{1}{p^2 + a^2}$	$p > 0$
$\cos at$	$\dfrac{p}{p^2 + a^2}$	$p > 0$
$\dfrac{t^{n-1}}{(n-1)!}$	$\dfrac{1}{p^n}$	$p > 0$
$\dfrac{e^{bt} \sin at}{a}$	$\dfrac{1}{(p-b)^2 + a^2}$	$p > 0$
$e^{bt} \cos at$	$\dfrac{p-b}{(p-b)^2 + a^2}$	$p > 0$
$\dfrac{t \sin at}{2a}$	$\dfrac{p}{(p^2 + a^2)^2}$	$p > 0$
$t \cos at$	$\dfrac{p^2 - a^2}{(p^2 + a^2)^2}$	$p > 0$

For this single-sided Laplace transform ranging from 0 to $+\infty$ some properties exist. We will present a few of them without giving any evidence.

$$L[a_1 f_1(t) + a_2 f_2(t)] = a_1 F_1(p) + a_2 F_2(p). \tag{3.18}$$

This is called the linearity property

$$L[e^{at}f(t)] = F(p-a). \tag{3.19}$$

This is called the first shifting property.

$$L[f(t-a)] = e^{-ap}F(p) \tag{3.20}$$

for

$$g(t) = f(t-a) \quad t > a$$
$$g(t) = 0 \quad\quad\quad t < a.$$

This is the second shifting property. If $L[f(t)] = F(p)$ then the Laplace transform of a derivative is

$$L\{f'(t)\} = pF(p) - f(0). \tag{3.21}$$

*   *   *

Example 3.5
If $f_1(t) = 3e^{-2t}$, $f_2(t) = 3 \cos 5t$ and $f_3(t) = (t^2+1)^2$ find the Laplace transform for each of these.

$$F_1(p) = 3/(s+2), \quad F_2(p) = 3p/(p^2+25) \quad \text{and} \quad F_3(p) = (p^4 + 4p^2 + 24)/p^3.$$

*   *   *

To transform backwards from the p domain into the time domain may require the partial-fraction method. In all cases this involves finding standard functions. We will give an example.

*   *   *

Example 3.6

$$F(p) = \frac{5s^2 - 15s - 11}{(s+1)(s-2)^3} = \frac{A}{s+1} + \frac{B}{(s-2)^3} + \frac{C}{(s-2)^2} + \frac{D}{s-2}.$$

Backward multiplication of the right-hand side of the equation to obtain the same denominator and equating the equivalent terms gives $A = -1/3$; $B = -7$, $C = 4$ and $D = 1/3$.

Then using the standard inverse transformations from table 3.1 we find for $f(t)$

$$L^{-1}[F(p)] = f(t) = -\frac{1}{3}e^{-t} - \frac{7}{2}t^2 e^{2t} + 4t e^{2t} + \frac{1}{3}e^{2t}.$$

*   *   *

3.2.3 Description of linear systems
In section 2.3 we have seen several possibilities for determining the characteristic behaviour of a system. If we want to describe this behaviour several tools are available. To start with, we assume that the systems to be considered are linear and

therefore fulfil the requirements as expressed in equations (2.6) and the conditions (a) to (c) following that equation. For convenience, the equation is repeated here.

$$a_0 S_i + a_1 \frac{dS_i}{dt} + a_2 \frac{d^2 S_i}{dt^2} + \ldots = b_0 S_o + b_1 \frac{dS_o}{dt} + b_2 \frac{d^2 S_o}{dt^2} + \ldots \quad (3.22)$$

If we assume that the input signal $S_i(t)$ is equal to $\exp(pt)$ where p is a complex quantity, defined as

$$p = \alpha + j\omega \quad (3.23)$$

then the differential equation (3.22) changes into an ordinary algebraic expression, because the derivative of an exponential expression is again the expression itself, apart from a possible constant. Note that if $\alpha < 0$ then $\exp(pt)$ represents a decreasing periodic signal, if $\alpha > 0$ then $\exp(pt)$ represents an increasing periodic signal and if $\alpha = 0$ the signal is purely periodic.

When we want to consider the ratio of the output quantity $S_o(p)$ and the input quantity $S_i(p)$, this ratio represents an analytic function in the complex frequency domain of p. This ratio, usually denoted by $A(p)$ or $H(p)$, is called the *transfer function* of the system. This ratio can be written

$$A(p) = H(p) = \frac{S_o(p)}{S_i(p)}. \quad (3.24)$$

Depending on which quantities are involved, this represents an impedance, an admittance or a dimensionless transfer function.

For a given circuit the desired expression in p is obtained when the appropriate algebraic equations are applied for every node and/or mesh. The substitution for any impedance of an inductance pL and for any capacitance $1/pC$ offers the desired expression in p. It is often more convenient to write the equations straightforwardly in p than to write the appropriate differential equations first. We observe that the term $j\omega$ corresponds to the term p in the p domain and that the p domain can be interpreted as an extension of the complex time domain. Further, it can be shown that the transfer function $A(p)$ can always be represented as the ratio of two polynomials, i.e. with the numerator and the denominator polynomial. If this is done, $A(p)$ can be written as

$$A(p) = \frac{a_n(p - z_1)(p - z_2) \ldots (p - z_n)}{b_m(p - p_1)(p - p_2) \ldots (p - p_m)}. \quad (3.25)$$

In the following example a description with the help of the Laplace transform is illustrated.

*   *   *

Example 3.7
An inductor of 2 H with a resistance of 16 Ω is series connected with a capacitor of 20 mF and a voltage source $U = 300$ V. $V_c(0) = 0$. Calculate the charge and current for $t > 0$.

Applying Kirchhoff's law we find for this series connection

$$2\frac{di}{dt} + 16i + \frac{q}{20 \times 10^{-3}} = 300.$$

Since $i = dq/dt$ it is more convenient to find a linear differential equation of second order:

$$2\frac{d^2q}{dt^2} + 16\frac{dq}{dt} + \frac{q}{20 \times 10^{-3}} = 300.$$

Taking the Laplace transform we find

$$2[p^2Q(p) - pq'(0) - q(0)] + 16[pQ(p) - q(0)] + \frac{Q(p)}{20 \times 10^{-3}} = \frac{300}{p}.$$

We know already that $q'(0) = q(0) = 0$; then applying the method of partial fractions we find

$$Q(p) = \frac{150}{p(p^2 + 8p + 25)} = \frac{6}{p} - \frac{6(p+4)}{(p+4)^2 + 9} - \frac{24}{(p+4)^2 + 9}.$$

Inverse transformation gives

$$q(t) = L^{-1}[Q(p)] = q(t) = (6 - 6\,e^{-4t}\cos 3t - 8\,e^{-4t\sin 3t})\varepsilon(t)$$

and for the current $i(t)$ we find

$$i(t) = \frac{dq(t)}{dt} = 50\,e^{-4t}\sin 3t\,\varepsilon(t).$$

Note that in $q(t)$ and $i(t)$ the factor $\varepsilon(t)$ indicates that both functions exist for $t > 0$.

$$* \quad * \quad *$$

3.2.3.1   *The polar and Bode plots*

Usually, we want to have a plot for the magnitude of $A(p)$ versus frequency and a phase plot versus frequency; then expression (3.25) becomes

$$|A(p)| = \frac{a_n|(p - z_1)\|(p - z_2)| \ldots |(p - z_n)|}{b_m|(p - p_1)\|(p - p_2)| \ldots |(p - p_m)|} \tag{3.26}$$

and a phase plot can be derived from

$$\arg[A(p)] = (\phi_1 + \phi_2 + \ldots + \phi_n) - (\theta_1 + \theta_2 + \ldots + \theta_m) \tag{3.27}$$

where

$$\phi_1 = \arg(p - z_1) \quad \phi_2 = \arg(p - z_2) \quad \text{etc} \quad \text{and} \quad \theta_1 = \arg(p - p_1) \quad \theta_2 = \arg(p - p_2) \text{ etc.}$$

For simplicity, we further assume that the complex frequency equals $p=j\omega$. Usually, a logarithmic decibel scale versus frequency is used in order to be able to depict several decades of frequency at once. Then it is easily seen from (3.26) that the following expression can be derived:

$$|A(j\omega)|_{dB} = 20 \log|A(j\omega)| = \sum_{k=1}^{n} 20 \log|j\omega - z_k| - \sum_{k=1}^{m} 20 \log|j\omega - p_k| + C \qquad (3.28)$$

where $C = 20 \log(a_n/b_m)$.

Of special interest are the cases where $j\omega$ varies if ω varies in the range $-\infty < \omega < \infty$. This implies that in the complex p domain the imaginary axis is fully covered. The plot of the transfer function in the p domain then describes the behaviour versus frequency of the system, for pure periodic *sinusoidal* signals. This graph is called a *polar plot* of $A(p)$ and completely describes the system's properties. Of course, in practical circumstances it makes sense only if $0 < \omega < \infty$. In the following example we shall demonstrate how a polar plot can be constructed.

*   *   *

Example 3. 8
Consider the simple circuit depicted in figure 3.3; then the transfer function $A(p)$ between $V_o(p)$ and $V_i(p)$ is

$$A(p) = \frac{V_o(p)}{V_i(p)} = \frac{1/pC}{R+1/pC} = \frac{1}{1+pRC} = \frac{1}{1+p\tau} \qquad (3.29)$$

where $\tau = RC$ is the so-called time constant (units in seconds). If we substitute for $p = j\omega$ and thus assume $\alpha = 0$ in (3. 23), then (3.29) becomes

$$A(j\omega) = \frac{1}{1+j\omega\tau}. \qquad (3.30)$$

In complex function theory it is shown that if ω varies from $-\infty$ to $+\infty$ and thus $p = j\omega$ follows a straight line in the complex frequency domain, then the conformal plot of a function of the type of $A(p)$ proves to be a (semi)circle. The remaining information required to draw this plot is three characteristic points. The following characteristic frequencies provide these three points: $\omega = 0$, $\omega = 1/\tau$ and $\omega = \infty$.

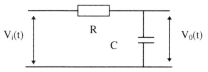

Figure 3.3 An *RC* circuit to illustrate the use of the polar and Bode plots.

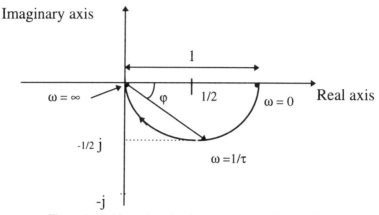

Figure 3.4   The polar plot for the circuit of figure 3.3.

Again from a physical point of view choosing $\omega = -\infty$ makes no sense at all. If $\omega = 0$ we see immediately that $A(j\omega) = 1$, if $\omega = 1/\tau$ then $A(j\omega)$ is given by

$$A(j\omega) = \frac{1}{1+j} = \frac{1}{2} - \frac{1}{2}j \qquad (3.31)$$

and if $\omega \to \infty$ then $A(j\omega)$ approaches zero via the negative imaginary axis. The result is drawn in figure 3.4.

*   *   *

We can make the following remarks concerning this result for the polar plot.

1   From equation (3.31) we can derive immediately that the plot is in the right half-plane below the *x*-axis.
2   For $\omega = 0$ (thus DC voltage) the transfer function is simply 1, which can be seen immediately by inspection of the circuit.
3   For $\omega \to \infty$ the circuit behaves as a short circuit, which will be the case for every circuit.
4   The frequency $\omega = 1/\tau$ is called the 3 dB frequency or the 0.707 frequency or the $1/\sqrt{2}$ frequency, which are all the same. This will become clear when we draw the Bode plot. See below.
5   The polar plot completely describes the behaviour of the circuit and shows the amplitude and phase of the transfer function simultaneously. Note that the phase is defined as the shift in radians between input and output signals versus frequency.
6   Polar plots are suitable for describing the response to sinusoidal input signals only.

*   *   *

From the polar plot, the magnitude of the amplitude $|A(p)|$, and phase $\arg[A(p)]$, versus frequency can be derived. This can be done by graphical methods, the amplitude and phase shift being measured at each frequency. This is illustrated for one frequency in figure 3.4. The combination of magnitude and phase plots versus ω is called a *Bode plot*.

We shall now use the same example to demonstrate the procedure for drawing a Bode plot. The Bode plot consists of two figures, and hence we have to derive the amplitude and the phase of $A(j\omega)$ versus frequency ω. Again in most cases it suffices to consider the same characteristic frequencies as in the previous example for the polar plot.

$$* \quad * \quad *$$

Example 3.9

In order to find the Bode plot we substitute the same frequency in (3.30) and determine the amplitude and phase of $A(j\omega)$. For $\omega = 0$ we find $|A(j\omega)| = 1$, for $\omega \to \infty$ we find $|A(j\omega)| \to -j/\infty$ and for $\omega = 1/\tau$ we find

$$|A(j\omega)| = \left|\frac{1}{1-j}\right| = \frac{|1|}{|1-j|} = \frac{1}{\sqrt{2}}.$$

Generally, in order to find the phase ϕ of $A(j\omega)$ it is convenient to describe (3.30) in the form $A + jB$ first. This yields

$$A(j\omega) = \frac{1-j\omega\tau}{1+\omega^2\tau^2}. \tag{3.32}$$

Then by definition $\phi = \arg[A(j\omega)]$ is given by

$$\arg[A(j\omega)] = \phi = \tan^{-1}\left\{\frac{\text{Im}[A(j\omega)]}{\text{Re}[A(j\omega)]}\right\} \tag{3.33}$$

or, in this example,

$$\phi(\omega) = \tan^{-1}\left(\frac{-\omega\tau}{1}\right).$$

Substituting the three characteristic frequencies we find for $\omega = 0$ that the phase shift equals 0 rad, for $\omega = 1/\tau$ the phase shift $\phi = -\pi/4$ rad and for $\omega \to \infty$ the shift is $-\pi/2$ rad. The results of both diagrams are plotted in figure 3.5.

$$* \quad * \quad *$$

We make the following remarks concerning the Bode plot.

1 The Bode plot provides information concerning the *amplitude* and the *phase shift* between input and output separately.

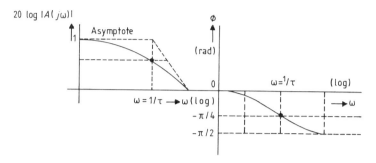

Figure 3.5   Bode plot for the circuit of figure 3.3.

2   In order to be able to cover a number of decades, especially in higher-order systems, the amplitude of $A(j\omega)$ and the frequency ω are usually depicted on logarithmic scales. Assume that this is done for $\omega = 1/\tau$ in the above example 3.9; then we find by applying the usually decibel notation that

$$20 \log|A(j\omega)||_{\omega=1/\tau} = 20 \log|1/\sqrt{2}| = -20 \times 1/2 \times 0.3010 = -3 \text{ dB}.$$

This is the already-mentioned dB frequency value and means that at that frequency the amplitude has become 3 dB less than the original amplitude with respect to the maximum value.

3   Often the Bode plot is indicated with its *asymptotes* only.
4   The minimum phase shift is 0 rad and the maximum phase shift is $-\pi/2$ rad. This is characteristic for a circuit with one time constant only, and this is usually called a first-order circuit.
5   The slope with which the amplitude of $A(j\omega)$ here decreases is -6 dB/octave.
6   This circuit is also known as a first-order *low-pass filter*, because frequencies higher than $f = 1/2\pi\tau$ are attenuated at 6 dB/octave.
7   Bode plots are only applicable for sinusoidal functions.

*   *   *

The results for both the polar plot and the Bode plot are summarized in table 3.2.
    Finally, one other type of plot is met, known as the Nichols chart, which represents the amplitude versus phase shift. This will not be discussed here in further detail.

Table 3.2   Summary of the results of polar and Bode plots for examples 3.8 and 3.9.

ω (rad s^{-1})	$A(j\omega)$	$|A(j\omega)|$	$20 \log|A(j\omega)|$	$\arg[A(j\omega)]$
$\omega = 0$	1	1	0	0
$\omega = 1/\tau$	$\frac{1}{2}(1-j)$	$1/\sqrt{2}$	-3	$-\pi/4$
$\omega \rightarrow \infty$	0	0	$-\infty$	$-\pi/2$

3.2.3.2 *Pole–zero plots*

A third and much more sophisticated method that can be used to describe the characteristics of a linear system is known as the pole–zero plot. Consider again expression (3.25) which we already know describes the behaviour of a linear system. It can be shown that, apart from a certain constant, the characteristic behaviour of the transfer function can be described with a limited number of points, usually called poles and zeros. It appears there exists a definite relationship between the position of the poles and zeros in the p domain and the qualitative and quantitative properties of the system. More precisely, the values of the independent variable p for which the numerator equals zero are called the zeros. The values of p for which the denominator equals zero are called the poles.

If the complex frequency domain is considered, it can be said that the zeros, $z_1, z_2, \ldots, z_n$ are the complex frequencies for which the transfer function $A(p)$ equals zero. The poles $p_1, p_2, \ldots, p_m$ are the complex frequencies for which the transfer function becomes infinite. This is easily verified by substitution in equation (3.25).

These poles and zeros in the complex domain can be plotted providing a series of points. Usually, the zeros are indicated with an 'o' and the poles with an 'x'. With this type of plot, the transfer function is also described completely apart from the constant a_n/b_m in equation (3.25). As will be seen, a pole–zero plot is more powerful than the polar or Bode plot. Note that the described method holds for linear systems only. In an abbreviated form we first present some properties of the pole–zero plots. Further, we assume that the systems are physically realizable.

- Owing to the real coefficients $a_0, \ldots, a_n$ and $b_1, \ldots, b_m$, we always have to deal with real transfer functions for real values of p. This is only possible for real values of poles and zeros, or those occurring as complex conjugate pairs. The function is then called a *conjugated analytical* function.
- Splitting the expression for the transfer function (3.25) into partial fractions, we obtain the following expression:

$$A(p) = \frac{V_o}{V_i} = \frac{a_n}{b_m}\left(\frac{A_1}{p-p_1} + \frac{A_2}{p-p_2} + \ldots + \frac{A_m}{p-p_m}\right) \tag{3.34}$$

or

$$V_o = V_i \frac{a_n}{b_m}\left(\frac{A_1}{p-p_1} + \frac{A_2}{p-p_2} + \ldots + \frac{A_m}{p-p_m}\right). \tag{3.35}$$

From this expression it appears that V_o can be interpreted as the superposition of a number of voltages of the form

$$V_o(k) = V_i \frac{A_k}{p-p_k} \tag{3.36}$$

or

$$V_o(p) = \sum_{k=1}^{m} V_i \frac{A_k}{p-p_k}. \tag{3.37}$$

Because p can be interpreted as the *Laplace operator* for d/dt, (3.36) corresponds to the differential equation in the time domain:

$$\frac{dV_o(k)}{dt} - p_k V_o(k) = A_k V_i. \tag{3.38}$$

If $V_i = 0$ the solution yields

$$V_o(k) = C \exp(p_k t) \tag{3.39}$$

where C is a constant.

Now note that, if p_k has a positive real part, $V_o(k)$ will increase in time, and the system is unstable and behaves like an oscillator. From this it may be concluded that the system is only stable when the poles of the transfer function have a real but *negative* part. The consequence is that for a stable system all poles have to be found in the left half-plane of the p domain. This condition is not required for the zeros. In principle, a single pole may be found on the imaginary axis.

- For a practical circuit, the number of poles is always greater than the number of zeros, because for $p = j\omega$ and $\omega \to \infty$ the modulus of the transfer function $|A(p)|$ $< \infty$. For $m = n$ the transfer function yields the constant $|A(p)| = a_n/b_m$. For every practical circuit, due to inherently present parasitic capacities, the transfer function reaches zero when $\omega \to \infty$.
- For passive two-ports it can be shown that the difference between the number of poles and zeros never exceeds unity. Note that, in passive two-ports, no active components such as transistors are involved. They consist of R, L and C components only. In these networks all the poles as well as the zeros have to be found in the left half-plane.
- For a series connection of unilateral circuits the transfer function equals the sum of the pole–zero plots of the independent systems. If a pole coincides with a zero they annihilate each other.

$$A = A_1 A_2 A_3 \dots . \tag{3.40}$$

The transfer function of a parallel connection of circuits contains the poles of the separate circuits but in general not the zeros of the separate circuits. We will illustrate this with a simple example.

*   *   *

Example 3.10

Let us consider again the circuit of figure 3.3. From (3.29) it can be derived that only one pole is involved. More specifically, we find $p_1 = -1/\tau$. This can be plotted

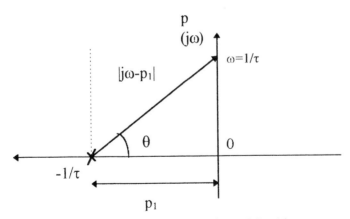

Figure 3.6 The pole–zero plot for figure 3.3, with $\tau = \tau_1$.

in the pole–zero plane as shown in figure 3.6. From this figure the following conclusions can be drawn:

1   The system is described with one point.
2   When the frequency ω varies from 0 to $+\infty$, p shifts in the positive direction of the imaginary y axis.
3   The angle θ indicates the phase shift between the input and output. Generally, the contribution of a pole is negative and that of a zero is positive. It is easily seen that, in this example, for $\omega = 0$ the phase shift is zero, and for $\omega \to \infty$ the phase shift is $-\pi/2$.
4   In this example the distance between a point on the $j\omega$ axis and the pole represents the amplitude of the transfer function at the indicated frequency. If more poles and zeros are involved, then the result is the ratio of the products of the corresponding distances for every zero and every pole.

*   *   *

This pole–zero method is also applicable to signals other than pure sinusoidal ones. If for instance a step response must be found, then it suffices to multiply the transfer function $A(p)$ by $1/p$, because the Laplace transform of the step function is $1/p$.

To complete this review let us consider the case where *feedback* is involved. This is illustrated in figure 3.7. The principle of feedback can physically be defined as a

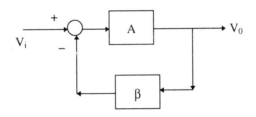

Figure 3.7 Block diagram of a feedback system.

fraction of the output (signal) that is coupled back into the input. We can distinguish between a positive feedback and a negative feedback.

As will be discussed in the next chapter, negative feedback can significantly improve the behaviour of an amplifier. To determine the pole–zero plot of such a system we assume that the poles and zeros of the separate systems are known. The transfer function of a system A_t, implying feedback, is

$$A_t = \frac{A}{1 - \beta A} \tag{3.41}$$

where

$$A(p) = a \frac{(p - z_1)(p - z_2) \ldots (p - z_n)}{(p - p_1)(p - p_2) \ldots (p - p_m)} = a \frac{N_1}{D_1} \tag{3.42}$$

and

$$\beta = b \frac{(p - z_1')(p - z_2') \ldots (p - z_n')}{(p - p_1')(p - p_2') \ldots (p - p_m')} = b \frac{N_2}{D_2}. \tag{3.43}$$

Substitution of (3.42) and (3.43) in (3.41) yields

$$A_t = \frac{a(N_1/D_1)}{1 - ab(N_1/D_1)(N_2/D_2)} \tag{3.44}$$

or

$$A_t = \frac{aN_1 D_2}{D_1 D_2 - abN_1 N_2}. \tag{3.45}$$

As can be seen from (3.45) the zeros of A_t are determined by the zeros of A and the poles of β. The poles of A_t can be derived from the equation $A\beta = 1$, where A_t becomes infinite, or

$$D_1 D_2 - abN_1 N_2 = 0 \tag{3.46}$$

or

$$\frac{N_1 N_2}{D_1 D_2} = \frac{1}{ab}. \tag{3.47}$$

The poles of A_t are determined by the equation (3.47). As stated before, the constants a and b are real so we have to look for values where equation (3.47) proves to have real values. For these values a graphical representation exists called *root loci*, showing the changes in pole position as ab varies. Concerning these root loci, without any proof, we mention some properties.

1 Root loci start in the poles of A and β.
2 Root loci stop in the zeros of A and β.
3 A mirror symmetry exists with respect to the x axis.
4 Generally a fraction of the real axis forms a part of the root loci.

5 For g poles and h zeros existing within the finite region, $g - h$ zeros are located in the infinite region.

Unfortunately it is not possible to measure poles and zeros. To determine whether a feedback system is stable, the Nyquist criterion is used. The Nyquist criterion is stated as follows: if the polar plot of $A\beta$ does not cover the point $+1$, the system is a stable one. This is equivalent to the condition that all poles of the stable feedback system are located inside the left half-plane.

3.3 ANALOGUE FUNCTIONS

Despite an increasing trend to digitize every electronic function where possible we are still confronted with many analogue functions in electronic measuring systems. As will be discussed in more detail in chapter 6, this is mainly due to the analogue nature of our environment. Think of a temperature, a pressure, a force, a voltage, radiation, magnetism, chemical reactions. They are all purely analogue in their nature.

When we convert these physical quantities into an electrical signal by an appropriate transducer, usually the signal level is very weak (microvolts). Before a digitization can be performed the signal has to be amplified to a suitable level. To amplify a very weak signal, often special precautions must be taken, concerning such aspects as offset and drift, noise levels and interference. All these aspects will be discussed in detail in later chapters.

However, other reasons also exist for the application of analogue circuitry. In some applications it is advantageous to apply analogue circuits, for instance in telecommunications and very-high-frequency (gigahertz) applications. Sometimes, in performing calculations, analogue circuitry is faster, more accurate, simpler and cheaper than digital circuitry. Other interesting applications are found in battery-operated equipment, such as pacemakers, blood flowmeters, auditory stimulators, hearing instruments, portable radios and so on. Low-voltage and low-power electronics are included here, which means operation from just one battery.

In general, we distinguish four different types of amplifier:

1 a voltage amplifier ($\Delta V_{out}/\Delta V_{in}$)
2 a current amplifier ($\Delta I_{out}/\Delta I_{in}$)
3 a transconductance amplifier ($\Delta I_{out}/\Delta V_{in}$)
4 a transimpedance amplifier ($\Delta V_{out}/\Delta I_{in}$).

Because of its general applicability one of the most popular building blocks in processing analogue signals is the operational amplifier. We shall present a collection of important functions realized by operational amplifiers.

3.3.1 The operational amplifier
An operational amplifier is a basic building block usually with a differential input and a single-ended output. By definition the amplifier has an infinitely high open-loop gain, zero output impedance, infinitely high input impedance, zero input current

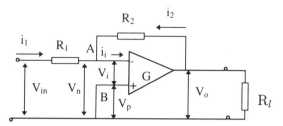

Figure 3.8 Operational amplifier with feedback loop.

and perfect symmetry in impedance for both inputs. The infinite gain requires feedback and as a consequence the input voltage $V_i = V_o/G$ will become zero. The operational amplifier with the usually applied notations is depicted in figure 3.8.

The theoretical case where $V_i = 0$ and $i_i = 0$ is called a *nullor*. It will be understood
that in practice this is not true, but it will be assumed to introduce a collection of
basic functions more easily. Practically the gain, G, for instance, can reach values of
10^5–10^9. To understand the function of the operational amplifier in its many applications, we will first look at its behaviour when used with a simple feedback loop as
is illustrated in figure 3.8. Compare this with figure 3.7 and regard the equivalence
of the two circuits. The input marked with a minus sign $(-)$ is called the *inverting*
input, as any signal applied to it will appear with opposite sign at the output. The
other input is called the *non-inverting* input.

As a consequence of the above-given definition the actual input voltage V_i between
the two amplifier inputs is zero, or in other words

$$V_p - V_n = 0 \tag{3.48}$$

where V_p is the non-inverting input voltage at node B with respect to common and
V_n is the inverting input voltage at node A with respect to the same common. Now
consider node A and apply Kirchhoff's first law to it. If an ideal amplifier is assumed,
the current equation can be written

$$\frac{V_{in} - V_n}{R_1} + \frac{V_o - V_n}{R_2} = 0 \tag{3.49}$$

or

$$V_o = \frac{R_1 + R_2}{R_1} V_n - \frac{R_2}{R_1} V_{in} \tag{3.50}$$

and as can be seen in figure 3.8 $V_p \equiv 0$; thus from (3.48) we see that $V_n = 0$ also. It
is said that V_n is virtually earthed, which does not imply that it is ground or earth,
but that it has the voltage level of ground. Then applying equation (3.48) we can
substitute $V_n = 0$ in (3.50), and this equation reduces to

$$V_o = -\frac{R_2}{R_1} V_{in}. \tag{3.51}$$

In practical circumstances for $V_o = 10$ V and $G = 10^6$, V_i equals $10\ \mu$V. Again, with the same assumptions as above it can be seen that the input impedance of the amplifier itself (node A) has become $R_1 \| R_2 / (1 - A\beta)$ where $G = A$ and with the load R_l we find $R_1 \| R_2 \| R_l$. To verify this, firstly note that the output impedance of the amplifier is zero and secondly the ideal voltage source's impedance equals zero. If the input impedance of the circuit is considered, or in other words the load impedance for the source V_{in}, this yields

$$R_{in} = R_1 = \frac{V_{in}}{i_1}. \tag{3.52}$$

It should be kept in mind that for minimum *offset error* the condition of perfect symmetry of the impedance at the input nodes A and B should be fulfilled. Offset is defined as the voltage to be applied at the input to make the output voltage zero when no signal voltage is present at the input. This offset voltage is related to the structure of the amplifier itself and will be discussed in more detail in chapter 7. Hence, we can distinguish the (a)symmetry of the circuitry at the amplifier's input and the (a)symmetry of the amplifier itself.

In the following sections we will discuss some frequently occurring important applications of the operational amplifier. Many variations exist, and they cannot all be discussed in this book. They are easily found in textbooks dedicated to applications of the operational amplifier.

3.3.2 Preamplifier with adjustable gain

Figure 3.9 shows an amplifier configuration with adjustable gain. In this configuration, only a fraction, V_t, of the output voltage V_o is used for feedback. This fraction V_t is given by

$$V_t = V_o \frac{R_4}{R_3 + R_4}. \tag{3.53}$$

The output voltage becomes

$$V_o = -\frac{R_3 + R_4}{R_4} \frac{R_2}{R_1} V_i. \tag{3.54}$$

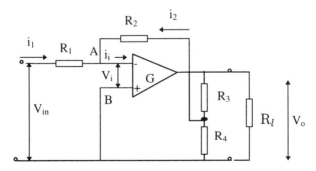

Figure 3.9 Operational amplifier with adjustable gain.

It can be concluded that attenuation of the output voltage before feedback is applied gives a higher gain. This has the special advantage that low-ohmic attenuators (some kiloohms) can be used, providing a reduction in noise and hum. A high R_2, combined with the always-present parasitic capacitors, will result in a large time constant, slowing down the speed performance of the amplifier.

Speed performance is often expressed by *slew rate*, which is defined as the maximum possible swing of the output voltage per unit of time, usually given in microvolts per second. (See also chapter 4.)

3.3.3 The non-inverting operational amplifier

The non-inverting amplifier is often used as a buffer amplifier to prevent any loading of a weak signal source. The configuration is depicted in figure 3.10. The voltage V_n at node A is given by

$$V_n = V_o \frac{R_1}{R_1 + R_2} \tag{3.55}$$

and with $V_p = V_n$ and $V_{in} = V_n$ we find

$$V_{in} = V_o \frac{R_1}{R_1 + R_2} \tag{3.56}$$

or

$$V_o = \left(1 + \frac{R_2}{R_1}\right) V_{in}. \tag{3.57}$$

The non-inverting input provides a high input impedance.

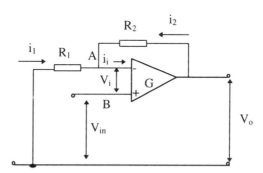

Figure 3.10   The non-inverting amplifier.

3.3.4 The differential amplifier configuration

As will be seen, the differential amplifier configuration is applied in cases where a difference voltage must be detected. Often the difference voltage of two signal sources is extremely small and superimposed on relatively high common-mode voltages. For instance, a bridge output voltage is between 5.1234 V and 5.1236 V which is only a

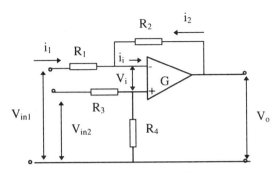

Figure 3.11 The differential amplifier configuration.

difference of $200\,\mu\text{V}$; then 5.1234 V is considered as the common-mode voltage. The configuration is able to discriminate between the common-mode signal and the difference signal and is shown in figure 3.11.

The general expression for this configuration is found by superposition of the expressions for the inverting and non-inverting amplifiers, or

$$V_o = \left(1 + \frac{R_2}{R_1}\right) \frac{R_4}{R_3 + R_4} V_{in2} - \frac{R_2}{R_1} V_{in1}. \tag{3.58}$$

Note that $R_4/(R_3 + R_4)$ is the fraction of V_{in2} that is available at node B. For minimum offset error the condition $R_1 \| R_2 = R_3 \| R_4$ must be fulfilled.

An exact differential expression for V_{in1} and V_{in2} is obtained by putting $R_1 = R_3$ and $R_2 = R_4$. The expression (3.58) becomes

$$V_o = \frac{R_2}{R_1} (V_{in2} - V_{in1}). \tag{3.59}$$

The above-described configuration is frequently applied for Wheatstone bridges. A drawback is the rather low input impedance, which is strongly improved in the so-called instrumentation amplifier. This configuration will be discussed in more detail in chapter 11.

3.3.5 The comparator

A comparator or zero detector is another application of an operational amplifier. It is frequently used to compare two signals surpassing a certain threshold value. This is illustrated in figure 3.12.

Because of the high gain of the amplifier a very small difference in input signal, of the order of $0.1\,\mu\text{V}$, will result in a large value of the amplifier's output voltage. Thus assume that if V_A is only some tenths of a microvolt lower than V_B the output will be fully positive, and in the opposite case, when V_A is slightly higher than V_B, the output will be fully negative. An arbitrary threshold voltage can be taken as a reference voltage.

When V_B is connected to earth the circuit compares V_A with zero, giving a sensitive zero detector which changes its output polarity when V_A passes zero. The error is

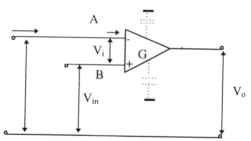

Figure 3.12   The comparator configuration.

only a small fraction of a microvolt. The comparators, especially the high-speed types, are very sensitive to oscillations near the (zero) crossing point. To prevent this, special precautions must be taken, for instance applying a capacitive decoupling of the supply voltages and both input terminals.

3.3.6 The integrator

With a capacitor in the feedback loop the operational amplifier can be used as an integrator. The principal circuit is shown in figure 3.13. As long as the capacitor C in the feedback loop is not fully charged it acts as a DC path and does not change the characteristics of an operational amplifier with a resistive feedback loop.

Applying again the current equation for the virtually earthed node A we find

$$\frac{V_{in}}{R_1} + \frac{dQ_c}{dt} = 0 \tag{3.60}$$

and with $Q_c = CV_o$ we find

$$\frac{V_{in}}{R_1} + \frac{d(CV_o)}{dt} = 0 \tag{3.61}$$

or

$$V_o = -\frac{1}{RC} \int V_{in} \, dt. \tag{3.62}$$

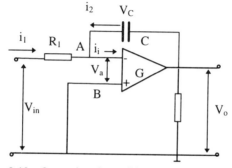

Figure 3.13   Operational amplifier used as an integrator.

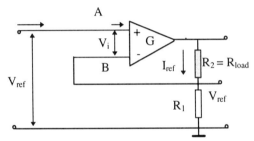

Figure 3.14 Operational amplifier used as a constant-current sink.

The moment the supply voltage across the capacitor is achieved the integration will stop. It can be seen from (3.62) that the output voltage is linear in t.

3.3.7 A constant-current source

With operational amplifiers excellent *current sources* and *sinks* can be realized. A simple example of a current sink is given in figure 3.14 where the resistance $R_2 = R_1$ is considered as the load resistance. The expression for the generated current is simply

$$I_o = I_{out} = \frac{V_{ref}}{R_1}. \tag{3.63}$$

Very low currents can be generated, but a drawback is that the load has to be floating. There are many variations on this theme. As shown, a current sink is involved, but when the functions of the inputs of the operational amplifier are reversed a *current source* is obtained. The accuracy is mainly determined by the voltage reference V_{ref} and the temperature coefficient of R_1.

Excellent voltage references are available. The best known are proportional to absolute temperature (PTAT) bandgap reference voltages offering a constant voltage of about 1.2 V. Also, temperature-compensated Zener diodes exist with a very low temperature coefficient of less than 5 ppm K^{-1}. Note that a simple Zener diode is a very temperature-sensitive device and not at all appropriate as an accurate voltage source.

3.3.8 The logarithmic amplifier

Occasionally a physical quantity has to be measured over a large range of values. This range is often called the *dynamic* range and is defined as the ratio between the largest and smallest signal values to be measured. In this case, a logarithmic amplifier can be used to compress the signal within smaller boundaries. The principle of the logarithmic amplifier is illustrated in figure 3.15.

In the feedback loop, a bipolar transistor is implemented. A bipolar transistor shows an excellent exponential relation between collector current and base–emitter voltage, V_{BE}. This relation is expressed in the following equation and can exist over

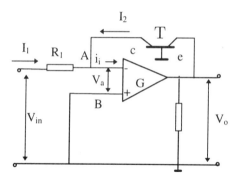

Figure 3.15 Circuit of a logarithmic amplifier. The bipolar transistor may be p or n type.

seven decades (10^7), with a good transistor, with acceptable accuracy:

$$I_C = I_{SS} \exp(qV_{BE}/kT) \tag{3.64}$$

where I_{SS} is the saturation current of the base–emitter diode, q the specific electric charge, V_{BE} the base–emitter voltage, k Boltzmann's constant and T the absolute temperature. If we apply the current equation to node A the result is

$$\frac{V_{in}}{R_1} + I_C = 0. \tag{3.65}$$

Substitution of (3.64) in (3.65) and use of $V_{BE} = V_o$ gives

$$\frac{V_{in}}{R_i} + I_{SS} \exp\left(\frac{qV_o}{kT}\right) = 0. \tag{3.66}$$

Rearranging (3.66) and taking logarithms we find

$$V_o = -\frac{kT}{q} \ln\left(\frac{V_{in}}{R_1 I_{SS}}\right). \tag{3.67}$$

Therefore the output voltage is proportional to the logarithm of the input voltage. A p-type or an n-type transistor can be applied, depending on the sign of the input signal V_i. Moreover, a parallel connection of a p-type and an n-type transistor is also applicable. From expression (3.67) for this circuit the weak points can be determined easily. There is a large temperature dependence. This can be overcome by compensation techniques, more sophisticated configurations being applied. If in figure 3.15 the functions of the transistor T and R_1 are interchanged, we obtain an exponential amplifier with which an *exponential expansion* can be realized. The relation then yields

$$V_o = R_1 I_{SS} \exp(qV_{in}/kT). \tag{3.68}$$

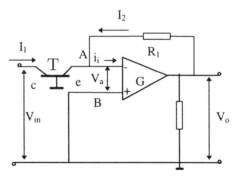

Figure 3.16 Exponential expansion of a signal. The transistor is either p type or n type.

This is illustrated in figure 3.16. Here also a large temperature dependence exists.

3.3.9 A current-to-voltage converter

Some transducers deliver an output current, for instance photodiodes and a variety of biological 'probes'. The currents produced are very small, of the order of nanoamperes. In particular, membranes of biological cells produce very low currents. In this case the currents must be converted into a voltage before they can be amplified. A feedback resistor R_1 and an operational amplifier provide this feature. This is depicted in figure 3.17. It is easy to see that the output voltage is governed by

$$V_{out} = -iR_1. \tag{3.69}$$

The input current of the operational amplifier causes an error, and hence amplifiers with a very high input impedance must be used. For instance, a FET input stage is excellent. Bias currents below 60 pA are reported for some of these types of amplifier. Also R_1 must be chosen to be high ($10^9 \, \Omega$).

3.3.10 An analogue multiplier

As we have already seen a logarithmic relation exists between the voltage and the current across a forward-biased pn junction This behaviour can be used to convert

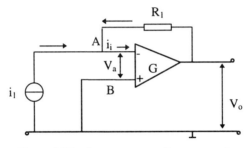

Figure 3.17 A current-to-voltage converter.

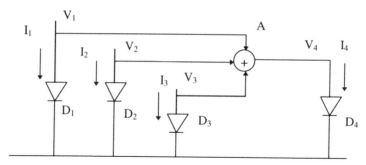

Figure 3.18   Principle of the analogue multiplier.

a multiplier into an adder. The principal circuit is depicted in figure 3.18. Assume that four currents $I_1, \ldots, I_4$ are flowing through their respective diodes $D_1, \ldots, D_4$. The diodes may be replaced by bipolar transistors; then the V_{BE} represent the different diode voltages. At point A, a summing element is implemented for which

$$V_4 = V_1 + V_2 + V_3 \tag{3.70}$$

and, substituting the relation

$$V = \frac{kT}{q} \ln\left(\frac{I_1}{I_{SS}}\right) \tag{3.71}$$

for the corresponding voltages, we find

$$V_4 = \frac{kT}{q} \ln\left(\frac{I_1 I_2 I_3}{I_{SS1} I_{SS2} I_{SS3}}\right) \tag{3.72}$$

or

$$V_4 = \frac{kT}{q} \ln(\alpha) \tag{3.73}$$

where $\alpha = I_1 I_2 I_3 / I_{SS1} I_{SS2} I_{SS3}$. Then from (3.73) we find

$$\alpha = \exp\left(\frac{q V_4}{kT}\right) = \frac{I_4}{I_{SS4}} \tag{3.74}$$

and, ultimately, we find for the product of currents

$$I_4 = I_{SS4}\, \alpha = I_{SS4}\, \frac{I_1 I_2 I_3}{I_{SS1} I_{SS2} I_{SS3}}. \tag{3.75}$$

Note that no temperature dependence is involved here. Several different variations on this principle exist, but they are all based on some current multiplication. Completely integrated analogue multiplier building blocks are available.

3.3.11 A peak detector

A peak detector can be used in AC measurements to determine for example the root mean square (RMS) value of a sine wave. For this purpose, to obtain the RMS value, the peak value must be multiplied by 0.707. The basic circuit of a peak detector is shown in figure 3.19.

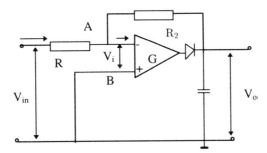

Figure 3.19 Principal circuit for a peak detector.

Since the diode is implemented in the feedback loop, the forward voltage drop is divided by the open-loop gain. The diode will conduct whenever the input is greater than the output, so the output will be equal to the peak value of the input voltage. Once the diode is conducting, the following relation exists between input and output voltages, where the capacitor will store this maximum value of V_{in}:

$$V_{out} = -\frac{R_2}{R_1} V_{in}. \tag{3.76}$$

If for instance the ratio of the resistances equals 0.707 the input voltage is multiplied by this value and the RMS value is obtained immediately.

3.4 FILTERS

Filters are used in widespread applications. Basically, filters are systems able to make a separation between different signals based on their frequency differences. Applications are found in antialiasing filters to confine the bandwidth to certain ranges within the specified frequency range of the system for accuracy and linearity. Other applications are found in improvement of the signal-to-noise ratio (S/N) and, for instance, to analyse the spectrum of the signal. Filters can be distinguished as passive and active filters. In active filters, active components are used which can reduce size considerably. To design and simulate filters excellent software is available, for instance with PSPICE. Other design tools for filters are also available in special design software for digital signal processors (DSPs) of Texas Instruments. With these DSPs all types of filter can be designed in software. Many different types of filter exist; however, we will mention here only five main types.

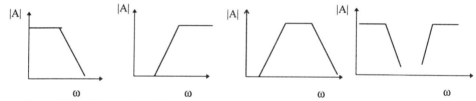

Figure 3.20 Illustration of four main categories of filters: (a) low-pass; (b) high-pass; (c) band-pass and (d) notch filter.

3.4.1 Types of filter

1 *Low-pass filters* have the characteristic that they pass all frequencies which are lower than the highest specified bandpass width of this filter.
2 *High-pass filters* have the characteristic that only frequencies which surpass a certain lowest limit are passed.
3 *Band-pass filters* allow only a certain predefined frequency band to pass.
4 *Band-stop filters* are the reverse of a band-pass filter. A predefined frequency bandwidth is stopped and cannot pass the filter. These filter types are also called *notch filters*.
5 *All-pass filters* can provide a predefined frequency-dependent phase shift or time delay.

Switched capacitor filters are manufactured on the principle that a resistor can be replaced by the equivalence of a capacitor. CMOS technology is well suited to the realization of switched capacitor circuits and hence is appropriate for large-scale integration. This is discussed in the next section.

Filters can be made as first-, second- or higher-order filters. The different types of filter are illustrated in figure 3.20. We will not go into further detail here, but reference is made to the literature, for instance the *Handbook of Active Filters* [8].

3.4.2 The switched capacitor

A special type of filter is found in switched capacitor filters. MOS technology offers the possibility to integrate very accurately these types of filter. The principle makes use of the possibility that capacitors can be used to simulate a resistor. To demonstrate this principle consider the circuit in figure 3.21. With the help of this circuit the capacitor C is alternately connected to V_1 and V_2, via Q_1 and Q_2. Then each

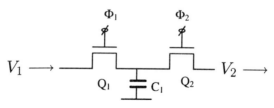

Figure 3.21 Principle of switched capacitors simulating a resistor.

time the charge flowing during the clock period T is given by

$$\Delta q = C(V_1 - V_2) \tag{3.77}$$

because the capacitor C is charged alternately to V_1 and V_2; then the average current $\bar{I}$ is given by

$$\bar{I} = \frac{\Delta q}{T} = \frac{C}{T}(V_1 - V_2). \tag{3.78}$$

From this equation the equivalent resistance can be determined:

$$R_{eq} = \frac{T}{C}. \tag{3.79}$$

It will be clear that, with the help of this similarity, very accurate resistors can be made, because capacitors and the time period T can be realized accurately in IC MOS technology. On the basis of this principle many applications such as filters are realized.

3.5 ANALOGUE-TO-DIGITAL AND DIGITAL-TO-ANALOGUE CONVERSIONS

3.5.1 Introduction

Analogue-to-digital and digital-to-analogue conversion, usually abbreviated to ADC and DAC, are frequently used in all kinds of digitized measurement equipment, but can also be recognized in many consumer electronics products owing to the ease of processing signals in a digital form. So, at the beginning of a system we can distinguish an A/D conversion as soon as the signal is amplified to the desired level at which conversion can be executed. At the end of a system, depending on the application, a D/A conversion can be found when analogue output is required.

In this section we will discuss both conversions but first we will present a short introduction to coding, because that is the essence of the conversion process. In other words, a given signal has to be transposed into another domain by an unambiguous representation and vice versa.

3.5.2 Coding

Coding can be defined as a kind of conversion of data from one representation into another by a certain function. Often a reference source code is predefined. For instance, a constant-voltage source can provide such a reference code. In the following we present a short introduction to coding of numbers.

Any real number can be represented by the expression

$$N = a_n g^n + a_{n-1} g^{n-1} + \ldots + a_1 g^{+1} + a_0 g^0 + a_{-1} g^{-1} + a_{-2} g^{-2} + \ldots + a_{-m} g^{-m}$$

or in an abbreviated way

$$N = \sum_{i=-m}^{n} a_i g^i \qquad (3.80)$$

where g is the base or radix and a is a number or symbol. If $m=0$ the number is an integer and if $n=0$ the number is a fraction. With $g=10$ we have the decimal system, if $g=2$ we have the binary system, if $g=8$ we have the octal system (symbols $0, \ldots, 7$), if $g=16$ (symbols $0, \ldots, 9$, A, B, C, D, E, F) we have the hexadecimal system, etc. The base number determines the number of possible symbols. A decimal number may also be represented by the following expression:

$$N_{10} = A \times 10^B \qquad (3.81)$$

where A is any number between zero and unity, or $0 \le A < 1$, and B is any positive or negative integer $\ldots, -2, -1, 0, 1, 2, \ldots$. In its turn A is defined as

$$A = a_1 10^{-1} + a_2 10^{-2} + a_3 10^{-3} + \ldots + a_n 10^{-n} \qquad (3.82)$$

where a_i is any positive integer between 0 and 9.

* * *

Example 3.11
Consider the number 98 102; we can write this number as $0.981\,02 \times 10^5$ so $A = 0.981\,02$ and $B=5$. We can write A as

$$A = 9 \times 10^{-1} + 8 \times 10^{-2} + 1 \times 10^{-3} + 0 \times 10^{-4} + 2 \times 10^{-5}.$$

In the same way we can write any number to base 2 as

$$N = (a_1 \times 2^{-1} + a_2 \times 2^{-2} + a_3 \times 2^{-3} + \ldots + a_n \times 2^{-n})2^B \qquad (3.83)$$

where here, of course, a_i can have the values 0 and 1 only. As will be seen, to indicate the sign of a number an extra bit has to be added.

* * *

Example 3.12
The binary representation of 5 can be written as 0101. When we write this number in the form of (3.80) this gives

$$N = (0 \times 2^{-1} + 1 \times 2^{-2} + 0 \times 2^{-3} + 1 \times 2^{-4})2^4.$$

To verify this, the expression can be rearranged back to the decimal representation. To avoid the multiplying factor in (3.80) a binary number can be represented as

$$N = a_n 2^{n-1} + \ldots + a_3 2^3 + a_2 2^2 + a_1 2^1 + a_0 2^0 \qquad (3.84)$$

where again a_i can have the values 0 and 1 only.

* * *

When an A/D conversion is required the first representation is preferred, because the binary number is always a fraction of some maximum reference value. Let us illustrate that again with an example.

* * *

Example 3.13
Assume the fraction $0.625 = 5/8$ must be represented in a binary form. Then, applying the notation of (3.80) this is simply 101 because

$$101 = (1 \times 2^{-1} + 0 \times 2^{-2} + 1 \times 2^{-3})2^0.$$

If we want to apply the two's complement method to assign negative numbers we just complement each bit and add 1 to it.

* * *

Example 3.14
Assume we want to represent the decimal number -0.625 in the binary form (3.80); then the procedure is

$+5/8$	0101
	1010 complementing
	1 adding 1
	___ +
$-5/8$	1011

The first bit represents the sign bit. When we add $5/8$ and $-5/8$ the result is zero with a loss of the carry bit, or

$5/8$	0101
$-5/8$	1011
___	___ +
0	10000

* * *

In ADCs and DACs the first bit is called the most-significant bit (MSB) and has a weight of one-half full scale of the converter. The second bit has a weight of one-quarter full scale, etc. For instance, with 8 bits we can count up to 256; the MSB represents the number 127, which is about half full scale. The last bit is called the least-significant bit (LSB), and has the weight $(\frac{1}{2})^n$ of the full-scale value.

The *resolution* of an ADC is defined as the number of bits assigned by the size of the LSB. For a DAC it is defined as the smallest possible incremental change in the output voltage (see also chapter 4). An alternative definition for the resolution of an ADC is the amount of input voltage change required to increment the output with one code change of the LSB. Resolution can be expressed in per cent of full scale or in binary bits.

* * *

Example 3.15

Consider a 12-bit ADC. The resolution is able to resolve 1 part in 2^{12}, i.e. 1 part in 4096. This equals 0.024 41% of full scale. So a converter with 10 V full scale can resolve an input change of $10/4096 = 2.441$ mV. In the same way this holds for a DAC; an increment of one LSB at the input produces a voltage change of 0.024 41 of full scale at the output. Note that resolution is not related to accuracy at all.

   The resolution of a converter can be related to the *dynamic range* in decibels. For example, a factor of 2 equals about 6 dB. Therefore the dynamic range can easily be given in decibels by the number of bits multiplied by 6. Thus a 12-bit converter has a dynamic range of $12 \times 6 = 72$ dB.

3.6 DIGITAL-TO-ANALOGUE CONVERTERS

As has been said already, in applications it can be necessary to change binary data into an analogue signal. Usually, this occurs at the end of a measuring chain after the signal has been processed in a digital form. A digital-to-analogue converter is defined as a device which converts a digital input into an analogue parameter, where the output is *quantized*, just like the input. Different parameters may be used for a DAC output; however, common ones are (1) a voltage, (2) a current or (3) a resistance. Two forms will be discussed, the fractionary binary and the ladder network.

3.6.1 The adder network

The diagram of the principle of DAC with the help of an adder network, also called *fractionary binary DAC*, is depicted in figure 3.22. For 2^n combinations at the input we find 2^n different levels at the output. We can write for the output level (3.82)

$$V_{out} = A = V_{ref} \sum_{i=1}^{n} b_i 2^{-i} \qquad (3.85)$$

where V_{ref} is a certain reference voltage, and b_i is the ith digital bit value with either a '1' or a '0' by definition. In figure 3.23 this principle is illustrated in more detail using this fractionary binary code. With respect to node A, this circuit behaves as a summing amplifier with weighted resistors in a *binary* fashion so that $R_1 : R_2 : R_3 : \ldots : R_n = 1 : 2 : 4 : 8 : \ldots$. They are all connected via electronic logic

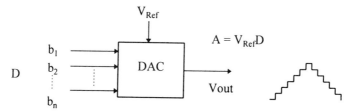

Figure 3.22 Diagram showing the principle of a DAC.

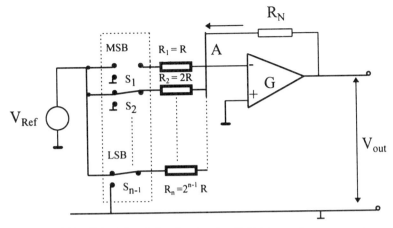

Figure 3.23 Principle of the circuit of a digital-to-analogue converter.

switches S_i to a reference voltage V_{ref}, or to ground, which is determined by the value of the corresponding bit b_i. A logic 1 applied to the appropriate switch connects the corresponding resistor to the reference voltage and increases the output voltage by the binary weighted increment involved. This process is expressed as

$$V_{out} = -R_N \left(\frac{1}{R} S_1 + \frac{1}{2R} S_2 + \ldots + \frac{1}{2^{n-1}R} S_{n-1} \right) V_{ref}. \qquad (3.86)$$

The switches $S_1, \ldots, S_{n-1}$ represent the binary multipliers b_i. Assuming $R_N = R/2$ we find for (3.86)

$$V_{out} = -\left(\frac{1}{2^0} S_1 + \frac{1}{2^1} S_2 + \ldots + \frac{1}{2^{n-1}} S_{n-1} \right) \frac{V_{ref}}{2} \qquad (3.87)$$

or

$$V_{out} = -\left(\frac{1}{2^1} S_1 + \frac{1}{2^2} S_2 + \ldots + \frac{1}{2^n} S_{n-1} \right) V_{ref}. \qquad (3.88)$$

This circuit has several disadvantages. The value of each input resistor is twice that of the preceding one so the absolute values become quite large. Moreover, matching, tolerances and temperature stability of the resistors influence the accuracy to a large extent. Finally, the input impedance of this circuit is low and changes continuously, producing offset errors.

3.6.2 The ladder network

A much better solution avoiding several of the problems is to make use of the R–$2R$ ladder network configuration. This circuit is depicted in figure 3.24. At each node $A_1, \ldots, A_n$ looking to the left, the same impedance R is seen and consequently the

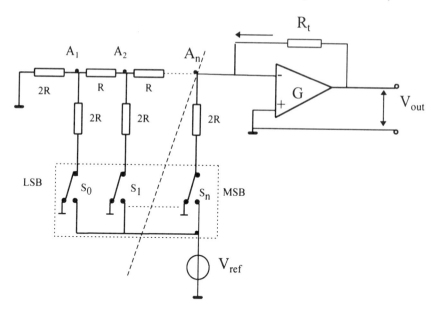

Figure 3.24   The ADC with an R–$2R$ ladder network.

current flowing from right to left, generated by the reference voltage V_{ref}, is divided in two at each junction. The input impedance of the operational amplifier has become a constant and independent of the binary word represented by the logical switch positions S_i, which are equivalent to a logic '0' or '1'.

For the conversion the expression appears to be

$$V_{out} = -V_{ref} \frac{R_t}{R} \sum_{i=0}^{n} S_i 2^{i-(n+1)} \tag{3.89}$$

where n is the number of bits used and S_i are the corresponding binary switches. The above two converter types are pure binary. It is possible to adapt both circuits to other code conversions just by implementing other resistor ratios or other kinds of resistor network. Finally, we mention serial DA converters, which may be preferred when the information is available in serial mode. Their conversion speed is less than that of parallel converters.

*   *   *

Example 3.16
Suppose a 4-bit DAC is used. Assume further that all bits are '1'; then the output voltage V_{out} is, using (3.86), given by

$$V_{out} = -\frac{R_t}{R}(1 \times 2^{-4} + 1 \times 2^{-3} + 1 \times 2^{-2} + 1 \times 2^{-1})V_{ref} = -\frac{R_t}{R} \times 0.9374 \times V_{ref}$$

and when $R_t = R$ we find

$$V_{out} = -0.9375 V_{ref}.$$

* * *

3.7 ANALOGUE-TO-DIGITAL CONVERTERS

The function of an analogue-to-digital converter is to convert the analogue input signal into a discrete number of normally binary steps that can easily be processed or presented in decimal form on a numerical display. By definition an analogue-to-digital converter produces a digital output of a finite number of discrete values, corresponding to the value of the signal applied to its analogue input. The difference between the value of the digital output and the analogue input is called the *quantization uncertainty*. The device can be unipolar or bipolar depending on the available polarity of the output.

The forms into which the analogue signal is converted may be of various natures, e.g. a voltage, a time or a charge, and depend on the type of ADC involved. As will be seen, in all cases each step should correspond to an accurately known voltage related to a very accurate stable reference voltage with which a comparison should be made.

If the step is a DC voltage the comparison is quite a simple matter, but when the step is of another nature, such as time or charge, the comparison can be made by converting the reference voltage into the same steps. For analogue-to-digital converters different principles of operation are available. The most widely used systems are the following.

1   Direct compensation
In this principle of operation, the input signal is compared with an internally generated voltage, which is increased in steps starting from zero. The number of steps needed to reach full compensation is counted and represents the desired output value.

2   Voltage-to-time conversion
Both input voltage and reference voltage are converted into a time of counts and the ratio of the two times is determined and represents the output value.

3   Voltage-to-frequency conversion
The input voltage is converted into a number of pulses and the number of pulses in a well defined constant time is counted and represents the output value. In the following sections, some widely known types of ADC are presented.

3.7.1   The dual-slope integration ADC
In this section we will discuss an application of the voltage-to-time conversion principle. The principle offers simple, accurate and very linear ADCs. The dual-slope integration ADC is one of the most well known and widely used types. The principle

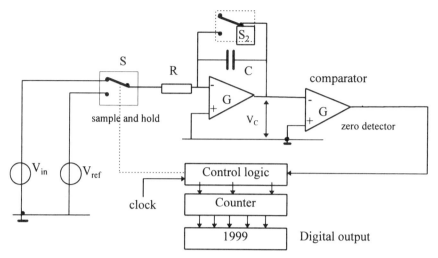

Figure 3.25 The principle of a dual-slope integration ADC. At the start the integrator is held discharged via S_2.

of operation is based on integration of the input voltage for a constant time or count number followed by the measurement of the time required to discharge the integrating capacitor with a constant current.

The principle is depicted in figure 3.25. At the start of a measuring period the logic circuit closes the switch S, and at the same time starts the timer, counting a fixed number of clock pulses corresponding to a fixed time T_1, e.g. during $N = 200\,000$ clock pulses. During the time T_1, the sampling period, V_{in} is connected to the input of the amplifier, and V_{in} will be integrated in the capacitor C.

It will be clear that at the end of this period V_C is proportional to V_{in}. At the input of the same period T_1 the logic disconnects the input V_{in} by the switch S, and connects the input to the reference voltage V_{ref} which should have the opposite polarity to the input V_{in}. The counter is reset to zero. At that moment, C starts to discharge with a constant current, because V_{ref} is a constant. The integrator integrates the reference voltage down to the comparator's threshold value at which time the counting is stopped. Assume then that the number of counts is n. This is depicted for two different input values V_i in figure 3.26.

The process can be described by an equation with respect to *charge* on the capacitor C, or

$$\int_0^{T_1} \frac{V_{in}}{R}\, dt - \int_{T_1}^{T_1 + T_2} \frac{V_{ref}}{R}\, dt = 0. \tag{3.90}$$

In the time interval T_1, V_{in} is supposed to be constant; then (3.90) can be written as

$$V_{in} T_1 = V_{ref} T_2 \tag{3.91}$$

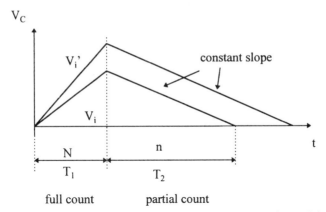

Figure 3.26 Graphical representation of the voltage V_C versus time of the dual-slope ADC for two different input values V'_i and V_i of V_{in}.

or

$$V_{in} = \frac{T_2}{T_1} V_{ref}. \tag{3.92}$$

If this equation is expressed in terms of the number of counts we obtain

$$V_{in} = \frac{n}{N} V_{ref}. \tag{3.93}$$

So V_{in} can be read immediately from the counter. The advantages of this type of ADC can be summarized as follows.

The errors of the integrator are present in the whole timing period T_1 and T_2 and will therefore cancel. The accuracy of the conversion will depend only on the accuracy of the reference voltage and not on the clock frequency. The only requirement for the clock is to be stable within the conversion period. The accuracy of the RC time is of less importance. For these ADCs an extremely high accuracy and linearity can be obtained. They are frequently used in measurement systems.

3.7.2 The sigma–delta modulator ADC

The sigma–delta modulator can be characterized as a continuously switching dual-slope ADC. The average charge in the capacitor C is kept at zero by a control loop. The principle is shown in figure 3.27.

The capacitor C stores the charge of the input signal which either has a current or a voltage character, I_{in} or V_{in}. In our discussion we will take voltage sources. Via the switch S the capacitor C is connected to two voltage reference sources different in polarity, charging or discharging the capacitor C. A control loop, consisting of the comparator, the D flip–flop and a clock generator f, determines the position of the switch S and therefore can be considered as a polarity detector. As a result of this control loop the average value of the capacitor is kept at zero. When no input

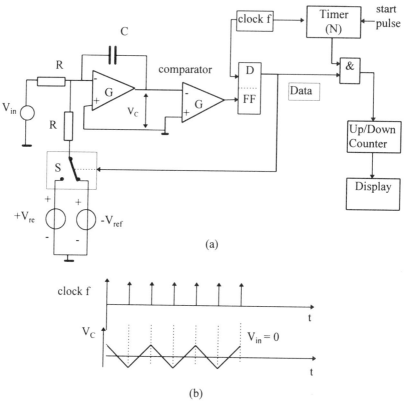

(a)

(b)

Figure 3.27 Principle of the sigma–delta modulator ADC and timing diagram for $V_{in} = 0$.

signal is present, the numbers of charge and discharge pulses are equal and just half of the number of pulses N in the *fixed* conversion time period, T. A triangular waveform signal (of several microvolts), synchronized with the clock f, will appear across the capacitor C (figure 3.27(b)).

When an input voltage is applied then a difference in the symmetrical up–down pulses will be found, which is proportional to the analogue input signal. Within the fixed conversion time T, the total number of pulses is N which is always given by

$$N = n_{up} + n_{down}. \tag{3.94}$$

During the conversion time T, the difference between the number of up–down pulses is counted and stored in the up/down counter; then after the conversion time the content of the up/down counter is

$$n = n_{up} - n_{down}. \tag{3.95}$$

Note that the maximum value of $n = \frac{1}{2}N$. Then if the charge equilibrium across the capacitor C, in the conversion period T, is considered, it can be shown that the ratio between the input voltage and the reference voltage is just the ratio between the stored pulses n in the up/down counter and the total number of pulses N, or

$$\frac{V_{in}}{V_{ref}} = \frac{n_{up} - n_{down}}{n_{up} + n_{down}} = \frac{n}{N} \tag{3.96}$$

or

$$V_{in} = V_{ref} \frac{n}{N}. \tag{3.97}$$

The number of counts N is fixed and equals the resolution. It is determined from the following expression

$$N = Tf \tag{3.98}$$

or the conversion speed is f/N. To obtain a high normal mode rejection ratio the clock frequency is adjusted to a multiple of the mains frequency.

3.7.3 Voltage-to-frequency converter (VFC)

Basically, a voltage-to-frequency converter is an oscillator whose output frequency is proportional to an input voltage. The frequency is measured with a normal counter. The conversion is realized by integrating the input voltage. As soon as a predetermined voltage level V_S is achieved, the level detector passes a command pulse to a switch in parallel with capacitor C of an integrator, causing the latter to be discharged by closing the switch. The switch then opens again and the cycle is repeated. Each time the switch is closed the level detector feeds a pulse to the counter. The circuit is depicted in figure 3.28.

The following relationship is valid:

$$V_{out} = \frac{1}{RC} \int_0^t V_{in} \, dt. \tag{3.99}$$

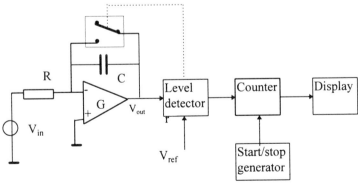

Figure 3.28 Principle of a voltage-to-frequency ADC.

When V_i is constant during the measuring time we may write

$$t = \frac{1}{V_i} \tau V_{out} \tag{3.100}$$

where $\tau = RC$ and the time required to make $V_{out} = V_{ref}$ is then related to V_i by

$$t = \alpha \frac{1}{V_i} \tag{3.101}$$

where $\alpha = \tau V_{out}$; then the frequency is given by

$$f = \frac{1}{t} = \frac{1}{\alpha} V_{in}. \tag{3.102}$$

By choosing the right values for R, C and V_{ref}, it is possible to give α an appropriate value to display a direct reading of the input voltage, V_{in}. The counter must be started and stopped by an internal generator in an exactly fixed known time interval T. The accuracy is limited because of the many parameters involved, R, C, V_{ref} and T. Note that the first part of the circuit in figure 3.26 acts as a voltage-controlled oscillator. This kind of VFC is particularly suitable for telemetry applications, since only a data channel is needed between the VFC and an applied counter–gate circuitry. However, this circuitry is also useful in interfacing applications for smart sensors.

3.7.4 Successive approximation ADC

The successive approximation method is another widely used ADC method. It combines high resolution and high speed. Basically it consists of a comparator, V_c, a digital-to-analogue converter D/A and a successive approximation register controlled by the comparator and the clock. The principle is shown in figure 3.29.

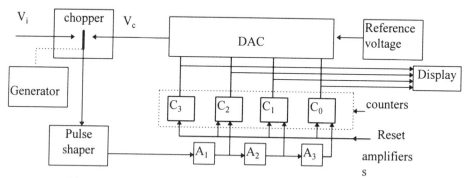

Figure 3.29   Block diagram of a successive approximation ADC.

The method is based on the comparison of the input voltage V_i with the output of a D/A converter, V_c, inserted in a feedback loop. This is performed using a kind of chopper circuit which alternately connects the pulse shaper circuit with V_i and V_c. At the start of the conversion cycle, the counters are set to zero and the D/A converter's MSB output is compared with the input. The MSB starts at half full scale. If it is smaller than the input, the MSB is left on and the next bit is tried. If the MSB is larger than the input, the MSB is turned off and the next bit is turned on. This process of comparison is continued until the output register contains the LSB of the complete output number. The output of the chopper will be a square wave of which the amplitude is a function of the difference between V_i and V_c. The pulse shaper converts the positive slope of this square wave into pulses with an amplitude related to that of the square wave and this signal is amplified by three amplifiers in series.

3.7.5 Parallel converter

To obtain a very high speed for example for video applications, a parallel converter can be used. Conversion rates of over 100 MHz are possible. The input uses a quantizer comprising a number $2^n - 1$ of comparators that is needed depending on the number of bits n. The output of the comparators must be converted into a digital mode with help of a decoder. Hence for an 8-bit converter 255 comparators are needed. These ADCs are very expensive becasue of the large number of components needed, but 16-bit converters are available.

3.8 DIGITAL FUNCTIONS

Digital functions have had a large impact on measuring systems and have changed the world of the performance of measuring systems both inside and outside the instrument. It is beyond the scope of this textbook to present a detailed treatment of digital electronics. In general it can be said that any analogue-implemented function has its digital counterpart. Also on a microscopic scale all digital functions are analogue. In this section we shall discuss a few examples of combinational circuits. Today, the most flexible digital components are programmable logic devices (UPLDs). This is a collective name of a large family of user-programmable components. The given examples frequently occur in electronic measurement systems. The discussion cannot be exhaustive and a basic grasp of knowledge of Boolean algebra and logic elements is assumed. The following items will be discussed:

- a decoder
- an encoder
- a multiplexer
- user-programmable logic devices (UPLDs).

3.8.1 A decoder

A decoder is defined as a logic building block that accepts a binary input and converts it into a certain output number one at a time. A simple example of a one-out-of-four decoder is shown in figure 3.30. The Boolean functions for the four output variables are

$$S_0 = \overline{A + B} \qquad S_1 = \overline{A + \overline{B}} \qquad S_2 = \overline{\overline{A} + B} \qquad S_3 = \overline{\overline{A} + \overline{B}}. \qquad (3.103)$$

A decoder is very useful for addressing purposes in memories, devices, equipment etc. The truth table is shown in the same figure.

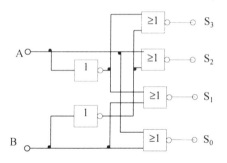

Input		Output			
A	B	S_0	S_1	S_2	S_3
0	0	1	0	0	0
1	0	0	1	0	0
0	1	0	0	1	0
1	1	0	0	0	1

Figure 3.30   Logic diagram of a one-out-of-four decoder.

3.8.2   An encoder

An encoder is a logic circuit that accepts any number of inputs and converts them into a binary code. As can be seen from this definition it has the reverse function of the decoder. An example of an encoder is the octal-to-binary encoder shown in figure 3.31. The truth table is again depicted in the same figure. Note that the names for a decoder and an encoder are often interchanged. Encoders are used to realize conversions between different codes, e.g. from one BCD (binary-coded decimal) to another BCD code.

From figure 3.31 the following Boolean functions are realized (3.101):

$$B_0 = D_1 + D_3 + D_5 + D_7$$

$$B_1 = D_2 + D_3 + D_6 + D_7 \qquad\qquad (3.104)$$

$$B_3 = D_4 + D_5 + D_6 + D_7.$$

From both these equations and the table itself it is easy to read the OR functions for the corresponding outputs B_i as a function of the different inputs D_i.

3.8.3   A multiplexer

The third group of combinational circuits is the multiplexers and demultiplexers. The word 'multiplexing' is derived from telecommunications techniques, where a large number of voice channels are transmitted over a small number of lines. In the same fashion, the digital multiplexer can be used to transmit a large number of digital signals over a few lines. Sometimes the multiplexer is termed a 'data selector'.

The reverse operation is called *demultiplexing*. A multiplexer can be defined as a combinational circuit that selects data from 2^n input lines and transmits them through a single output line. This definition closely resembles the definition of the encoder, but the difference is the presence of a number of additional input lines called the

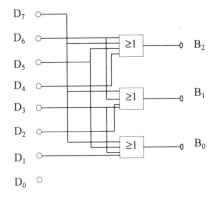

Input								Output		
Octal digits								Binary Digits		
D_7	D_6	D_5	D_4	D_3	D_2	D_1	D_0	B_2	B_1	B_0
							1			
						1				1
					1				1	
				1					1	1
			1					1		
		1						1		1
	1							1	1	
1								1	1	1

Figure 3.31 Logic diagram and truth table of an octal-to-binary encoder. In the table all empty cells must be read as zeros.

address or *selector* lines. An example of an eight-input digital multiplexer is shown in figure 3.32.

The eight input lines are connected to eight four-input AND gates which are controlled by the address inputs S_0, S_1 and S_2. It can be derived from the truth table that only one input at a time is connected to the output via the eight-input OR gate. Often the AND gates have a fifth input to provide an *enable* input, by which all inputs can be switched on and off from the output simultaneously.

As said previously, the reverse operation is called demultiplexing and an example of a four-output digital demultiplexer is given in figure 3.33. This circuit is self-explanatory. Numerous different digital components exist and for these parts reference is made to the relevant literature.

3.9 USER-PROGRAMMABLE LOGIC DEVICES

User-programmable logic devices (UPLDs) is a collective name for a large family of user-programmable components. A distinction is made between *combinational* circuits and *sequential* circuits. Combinational circuits are circuits without a memory function. In other words, combinational circuits can be defined as systems in which

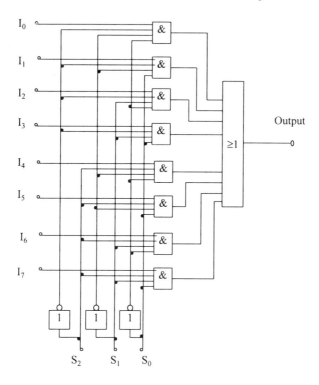

Address			Selected
S_2	S_1	S_0	line
0	0	0	0
0	0	1	1
0	1	0	2
0	1	1	3
1	0	0	4
1	0	1	5
1	1	0	6
1	1	1	7

Figure 3.32 An eight-input multiplexer and the related truth table.

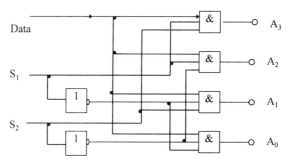

Figure 3.33 A four-output digital demultiplexer.

the output is a function of the present state of the input values only. In contrast, sequential circuits are systems in which the output is also a function of the previous values of the input values, and hence in sequential systems a memory function must be present. It is not possible to present a full extensive treatment of UPLDs, because of space limitations. Therefore, we have to confine ourselves to a brief introduction of the main concepts.

Every logic function can be built with standard parts, e.g. with help of the TTL 7400 series, or in the CMOS 5400 series, which offer a large variety of functional parts, such as AND, OR, NOT, NAND, NOR, XOR. However, the greater the number of separate parts available, the more random a choice is made and often a non-structured design results, which gives poor testability and reliability, owing to too much interconnecting wiring. Also, the needed printed circuit board (PCB) will be complicated and hence expensive. For programmable arrays these problems are strongly reduced. The only disadvantage may be that special programming equipment is needed to blow the fuses, or to program the PLDs of the logic array used, to realize the desired function. As said previously, there is a fundamental difference involved; therefore we will split our discussion into two sections, one for combinational circuits and one for sequential circuits.

3.9.1 Combinational logic

We state that any switching function of n variables $f(x_1, x_2, \ldots, x_n)$ may be expressed as a sum of products, where in each product every input variable appears in either its true or its complemented form. The form is called a standard normal form, full disjunctive form or sum of minterms. It is generally expressed as

$$F = \sum_i g_i m_i \qquad (3.105)$$

where m_i is a minterm and $g_i = 0, 1$.

* * *

Example 3.17
The following expression is an example of a sum of minterms

$$F(x_1, x_2, x_3, x_4) = \bar{x}_1 x_2 \bar{x}_3 \bar{x}_4 + \bar{x}_1 x_2 \bar{x}_3 x_4 + \bar{x}_1 \bar{x}_2 x_3 \bar{x}_4 + \bar{x}_1 x_2 x_3 \bar{x}_4$$
$$+ x_1 \bar{x}_2 x_3 \bar{x}_4 + x_1 x_2 x_3 \bar{x}_4.$$

* * *

In the same way, on the basis of the duality principle, any switching functions may also be expressed as a product of sums, where in each sum every input variable appears either in its true or in its complementary form. This form is known as a standard product of sums, full conjunctive normal form or product of maxterms. It is generally written as

$$F = \prod_i G_i M_i \qquad (3.106)$$

where M_i is a maxterm and $G_i = 0, 1$. Note that these standard forms do not represent the minimal form but the number of variables in these product terms or sum terms may be reduced. The word *minterm* is derived from the idea that a minterm covers a minimum area of a Karnaugh diagram; a *maxterm* covers a maximum surface area of a Karnaugh diagram.

Boolean algebra is applied to obtain the reduced forms.

*   *   *

Example 3.18
The example given previously may also be written as

$$F(x_1, x_2, x_3, x_4) = (x_1 + x_2 + x_3 + x_4)(x_1 + x_2 + x_3 + \bar{x}_4)(\bar{x}_1 + \bar{x}_2 + x_3 + x_4)$$

$$\times (\bar{x}_1 + \bar{x}_2 + x_3 + \bar{x}_4)(\bar{x}_1 + x_2 + x_3 + x_4)(\bar{x}_1 + x_2 + x_3 + \bar{x}_4)$$

$$\times (x_1 + x_2 + \bar{x}_3 + \bar{x}_4)(x_1 + \bar{x}_2 + \bar{x}_3 + \bar{x}_4)(\bar{x}_1 + x_2 + \hat{x}_3 + \bar{x}_4)$$

$$\times (\bar{x}_1 + \bar{x}_2 + \bar{x}_3 + \bar{x}_4).$$

*   *   *

When we apply Boolean algebra we find for both examples the reduced forms

$$F(x_1, x_2, x_3, x_4) = \bar{x}_1 x_2 \bar{x}_3 + x_3 \bar{x}_4$$

or

$$F(x_1, x_2, x_3, x_4) = (\bar{x}_1 + x_3)(x_2 + x_3)(\bar{x}_3 + \bar{x}_4).$$

Now it can be shown that any switching function can be realized with AND, OR and NOT gates. For instance, the example presented can be realized in the two different forms as depicted in figure 3.34.

3.9.2 Realizations of programmable arrays

On the basis of these concepts complete PLDs are realized using stages of arrays of NOTs (inverters), ANDs and ORs, which can be connected or disconnected via selectable switches. In general, in PLDs three stages can be distinguished. The first stage generates the complemented form where needed; the second stage consists of an array of AND functions, producing AND and NAND combinations to generate product terms. The third stage consists of an OR function producing the output form. Compare, for example, figure 3.34(a). Firstly, we will see how a product or AND term is realized. This is illustrated in figure 3.35.

The fuses can be blown with the help of a special programming apparatus. The product term is determined by the fuses *not* blown. In this diagram the diodes act as electronic switches. These switches can be programmed by a special programming apparatus, which is able to blow a thin aluminium interconnecting wiring on the chip itself. With the help of these basic AND arrays it is easy to implement whole arrays of product terms. The fuses and diodes are replaced by a dot. This is indicated in figure 3.36. When we extend this diagram with inverters, any product term can be made. In figure 3.37 the implementation is shown. In the same way as for an

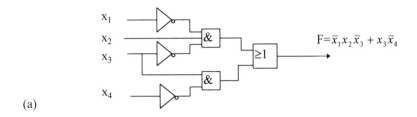

(a)

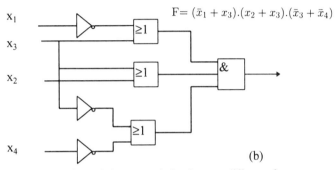

(b)

Figure 3.34 Realization of the example in the two different forms.

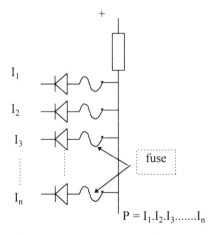

Figure 3.35 Principle of a programmable n-input AND gate.

AND array, an OR array can be made. This is illustrated in figure 3.38. This OR diagram can be repeated and an array of ORs is obtained such as is illustrated in figure 3.39. We have again replaced the fuses and switches (diodes) by dots. Finally, we can complete the programmable array by gluing together the AND and OR

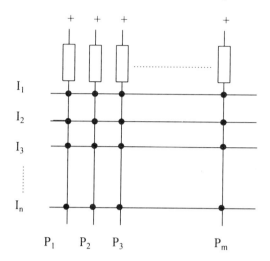

Figure 3.36 Programmable AND array with different product terms, P_i. The fuses and switches are replaced by a dot. Where no dot is present the fuse is blown. Some manufacturers use a cross instead of a dot for a programmable switch.

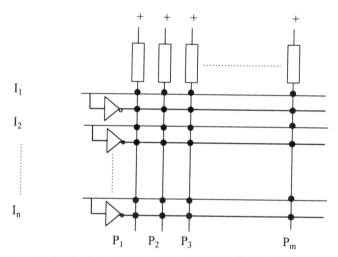

Figure 3.37 Implementation of an array of product terms.

arrays. Note that buffering is required to reduce the load of both arrays. This is illustrated in figure 3.40.

On the basis of this concept a large number of families have been developed, by the application of different technologies and different types of combination of arrays for the logic components, the fuses and the switches. Four major types of programmable array exist, which will be discussed in brief:

1   (P)ROM, (programmable) read-only memory
2   (F)PLA, (field-) programmable logic array

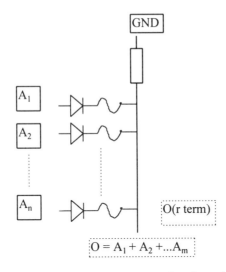

Figure 3.38 Implementation of an OR function with diodes.

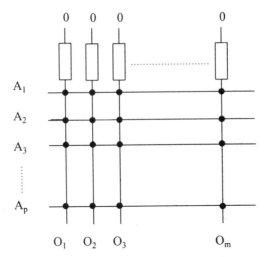

Figure 3.39 The implementation of an OR array.

3 (F)PGA, (field-) programmable gate array
4 PAL/HAL, programmable array logic/hard array logic.

Programmable read-only memory
A programmable read-only memory is a memory device in which the selection of the terms is realized using all minterms of a given number of input variables. The minterms are realized in the fixed AND field. The minterms are connected to a programmable OR array. When the OR array is programmed by the manufacturer we speak of a ROM. A PROM may be programmed by the user. Applications are found in memory devices and in XOR-type functions.

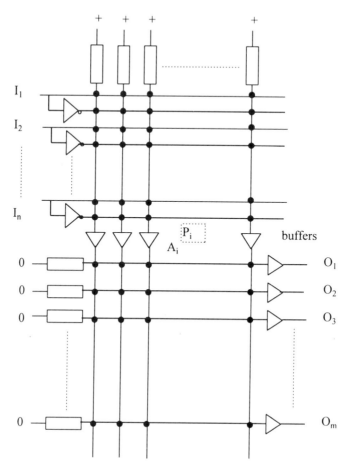

Figure 3.40   Complete programmable array. Complex PLDs contain over 20 000 gates.

Field-programmable logic array
In field-programmable logic arrays the AND and OR field are both programmable. PLAs are programmed by the manufacturer, whereas FPLAs may be programmed by the user. The different types differ in the number of pins. Often the output functions can be programmed independently using XORs available at the output. In some types also tristate outputs are available, which makes them well suited for bus systems. Other features may be buffers for the non-inverted input lines.

Field-programmable gate array
Field-programmable gate arrays have no OR field and can perform single-level AND functions. The PGA is the mask-programmable device. Applications can be found in machine state decoders, fault monitors, code detectors, peripheral selectors.

Programmable array logic (PALs)/hard array logic (HALs)
In programmable array logic/hard array logic devices the AND array is programmable and the OR array is not. This reduces flexibility to a large extent, but a whole

family of different PALs is designed to overcome this drawback. The HAL version is the mask-programmable family which is programmed by the manufacturer. We summarize this in table 3.3 and we have added some other known PLDs. In this respect we also have to mention here the application-specific integrated circuits (ASICs). They can be understood as completely customized ICs designed for specific applications and they are usually manufactured in limited quantities. Special design houses deliver all the support needed by the customer.

3.9.3 Sequential logic

In sequential logic a memory function is present. The output is a function of the input variables and the previous values of the input gates. The principle of a sequential machine is depicted in figure 3.41.

Basically, the system consists of a combinational part and a memory part. The memory contains information from s binary outputs of 2^s combinations of possible outputs of $y_1, \ldots, y_s$. Each combination is called a state variable of past events. Then the combinational logic has the inputs $i_0, \ldots, i_n$ and the state variables $y_1, \ldots, y_s$. The output generates m outputs $F_0, \ldots, F_m$ and p excitation variables $Y_1, \ldots, Y_p$. The function F_i can be written

$$F_i = h_i(i_1, i_2, \ldots, i_n, y_1, y_2, \ldots, y_s) \text{ with } i = 1, 2, \ldots, m$$

and

$$Y_j = g_j(i_1, i_2, \ldots, i_n, y_1, y_2, \ldots, y_s) \text{ with } j = 1, 2, \ldots, p.$$

Note that these functions are time-dependent equations and that they are valid only when a stable state is obtained. For the memory function a type of flip–flop (FF) may be used. The FFs may be clocked in an asynchronous way and in a synchronous way when one clock controls all the actions.

Table 3.3 Review of some families of user programmable devices (UPLDs).

Name	Full name description	Main features
(P)ROM	Programmable read-only memory	Programmable OR array only; fixed AND array
(F)PLA	Field-programmable logic array	Programmable AND and OR arrays
(F)PGA	Field-programmable gate array	Only programmable AND functions present; no OR field present
PAL/HAL	Programmable array logic/ hard array logic	Programmable AND array only; fixed OR matrix
LCA	Logic cell array	Makes use of static memory technology, offering reprogrammable devices
EPLD	Erasable programmable logic device	Contains an output macro cell (OMC) offering sequential logic circuitry.
PEEL	Programmable electrically erasable logic	Contains an output macro cell (OMC) offering sequential logic circuitry
GAL	Generic array logic	Contains an output macro cell (OMC) offering sequential logic circuitry

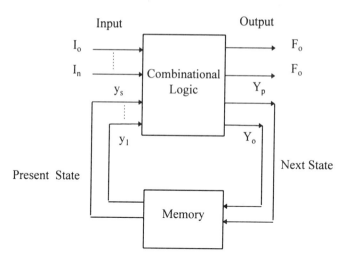

Figure 3.41   Principle of a sequential system.

The PLDs which can provide these sequential circuits have a kind of register built in. Two basic types are distinguished (see also table 3.3):

1   (F)PLS, (field-) programmable logic sequencer
2   PAL/HAL, programmable array logic/hard array logic.

A further detailed discussion requires more space. For further details reference is made to the literature and the documentation of the different manufacturers, which may be obtained from Texas Instruments, National Semiconductor and/or Philips. It must also be noticed that very powerful software packages are available for the design and implementation of of UPLDs. Some well known software packages from independent CAE distributors are ISDATA and Data I/O. For instance, after the design and specification of the required system functions a completely automated generation of test vectors and a testing cycle follow. A further recommendation for the most optimal type of PLD is also indicated. When the suggested PLD is connected to a special programming apparatus the software can be loaded.

3.10   MISCELLANEOUS

In this section we will discuss the well known Wheatstone bridge configuration often applied at the input of a measuring chain. Basically, this circuit can be interpreted as a type of interfacing circuitry between the outside world and the electronics part. The relevance of a Wheatstone bridge is not easily overestimated as interfacing circuitry. Because silicon integration technology has made possible completely new implementations of these bridges, with new features and characteristics, we will present a discussion of an integrated bridge configuration. We will start with the conventional concept.

3.10.1 The conventional Wheatstone bridge configuration

In many cases, the Wheatstone bridge provides the best configuration for input transduction to obtain maximum sensitivity for a certain physical input parameter, be it a temperature, a pressure or any other physical quantity. In particular, when resistive elements are used as the sensing elements to perform the transduction, a Wheatstone bridge configuration is highly preferred. For modulating transducers, as will be seen in more detail in chapter 6, an auxiliary energy source is required. This implies that, for such a type of transducer, a so-called common-mode signal is present. Applying a Wheatstone bridge can suppress this common-mode signal to a large extent. The bridge composed of four resistive elements is illustrated in figure 3.42. Basically, the four branches can be any type of element sensitive to the physical quantity to be measured. A balanced bridge configuration shows an inherent insensitivity to the bridge supply voltage and the impedance of the supply voltage and the detector. This can be derived from the equilibrium condition for a bridge configuration. In general, the output voltage V_o is given by

$$V_o = V_{BB}\left(\frac{R_3}{R_1 + R_3} - \frac{R_4}{R_2 + R_4}\right) \tag{3.107}$$

where V_{BB} is the supply voltage, and the equilibrium condition gives

$$\frac{R_3}{R_1 + R_3} - \frac{R_4}{R_2 + R_4} = 0 \tag{3.108}$$

or

$$R_1 R_4 = R_2 R_3. \tag{3.109}$$

This expression is known as the equilibrium condition. In this fashion, a bridge configuration may be considered as a passive *nulling* element, because the measurand is that output value of a branch setting that satisfies (3.109). In that case, the influence of the supply voltage V_{BB} is put aside. Compare this with the nullor concept of an operational amplifier discussed in section 3.3.1.

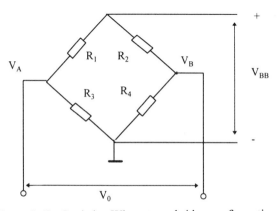

Figure 3.42 Resistive Wheatstone bridge configuration.

It will be obvious that when the *deflection* method is applied the output signal is linearly dependent on the supply voltage, V_{BB}, and so the measurand is in accordance with equation (3.104). For the Wheatstone bridge the following parameters are important.

1   The *differential sensitivity* is defined as the ratio of the change in output voltage due to a change in one of the sensor (input) elements; applying this for a change in R_1 and neglecting the sign we find

$$S_{diff} = \frac{dV_o}{dR_1} = V_{BB} \frac{R_3}{(R_1 + R_3)^2}. \qquad (3.110)$$

2   More important for a bridge configuration is the *relative sensitivity* defined as

$$S_{rel} = \frac{dV_o}{dR_1/R_1} = \frac{dV_o}{dR_1} R_1 = V_{BB} \frac{R_3 R_1}{(R_1 + R_3)^2}. \qquad (3.111)$$

Substitution of $F = R_1/R_3$ in equation (3.108) yields

$$S_{rel} = V_{BB} \frac{F}{(1 + F)^2}. \qquad (3.112)$$

This expression appears to have a maximum for $F = 1$, or for $R_1 = R_3$, which then results in

$$S_{rel} = \frac{1}{4} V_{BB}. \qquad (3.113)$$

Thus a maximum sensitivity is achieved as the transducer elements are pairwise equal, or

$$R_1 = R_3 \qquad R_2 = R_4. \qquad (3.114)$$

By inspection of (3.109) it will be clear that a high supply voltage will also increase the sensitivity. The concept of *resolution* as defined in section 3.4 can also be specified here. Analogously to this definition, it is defined as the smallest change which can be processed by the system.

Numerous examples of interfacing circuitry exists for Wheatstone bridges. They will be discussed in detail in chapter 6.

3.10.2   An integrated Wheatstone bridge

When a resistive bridge configuration is integrated in silicon a perfect matching and temperature tracking for all branch resistors may be expected. This is because the four resistors are integrated on one chip in one batch process. The temperature coefficients are equal and a perfect temperature tracking in silicon is guaranteed. An

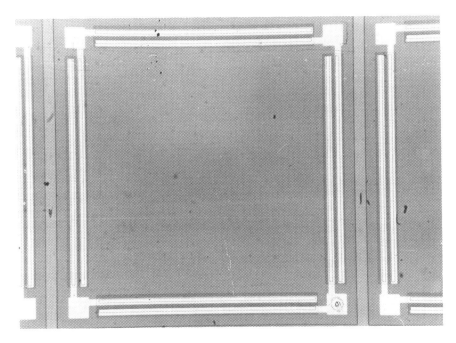

Figure 3.43 Picture of an integrated silicon Wheatstone bridge configuration. Chip size is 4 mm^2.

example of such a bridge is depicted in figure 3.43. The four resistors are integrated on a 4 mm^2 silicon chip and have equal resistances.

In practice, the ideal behaviour is not always achieved and small deviations will occur. A considerable effort is being made in the manufacturing process of all types of integrated bridge transducer to improve this. Often a kind of trimming is executed to meet the specifications with respect to temperature behaviour. Temperature effects are still one of the main reasons for inaccuracy in bridge transducers. These errors are normally known as drift phenomena and will be discussed in chapter 7 in more detail. In particular, the long-term behaviour is often not specified at all.

In addition to trimming techniques, compensation techniques are available. An interesting development for bridge transducers proposes to eliminate the influence of temperature with a thermal feedback principle. Moreover, this guarantees constant sensor parameters. This is discussed in the following section.

3.10.3 The principle of the constant-voltage, constant-current bridge configuration

Generally the bridge output voltage V_o is a function of more than one parameter. We have seen already in (3.107) for the influence of the supply voltage V_{BB} that a change in one of the transducer elements R_i can also be caused by other physical parameters than the desired one alone. Assume that the output voltage V_o is a function of four parameters, which might be related to each other; in other words, they cannot be considered as strictly independent. In the following equation the suggested parameters are the supply voltage, V_{BB}, the supply current, I_{BB}, one or more of the resistive sensing elements R_i and the ambient temperature, T_{amb}:

$$V_o = f\{V_{BB}, R_i, I_{BB}, T_{BB}\}. \tag{3.115}$$

Of course the preferred output voltage is obtained when the output voltage is a function of a single measurand affecting R_i only. When V_{BB}, I_{BB} and T_{amb} have become constants, then this condition is fulfilled and we obtain the following expression:

$$V_o = f(R_i). \tag{3.116}$$

In many cases it appears to be that the ambient temperature changes are one of the most disturbing factors affecting all other parameters. Often a number of different time constants are related to them, which complicates matters again. In that case, it suffices to maintain a constant environmental temperature, T_{amb}. Now consider again the integrated bridge configuration of figure 3.42 and assume that the four equal resistive elements R_i have a positive and linear temperature coefficient. It can then be argued that, if the ratio of V_{BB} and I_{BB} is kept constant, the average level of the magnitude of the four resistor values is constant as well. A constant ratio of the average value of all resistors requires that both the supply voltage and the supply current in the bridge are kept constant. This can be realized by integrating a second bridge and applying a type of thermal feedback in the chip. This double-bridge

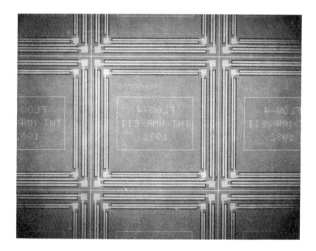

Figure 3.44 Integrated silicon double-bridge configuration. The two bridges have equal resistances because of the constant length-to-width ratio in all eight branch resistors. Chip size is 4×4 mm^2.

configuration is depicted in figure 3.44. The two bridges have equal impedances owing to the constant length-to-width ratio of all eight branch resistors. Because of the relative high thermal conductivity of silicon (80–150 W mK^{-1} depending on the doping concentration) and short intermediate distance of several micrometres, the

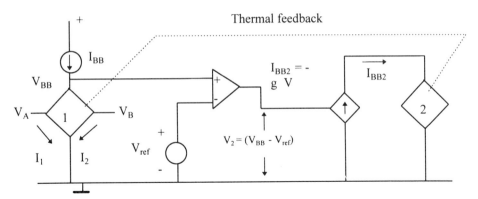

Figure 3.45 Double-bridge circuit for a constant bridge supply voltage V_{BB} and current I_{BB}. Bridges 1 and 2 are integrated on the same silicon substrate and tightly thermally coupled.

two bridges are coupled thermally by the substrate and this guarantees a complete thermal connection.

The circuit needed to produce a constant V_{BB} and I_{BB} is depicted in figure 3.45. The first, outer, bridge is connected to a current source delivering I_{BB}. The second, inner, bridge 2 is connected to a voltage-controlled current source. Its voltage level is determined by

$$I_{BB2} = g_m G(V_{BB} - V_{ref}) \tag{3.117}$$

where g_m is the transconductance of the voltage-controlled current source. Each deviation of V_{BB} from the reference voltage V_{ref} is amplified via the 'comparator' configuration A and fed into the second bridge by the voltage-controlled current source. Hence each deviation results in a change in power dissipation in the second bridge, causing a temperature change in the chip and thus in a change in the first bridge. This change in temperature will continue until V_{BB} is exactly equal to V_{ref}. Consequently, the supply voltage and current of the outer bridge are constant under all measuring conditions when negative feedback is guaranteed.

With this configuration the measurand developed in the outer bridge will be a function of the current division I_1 and I_2 alone, where the sum of the two currents is constant.

This principle is called *thermal feedback*, because a fixed operating point for the sensor is obtained for the supply voltage and current, the ratio of V_{BB} and I_{BB}, which equals the average sensor resistance R_S, and hence the sensor temperature T_S.

It is not strictly necessary that the two bridges are equal, or that the inner circuit is a bridge configuration, but the applied configuration guarantees maximum performance in response time and thermal coupling. Further improvements in response time can be made by micromachining. See also chapter 6. It can be concluded that this principle of operation is well suited for many different input transducer configurations manufactured in silicon technology.

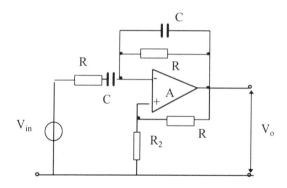

Figure 3.46 Circuit for problems 3.1 and 3.3.

PROBLEMS

3.1 Determine the transfer function $H(j\omega) = V_o/V_{in}$ for the circuit depicted in figure 3.46.

3.2 In figure 3.47 an active filter is shown. (a) Determine the transfer function $H(j\omega) = V_o/V_{in}$. (b) For which ω is $H = 0$? (c) For which ω does $H(j\omega)$ have a real value?

3.3 Determine the Bode plot and the polar plot for the circuit depicted in figure 3.46.

3.4 Determine the pole–zero plot for the circuit depicted in figure 3.47.

3.5 Give a circuit for the non-inverting summation of three voltages V_1, V_2, V_3 and give an expression for the output voltage. The gain factor should equal 12. Design the circuit for minimum offset error.

3.6 Design a circuit to determine the root mean square value of a full sine wave function. An arbitrary AC signal is given; which solution is preferred?

3.7 A precision current source of 50 mA is required. The available operational amplifier can deliver 15 mA. Design a circuit and derive an expression for the current source. Discuss factors which determine the (in)accuracy.

3.8 Design a voltage regulator for 2 V and 1 A. The supply voltage is 3.5–15 V. Try to avoid the implementation of a Zener diode.

3.9 Design a digital multiple-input system which can measure 32 different temperatures with an inaccuracy of <1%. Every 24 h 2048 measurements have to be carried out. The

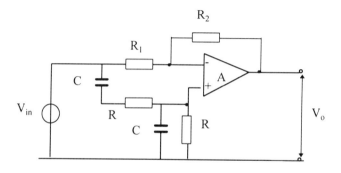

Figure 3.47 Circuit for problem 3.4.

measurements are temporally stored in a memory and have to be read the next day. What is the required memory capacity and how many memory address lines are needed? If a digital multiplexer is applied how many mux control lines are then required?

3.10 What is the main advantage of a successive-approximation ADC? What is the main advantage of the dual-slope ADC?

3.11 What are the main differences between an ASIC and a UPLD?

BIBLIOGRAPHY

[1] Sardijn W 1994 The design of low-power analog integrated circuits and their applications in hearing instruments *PhD Thesis* (*Delft*: *DUP*)

[2] Lewin D 1986 *Design of Logic Systems*, (New York: Van Nostrand Reinhold)

[3] van den Eijnden P M C M 1985 *A Course on Field Programmable Logic* (Eindhoven: Eindhoven University)

[4] Sudar M H, Van Erk M H and Onstee H G 1978 *Digital Instrument Course* part 3 (Eindhoven: Philips)

[5] Regtien P L L (ed) 1978 *Modern Electronic Measuring Systems* (Delft: DUP)

[6] van Putten A F P 1988 *Sensors Actuators* **13** 103–15

[7] van Putten A F P 1990 *IEEE Trans. Instrum. Meas.* **39**

[8] Johnson D E, Johnson J R and Moore H P 1980 *A Handbook of Active Filters* (Englewood Cliffs, NJ: Prentice Hall)

[9] Haskard M R and May I C 1988 *Analog VLSI Design, nMOS and CMOS* (Englewood Cliffs, NJ: Prentice Hall)

Four

SYSTEM SPECIFICATIONS

4.1 INTRODUCTION

In the previous chapters, we have discussed several aspects of measurement set-ups. Recent developments show an increasing impact of and dependence on electronic measurement systems. This occurs in the private and the industrial environments, as well as in the laboratory field. The electronic impact in our society is illustrated in more detail in table 1.1. It will be obvious that maintaining a high standard of quality during a manufacturing process makes it obligatory to perform extensive measurements. In general the overall specifications of a measurement system are highly application and marketing dependent. In all cases we want to know whether the measurement system complies with the claimed specifications. If the system either fails to meet or goes beyond its *predefined* specifications, it is said that the system fails.

Several system specifications are in common use. Very often the technical specifications of a system are badly specified and confusing. Here we attempt to clarify this matter. In general, a global description of a system involves such aspects as

- technical specifications
- quality and reliability
- ergonomic aspects
- environmental constraints
- price–performance ratio
- maintenance requirements.

Most of these aspects will be discussed more fully in chapter 5, concerning reliability, where their interrelations will be shown as well. Some aspects of ergonomics are discussed in chapter 12. In this chapter we shall discuss the most important specifications to which every technical system can be subjected, whether it is a sensor, a part, a component or a complete system. To avoid or reduce certain errors, a discussion of some improving measures will also be presented.

4.2 TECHNICAL SPECIFICATIONS

System specifications characterize the system's behaviour under predefined conditions and are usually provided by the manufacturer. Within certain well defined limits deviations are allowed. These accepted deviations are called *tolerances*. Specifications can cover large and different areas of conditions and in order to be able to obtain the desired information, a detailed knowledge of the system behaviour is required.

4.2.1 Application area

This concerns the physical quantity or quantities to be measured or process to be monitored and/or measured. Other aspects include the required supply voltages such as AC or DC, the environmental conditions and rangings. This is illustrated in the following example.

*   *   *

Example 4.1

A digital temperature measurement system is specified as follows:

- temperature range from −50 to +150 °C
- operating temperature from −10 to +40 °C
- storage temperature from −40 to +75 °C
- power 100 V AC ± 10%, 120 V AC ± 10%, 240 V AC ± 4%, selected by internal switches, 50–400 Hz, 5 W maximum
- display 7.5 mm high, $3\frac{1}{2}$-digit, 7-segment LCD, with decimal point location and temperature indication
- analogue output, 0–10 V for complete range and load > 1 kΩ, 0–20 mA or 4–20 mA, load < 600 Ω
- relative humidity < 90%.

All other specifications are only applicable within the specified range.

*   *   *

Example 4.2

Of a digital voltmeter FLUKE 8050 A the following (short) list of specifications is known. All electrical specifications given apply for an operating temperature of 18 °C–28 °C. Functions are DC volts, AC volts linear and decibels, DC and AC, resistance diode test, conductance, relative. The following table for DC measurement may be given.

Range	Resolution	Accuracy for 1 year
±200 mV	10 μV	
±2 V	100 μV	
±20 V	1 mV	±0.03% of reading + 2 digits
±200 V	10 mV	
±1000 V	100 mV	

- input impedance, 10 MΩ in parallel with <100 pF, all ranges
- common-mode rejection ratio, >90 dB at DC, 50 Hz or 60 Hz, 1 kΩ unbalanced
- common-mode voltage maximum, 500 DC or peak AC
- response time to rated accuracy, 1 s maximum
- maximum input, 1000 V DC or peak AC or continuous, less than 10 s.

For AC measurements the following specifications are valid (AC volts (true RMS responding, AC coupled); voltage readout accuracy, ±(percentage of reading + number of digits), between 5% of range and full range).

Input voltage	Resolution	Range	20 Hz†	45 Hz	1 kHz	10 kHz	20 kHz	50 kHz
10 mV–200 mV	10 μV	200 mV						
0.1 V–2 V	100 μV	2 V	1% + 10		5% + 10	1% + 10	5% + 10	
1 V–20 V	1 mV	20 V						
10 V–200 V	10 mV	200 V						
100 V–750 V	100 mV	750 V			Not specified			

†Typically, 3 to 5 digits of rattle will be observed at full scale at 20 Hz.

dB Voltage (dB, dBm, dBW and dBV)

Voltage measurements are allowed in decibels referenced to any user-selected reference level or any selectable reference impedances.

Current measurements

Standard current measurements can be made from 10 nA to 2 A DC and 10 μA to 2 A AC true RMS.

Resistance measurements

Standard resistance measurements from 10 mΩ to 20 MΩ are possible.

Conductance (1/Ω)

Fast noise-free resistance measurements can be made up to 100 000 MΩ.

Relative measurements

Any input applied when the RELATIVE button is depressed to the ON position is taken as the zero reference point.

Other specifications are often related to the environmental (operating) conditions, such as operating temperature, data storage temperature and relative humidity.

Electromagnetic compatibility

For every system electromagnetic compatibility (EMC) specifications are also of increasing importance. These EMC requirements are related to the susceptibility and the emissivity of interference. In January 1996 EMC legislation came into effect for the European market. This will be discussed in more detail in chapter 8. For complete details the manufacturers' manuals should always be consulted.

*   *   *

In these two examples, we have seen quite a number of technical terms and specifications (table 4.1). We start now with a overview of the most frequently used parameters and specifications. Then we will discuss these in more detail.

4.2.2 Resolution

We have already discussed resolution in the previous chapter, but for uniformity the definition is repeated here in a slightly different way. Resolution is defined as the smallest change at the input which can be processed by the system. Resolution is defined in different fashions, e.g.

1 the smallest change of the input value, δx_{min}
2 the smallest change of the input value related to the maximum input value, $\delta x_{min}/x_{max}$
3 the reverse of the above definition, or $(\delta x_{min}/x_{max})^{-1}$.

The last two definitions are also called relative resolutions. There is no conformity in the application of these definitions.

*   *   *

Example 4.3
Firstly, consider a four-digit decimal display with a fixed decimal point, one position from the right. The specified resolution is 0.1 unit. The maximum value to be displayed is 999.9, and consequently the relative resolution is about 10^{-4} or 10^{+4}.

Secondly, we have to determine the relative resolution of a digital voltmeter (DVM) in voltage mode, in current mode and in resistance mode. From the specifications delivered by the manufacturer, we can read the resolutions in the different modes. Assume they are specified as follows:

$$10\text{ mV at } \pm 200\text{ V} \qquad 0.1\ \mu\text{A at } 2\text{ mA} \qquad 0.1\ \Omega \text{ at } 2\text{ k}\Omega.$$

Then the respective relative resolutions are as follows:

in voltage mode $200/10 \times 10^{-3} = 2 \times 10^4$
in current mode $2 \times 10^{-3}/0.1 \times 10^{-6} = 2 \times 10^4$
in resistance mode $2 \times 10^3/0.1 = 2 \times 10^4$.

Note that in each mode the same relative resolutions are obtained and that just a dimensionless number for the resolution is obtained.

* * *

4.2.3 Accuracy and inaccuracy

The *accuracy* of a system is a figure of merit which describes the *probability* that the measurand is correct. A lot of confusion exists in this statement, because often the specified accuracy of a system is related to its *inaccuracy*. If it is claimed the accuracy is 1%, we certainly do not mean the accuracy is 1%, but the inaccuracy is 1% and hence the accuracy is 99%. Therefore we say the accuracy equals 1 − inaccuracy and so they are the complements of each other.

Table 4.1 Review of frequently used part and system parameters.

Parameter	Expression and/or symbol	Description
Absolute resolution	δx_{min}	The smallest change at the input which can be detected at the output
Relative resolution	$\delta x_{min}/x_{max}$	The ratio of the smallest change at the input which can be processed and the maximum allowed change at the input
Absolute sensitivity	$S = \delta y/\delta x$	The ratio of the change at the output and the change of the measurand at the input, sometimes called differential sensitivity
Relative sensitivity	$S = \delta y/\delta x\vert_{x=0}$	*Standardized* absolute sensitivity with respect to output value when the input value is zero
Cross-sensitivity (parasitic)	$S = \delta y/\delta x_i$ $i = 1, 2, 3\ldots$	The change of the output value caused by more than one measurand
Directional sensitivity	$S = f(\varphi) = \delta y/\delta \varphi_i$	The sensitivity as a function of the incident angle
Accuracy	—	The probability that the measurand is in agreement with an accepted standard value
Inaccuracy or uncertainty	—	$1 -$ the accuracy
Absolute error	$m - \mathrm{tv}$	The measurand, m, minus the true value, tv
Absolute value of the error	$\vert m - \mathrm{tv}\vert$	The absolute value of the measurand minus the true value
Fractional absolute error	$(m - \mathrm{tv})/\mathrm{fsd}$	The ratio of the absolute error and the value of full-scale deflection (fsd)
Fractional error	$\vert m - \mathrm{tv}\vert/m$	The ratio of the absolute value of the error and the measurand
Non-linearity	Δx_{max}	The maximum deviation of a specified approximated linearity curve
Relative non-linearity (static)	$\Delta x_{max}/x_{max}$	The ratio of the maximum deviation of a specified approximated linearity (best-fit) curve and the maximum value of that curve (jamming, clipping, saturation, dead zone)
Slew rate (dynamic)	$\mathrm{V}\,\mu\mathrm{s}^{-1}$	The time rate of change of the output voltage for a voltage step applied at the input
Slew rate (for sinusoidal signals only)	f_{max}	The maximum frequency at which the output voltage swing can be obtained without distortion
Bandwidth	$B = f_{nom} - f_{3dB}$	The difference in frequencies between the nominal value of power, voltage or current and their corresponding 3 dB dropped value
Hysteresis	—	The deviation of the output signal as a function of the input signal regardless of the direction
Offset	mV, μV	The value of the output signal when the input is zero (sometimes referred to the input)
Drift	$\mu\mathrm{V}\,\mathrm{K}^{-1}$ or $\mu\mathrm{V}\,\mathrm{month}^{-1}$	A slow variation of the output signal as a function of temperature or time (hours, months or years)
Dynamic range	x_{max}/x_{min}	The ratio of the maximum and minimum signal level which can be handled within a specified accuracy
Temperature range	$T_{min} - T_{max}$ (°C)	The temperature range in which the system can operate within its specifications
Reliability	$R(t)$	The probability that the system will function within its predefined specifications after a predefined period of time
Mean time between failure	MTBF (h)	For a repairable system the average time between two failures over its total life cycle
Power supply rejection ratio	PSRR (dB)	The ratio of the change in supply voltage and a differential mode voltage producing the same output voltage
Common-mode rejection ratio	CMRR (dB)	The ratio of a common-mode voltage and a differential-mode voltage producing the same output voltage

The inaccuracy of a measuring system is strongly dependent on the way the information is displayed, because measuring ultimately means displaying a quantity on a certain symbolic scale. It will be obvious that inaccuracies are always involved in every system. Causes of inaccuracies include calibration faults, tolerances of components, interference, mismatching of stages, applying the wrong method and environmental conditions and influences. We distinguish two types of error: (1) the absolute error and (2) the fractional error.

1 *Absolute error*
(a) The absolute error can be defined as the difference between the measurand, *m*, and the true value, tv, or

$$\text{absolute error} = m - \text{tv}. \tag{4.1}$$

(b) The absolute error value is defined as the absolute error, neglecting the sign, or

$$\text{absolute error value} = |m - \text{tv}|. \tag{4.2}$$

(c) The fractional absolute error is defined as the ratio of the absolute error and the value of full-scale deflection fsd, often expressed as a percentage. In mathematical form

$$\text{fractional absolute error} = \frac{m - \text{tv}}{\text{fsd}} \times 100\%. \tag{4.3}$$

*   *   *

Example 4.4
Suppose the measured value is 19.8 V, the true value is 19.9 V and full-scale deflection is 100 V; then

$$\text{absolute error} = 19.8 \text{ V} - 19.9 \text{ V} = -0.1 \text{ V}$$
$$\text{absolute value of error} = |-0.1 \text{ V}| = 0.1 \text{ V}$$
$$\text{fractional absolute error} = (0.1/100) \times 100\% = 0.1\%.$$

2 *The fractional error*
The fractional error is defined as the ratio of the absolute error value and the measurand, often given as a percentage value; hence

$$\text{fractional error} = \frac{|m - \text{tv}|}{m} \times 100\%. \tag{4.4}$$

*   *   *

Example 4.5
Applying the same values as in example 4.4 we find

$$\text{fractional error} = |(19.8 - 19.9)/19.8| \times 100\% = 0.5\%.$$

*   *   *

The manufacturer has to provide the inaccuracy of the measuring system; in its turn every measurement must be accompanied by its inaccuracy, otherwise its value is meaningless.

Constant *independent* errors can be considered as additive errors, while errors depending on the measured value can be considered as multiplying errors. For instance, a measured voltage can be expressed in the following relationship in which γ is a multiplicative factor and $V_{\text{error}}(E_1, \ldots, E_n)$ includes the additive errors, which do not depend on the measurand measurement:

$$V_{\text{out}} = \gamma V(S) + V_{\text{error}}(E_1, E_2, \ldots, E_n). \tag{4.5}$$

Multiplying errors often deliver a constant fractional error and are strongly related to the inaccuracy of the display method applied. This is illustrated in figure 4.1 where three types of error are depicted with their functional behaviour and their absolute and fractional behaviour as a function of the deflection.

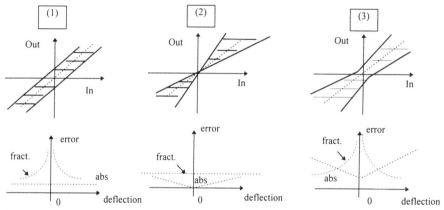

Figure 4.1 The differences between additive and multiplicative errors: (1) additive error with constant absolute error; (2) multiplicative error with constant fractional error; (3) total inaccuracy.

4.2.4 Sensitivity

Concerning the Wheatstone bridge in section 3.6 we have already met a definition of sensitivity. Here we shall define sensitivity in a broader scope. For every measuring system the sensitivity can be defined as the ratio of a change y at the output, caused by a change x at the input, or

$$S = \frac{y}{x} \quad \text{or} \quad S = \frac{\delta V_{\text{out}}}{\delta x_i} \tag{4.6}$$

where, for instance, V_{out} is an output voltage and x_i is an arbitrary input value. When a constant sensitivity is involved, then a single number suffices; otherwise, the sensitivity changes and must be specified for each point. Sensitivity can be expressed in volts per ampere, centimetres per volt or volts per division, for example, for a current-to-voltage converter, a recorder or an oscilloscope respectively. In most cases, a measuring system also appears to be sensitive to other parameters in addition to the desired one. These sensitivities are called parasitic sensitivities. They are sometimes specified with respect to the sensitivity of the quantity to be measured. In table 4.1 two other sensitivities are mentioned, i.e. the cross-sensitivity and the directional sensitivity. Usually, the cross-sensitivity is a parasitic effect. In particular in sensors cross-sensitivity is a disturbing effect affecting the output signal with undesired physical parameters.

$$* \quad * \quad *$$

Example 4.6
Consider a piezoelectric positioner with a voltage sensitivity of $1 \, \text{mV} \, \mu\text{m}^{-1}$ and a temperature sensitivity of $-10 \, \mu\text{V} \, \text{K}^{-1}$. From this it can be derived by division that a change of $10 \, \mu\text{V}$ corresponds to a change in position of 10 nm. Thus the parasitic temperature sensitivity is $-10 \, \text{nm} \, \text{K}^{-1}$.

$$* \quad * \quad *$$

Example 4.7
Consider a simple analogue temperature measurement system with a sensitivity of $100 \, \text{mV} \, \text{K}^{-1}$. In the system an amplifier is involved which appears to have a temperature sensitivity resulting in an output change of $0.1 \, \text{mV} \, \text{K}^{-1}$. The parasitic temperature sensitivity can be expressed as $1/1000 \, \text{K} \, \text{K}^{-1}$ or $1 \, \text{mK} \, \text{K}^{-1}$. In other words, a temperature change in another part of the system than the temperature sensor itself can affect the output.

$$* \quad * \quad *$$

Ultimately, the sensitivity can be made extremely high by electronic means, theoretically without any limit. However, going beyond certain limits no longer makes sense, because the real physical limit in sensitivity is *noise*. Noise is just a collective noun for a number of mechanisms which are inherently present in every component. The subject of noise is dealt with in chapters 9 and 10. We stress already here that noise and interference have different causes and will be treated separately. Many authors confuse these two mechanisms.

4.2.5 Linearity
As can be seen from (4.6) if a strict linear behaviour of the transfer characteristic is involved, the sensitivity is independent of the x value. If the system is non-linear then the sensitivity is a function of x and y. It will be clear that in practical circumstances a linear behaviour is preferred, e.g. because the sensitivity can be specified with one number. This is illustrated in a exaggerated form in figure 4.2.

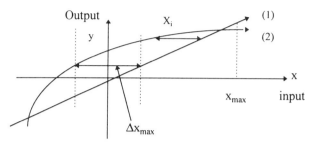

Figure 4.2 Plot of the output versus input, $y = f(x)$; curve (2) is the real transfer characteristic and curve (1) the approximated transfer characteristic.

The *non-linearity* of a system can be defined as the maximum deviation of the specified approximated linearity. In figure 4.2 the non-linearity can be denoted by Δx_{max}, or expressed as *relative non-linearity*:

$$\text{relative non-linearity} = \frac{\Delta x_{max}}{x_{max}}. \tag{4.7}$$

This value can be expressed in a percentage. In figure 4.3 other examples of frequently occurring non-linearity mechanisms are depicted, i.e. (a) saturation, e.g. in a transistor characteristic, (b) jamming, e.g. the maximum supply level is achieved, and (c) dead zone, e.g. when a hysteresis is involved. These non-linearities are called *static*, because they can occur when slowly varying signals are involved. When rapid changes are involved another type of non-linearity comes into scope which is called *slew rate*. Slew rate is defined as the time rate of change of the output voltage for a voltage step applied at the input. For instance, the slew rate of an amplifier is specified as $100 \text{ V } \mu\text{s}^{-1}$. For sine-wave-shaped signals an alternative definition is used: it is the maximum frequency at which the peak output voltage swing can be obtained without distortion. The effect of slew rate is shown in figure 4.4. As is illustrated, changes in the input are time limited at the output; in other words the slope is always less than $\pi/2$.

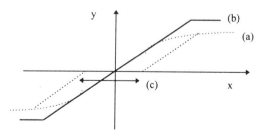

Figure 4.3 Examples of static non-linearities: (a) saturation; (b) jamming; (c) a dead zone.

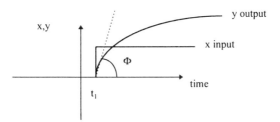

Figure 4.4   The effect of slew rate of an amplifier.

4.2.6   Offset and drift

Offset can be defined as the deviation of the output signal from zero when the input is zero. A difference is made between input offset and output offset. Drift can be defined as a slow variation of a zero setting. We will not go in detail here, because a comprehensive discussion of offset and drift is presented in chapter 7. Also some precautions to eliminate drift in sensors will be discussed. Here we provide but an example of the specifications of an operational amplifier.

*   *   *

Example 4.8

In every databook of operational amplifiers specifications can be found concerning offset and drift of a specific amplifier. For instance for instrumentation amplifiers the following specifications are known.

Specification	Typical	Maximum
Input offset voltage	$\pm 40.0\,\mu\mathrm{V}$	$100.0\,\mu\mathrm{V}$
Drift	$\pm 0.4\,\mu\mathrm{V}\,\mathrm{K}^{-1}$	$1.8\,\mu\mathrm{V}\,\mathrm{K}^{-1}$
Long-term stability	$0.4\,\mu\mathrm{V}\,\mathrm{month}^{-1}$	$2.0\,\mu\mathrm{V}\,\mathrm{month}^{-1}$
Offset current	$10.0\,\mathrm{nA}$	$75.0\,\mathrm{nA}$
Slew rate	$1.7\,\mathrm{V}\,\mu\mathrm{s}^{-1}$	$1.9\,\mathrm{V}\,\mu\mathrm{s}^{-1}$

*   *   *

4.2.7   Rejection factor

In section 3.3.1 we introduced the concept of the differential amplifier in a theoretically ideal form with infinitely high open-loop gain, infinitely high input impedance, zero input current and zero output impedance. As a building block it is usually called an operational amplifier. Of course, a real operational amplifier does not have these ideal properties and to characterize the real behaviour some characteristic figures of merit are often used.

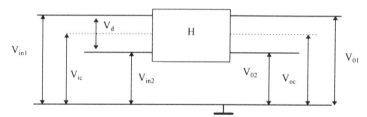

Figure 4.5 An operational amplifier with a double-ended output as a black box.

We have also seen that one of the most powerful characteristics of a differential amplifier is its ability to amplify a very small voltage difference between two relatively large signals, without amplifying the two large signals. The advantage of such a building block will be quite clear, especially in circumstances where a large common-mode signal is present at both inputs, no matter whether it is a signal level or interference.

Let us now consider a differential amplifier with a differential input and output, as a black box. This is illustrated in figure 4.5 and it is the most general case. For the input voltages the following definitions will be applied:

$$V_{id} = V_{in1} - V_{in2}.$$ (4.8)

which is called the differential-mode input voltage. The common-mode input voltage is defined as

$$V_{ic} = (1/2)(V_{in1} + V_{in2}).$$ (4.9)

When the amplifier does not have perfect symmetry, for whatever reason, we can state that a common-mode voltage at the input generates a differential-mode voltage at the input and hence a common-mode voltage at the output. In the same way a differential-mode signal at the input can generate a differential-mode and a common-mode signal at the output, or

$$V_{od} = A_{dd} V_{id} + A_{cd} V_{ic} \qquad V_{oc} = A_{do} V_{id} + A_{cc} V_{ic}$$ (4.10)

where

$$V_{od} = V_{o1} - V_{o2} \qquad V_{oc} = (1/2)(V_{o1} + V_{o2})$$ (4.11)

and by definition A_{dd} is the differential to differential gain, A_{dc} is the differential to common-mode gain, A_{cd} is the common-mode to differential gain and A_{cc} is the common-mode to common-mode gain; by definition we have also

$$A_{dd} = \frac{V_{od}}{V_{id}} \text{ at } V_{ic} = 0 \qquad A_{dc} = \frac{V_{oc}}{V_{id}} \text{ at } V_{ic} = 0$$

$$A_{cd} = \frac{V_{od}}{V_{ic}} \text{ at } V_{id} = 0 \qquad A_{cc} = \frac{V_{oc}}{V_{ic}} \text{ at } V_{id} = 0.$$ (4.12)

From (4.10) we see that, generally, the differential-mode output voltage and the common-mode output voltage are the superposition of two contributions. The ideal situation exists when A_{cd} equals zero. Note that the gain factors A_{dc} and A_{cc} are of less importance, although both should be as small as possible. With help of (4.10) we can now define the following two figures of merit for an operational amplifier:

1 The *discrimination factor*, F, is defined as the ratio of the differential to differential gain and the common-mode to common-mode gain, or

$$F \overset{\text{def}}{=} \frac{A_{dd}}{A_{cc}}. \tag{4.13}$$

The discrimination factor, F, is of importance only when a double-ended output is involved, as is illustrated in figure 4.5. The factor F makes no sense when no common-mode signal at the output is present. However, a common-mode voltage at the input can always generate a differential-mode signal at a single-ended output. The most common operational amplifier configurations have a single-ended output.

2 The *rejection factor*, or more precisely the *common-mode rejection ratio* (CMRR), is defined as the ratio of the differential to differential gain and the common-mode to differential gain, or

$$H \overset{\text{def}}{=} \frac{A_{dd}}{A_{cd}}\bigg|_{V_{\text{out}} = \text{const.}} \tag{4.14}$$

or in decibels

$$20 \log|H| = 20 \log \left| \frac{A_{dd}}{A_{cd}} \right|. \tag{4.15}$$

The CMRR is normally expressed in decibels and is the most crucial factor of a differential amplifier. It describes the ability of the amplifier to eliminate a common-mode signal connected at the input. This is strongly related to the *symmetry* of the amplifier itself and its input configuration. It will be clear that the CMRR should have a large value, e.g. 120 dB is a good value, and 70 dB is a rather poor value. Sometimes the reciprocal value of H is used.

Finally, it will be clear that the differential to differential gain A_{dd} must be high in contrast to A_{cc}, which must have an extremely low value. Thus a high value for F is preferred. It should be noted that F can always be improved, while the rejection factor H as a system-dependent quantity is a fixed number. The CMRR H is always provided by the manufacturer, whereas F is not.

It is noticed here that the CMRR can also be influenced by the symmetry of the input circuitry connected to the amplifier, temperature effects of the input circuitry and power supply variations. We then talk about temperature rejection ratio (TRR) and power supply rejection ratio (PSRR).

When the CMRR H is considered under conditions of the same output voltages for both A_{dd} and A_{cd}, then the CMRR can be expressed directly in voltages. By substitution of (4.11) we then find for H with $A_{dd} = V_{od}/V_{id}$ and $A_{cd} = V_{od}/V_{ic}$

$$H = \frac{V_{ic}}{V_{id}}\bigg|_{V_{out}=const.} \tag{4.16}$$

We state here that any common-mode influence which generates a differential-mode voltage at the output can be expressed in a specific type of rejection factor, whatever the cause may be.

As will be shown in chapter 8, it is important that the conditions under which the CMRR is measured are specified. In particular, the frequency will affect the CMRR. In that chapter we shall repeat some important aspects of the differential amplifier. An example of a series system will illustrate the above-given concepts in some more detail.

*   *   *

Example 4.9
In figure 4.6 a series system is drawn for which the CMRR must be determined. Assume that the first system has a CMRR of H_1, a discrimination factor F_1 and a gain factor A_{dd1}. The second system has a single-ended output and has a CMRR H_2 and a gain A_{dd2}. Because of the single-ended output there is no F_2 involved.

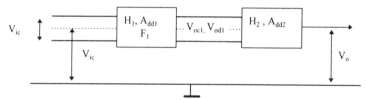

Figure 4.6 A series connection of two systems.

Applying (4.10) to the differential output voltage of the amplifier we can write

$$V_{od1} = A_{dd1}\,V_{id1} + A_{cd1}\,V_{ic} \tag{4.17}$$

or, because $1/H_1 = A_{dd1}/A_{cd1}$,

$$V_{od1} = A_{dd1}\left(V_{id1} + \frac{V_{ic}}{H_1}\right). \tag{4.18}$$

For V_{oc1} we find with (4.10)

$$V_{oc1} = A_{dc1}\,V_{id1} + A_{cc1}\,V_{ic}$$

or

$$V_{oc1} \approx A_{cc1}\,V_{ic} \tag{4.19}$$

because $A_{\text{dc1}} V_{\text{id1}} \ll A_{\text{cc1}} V_{\text{ic}}$. For the second stage we find

$$V_{\text{o2}} = A_{\text{dd2}} V_{\text{od1}} + A_{\text{cd2}} V_{\text{oc1}} . \tag{4.20}$$

Substituting (4.18) and (4.19) in (4.20) and rearranging, we find for the output voltage V_{o2} of the second stage

$$\frac{1}{H_{\text{t}}} = \frac{1}{H_1} + \frac{1}{H_2 F_1} . \tag{4.21}$$

The influence of H_1 is largest; the influence of H_2 is reduced by a factor of F_1.

* * *

Example 4.10
A system with a specified CMRR of 10^4 is used to compare two voltages of about 10 V. Then the lowest voltage difference which can be detected with this system is determined by the rejection factor. This implies that a common-mode input of 10 V results in an output of ± 1 mV. In other words, the accuracy is guaranteed to $10/10^4 = 1$ mV. If the same system is applied to measure a voltage difference of 10 mV with a common-mode signal of again 10 V, then the inaccuracy is again 1 mV, or the inaccuracy as a percentage can be derived from

$$V_{\text{in1}} - V_{\text{in2}} \pm \frac{V_{\text{CM}}}{H} = 10 \pm 1 \text{ mV}$$

and this equals a rather poor accuracy of 90% (inaccuracy of 10%).

* * *

4.2.8 Dynamic range
The (dynamic) range of a system is defined as the ratio of the maximum and minimum signal levels which can be handled by the system within a specified accuracy.

The maximum signal level will be restrained by the operational limits in which the transfer of the signal is defined. For instance, the supply voltage will form a natural limit. The minimum signal level to be handled is limited by the so-called boundary sensitivity again within specified accuracy. Natural limits here are for instance thermal voltages, noise, drift and hysteresis. Usually, the dynamic range is expressed in a single number, in a percentage, or in decibels, for instance 10^3 equals 60 dB. Some aspects of the dynamic range can be clarified best by discussing an example in which the measurement of a DC voltage is performed.

* * *

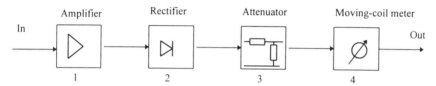

Figure 4.7 Measurement set-up for determining the dynamic range of DC voltages (courtesy of Delft University Press).

Example 4.11

For the DC voltage measurement system shown in figure 4.7, with a required fractional inaccuracy of 4%, the dynamic range must be determined. The system consists of four stages.

1 The amplifier stage has a gain setting between 1 and 1000 times. The equivalent noise voltage is 10 μV, and the maximum voltage range for both input and output is 10 V.
2 The bridge rectifier consists of four diodes with a dead zone of 0.1 V. The maximum allowed reverse voltage is 100 V.
3 The attenuator is composed of resistors to match the moving-coil meter to the system.
4 The moving-coil meter has an error class of 0.5, which shows full-scale deflection at 100 mV and 100 μA.

The claimed fractional inaccuracy requires an inaccuracy of less than 1% for every stage involved. We will consider every stage in detail and hence presume that the inaccuracy is equally divided over all four stages, i.e. 1% each.

For the first stage we can derive from the data that a required inaccuracy of 1% equals an inaccuracy of 10 μV and hence the lowest acceptable value to be measured is 100 times larger, or 1 mV. When the gain setting is 1000 the output voltage is 1 V and this value compared with the allowed 10 V maximum for input and output delivers a dynamic range for the amplifier's stage of merely $10/1 = 10$. When the gain setting is only 1 we have a dynamic range of $10/10^{-3} = 10^4$.

The bridge rectifier in the second stage has a dead zone of 0.1 V; if here again a 1% inaccuracy is claimed, then this dead zone equals the 1% inaccuracy and the lowest acceptable value therefore is 10 V. The maximum allowed reverse voltage is 100 V and so the dynamic range of this stage is $100/10 = 10$.

As can be seen already from these two stages the voltage matching fails to be completely corrected. This is illustrated in figure 4.8. To solve this matching problem a second amplifier stage might be introduced, but this requires a supply voltage of 100 V and so is not very attractive. A better solution is to diminish the rectifier's dead zone by a factor of 10. This can be realized by introducing a small voltage source of 0.1 V series connected with the rectifier's output. The new threshold value of the rectifier becomes 0.01 V which then forms the 1% inaccuracy. This delivers a new acceptable lowest value of 1 V and so a better voltage matching is created.

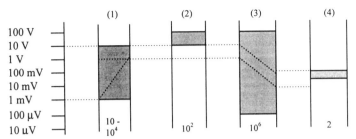

Figure 4.8 Schematic diagram of the dynamic ranges of the four stages.

The third stage offers no problems at all and is merely required to match the moving-coil meter. The lowest value will be determined by thermal voltages which have an order of magnitude of $1\,\mu V$ (1%); hence, $100\,\mu V$ is the lowest acceptable value with an inaccuracy of 1%. The upper limit can be claimed to be 100 V, and then the dynamic range is $100/100 \times 10^{-6} = 10^{6}$. In this stage, complete voltage matching is easily obtained.

The last stage is the moving-coil meter and this appeared to be the most crucial one. An error class of 0.5 implies an absolute inaccuracy of 0.5% of full-scale deflection, which equals here 0.5 mV. Then the lowest acceptable value within the claimed inaccuracy is 100 times larger, or 50 mV. This means a dynamic range of only $100/50 = 2$. It will be obvious that the last stage determines the overall accuracy. This is also shown in figure 4.8. For clarity the results are summarized in table 4.2.

Of course we can allow greater inaccuracy to improve the dynamic range, but that is not the right way to handle things. To solve this problem, logarithmic compression can be applied, which requires a type of logarithmic amplifier. The principle of operation is based on diminishing the ratio of the maximum and minimum available signal, as discussed already in chapter 3.

For instance, the transfer function of a logarithmic amplifier is described by

$$V_{o} = \log_{10}\left|\frac{V_{in}}{V_{1}}\right| \tag{4.22}$$

in which V_{in} is the input voltage and V_{1} is the lowest acceptable input voltage for accuracy reasons. Then it can easily be seen that the ratio of maximum and minimum voltages is logarithmically reduced.

* * *

Table 4.2 Results from example 4.11.

Stage	Error (1%)	Lowest acceptable value (100 × error)	Upper acceptable value	Dynamic range
Amplifier	$10\,\mu V$	1 mV	10 V	10 (gain, 1000)
Amplifier	$10\,\mu V$	1 mV	10 V	10^{4} (gain, 1)
Rectifier	0.1 V	10 V	100 V	10
Attenuator	$1\,\mu V$	$100\,\mu V$	100 V	10^{6}
Moving-coil meter	0.5 mV	50 mV	100 mV	2

Example 4.12
Suppose the amplifier in example 4.11 is replaced by a logarithmic amplifier with transfer function (4.22); then the output voltage changes from 0 to 4 V. This is in contrast to the input which can change over four decades, and hence a dynamic range of 10^4 is converted into 4. Exponential expansion can be implemented, after logarithmic compression, when required.

$$* \qquad * \qquad *$$

4.2.9 Reliability
Reliability, $R(t)$, is defined as the probability that a system will function after a predefined number of operating hours within its predefined specifications. For parts, components and systems it is of increasing importance that figures can be given (and often are demanded by contract) about the reliability of systems. Our whole society depends on the reliability of systems. Therefore, chapter 5 is devoted to reliability of technical systems. As will be shown, the mean time between failure (MTBF) in hours will usually be given.

4.2.10 The number of measurements per unit time
This topic is the main reason why it is becoming common to implement computing power in a measurement system. In processes often hundreds of measurements have to be made in a very short time. Collected data have to be edited, calculated, stored, compared with set-points and, after conversion, have to be sent out again. Finally, calibration, automatic nulling and autoranging are often highly desirable features to implement in a measurement system. It will be quite clear that this is impossible to perform by hand. There is an increasing need for fully automated multichannel measurement systems. Aspects of these systems are discussed in chapters 2 and 3.

To emphasize the power of a modern measurement system we compare a classical configuration as illustrated in figure 4.9 with a sophisticated system as depicted in figure 4.10. The differences are summarized in table 4.3.

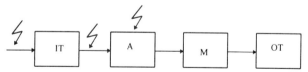

Figure 4.9 Classical measurement set-up. IT is input transducer, A is amplifier, M is modulator, OT is output transducer. The zig-zag arrows represent interference.

The classical measurement system consists of a one-channel configuration and is composed of at least three stages. The first stage represents a type of input transducer; the second stage consists of a signal modifier or conditioner which for instance implies a filter, and an amplifier; the third stage is a type of output transducer or display.

In figure 4.10 the multichannel arrangement shows n different input transducers, n amplifiers, an interface, a multiplexer, or scanner, a low-pass filter, a sample-and-hold function, an A/D converter, a hard or soft disk drive unit, a printer and plotter,

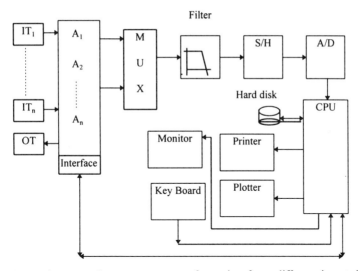

Figure 4.10 Automated measurement configuration for *n* different input signals.

a keyboard, a monitor and a computer. The amplifiers can be made of so-called modular blocks and are easily adapted to the quantity to be measured. Usually, the amplifiers and filters are software programmable, this means that the number of used channels, the gain factors and the cut-off frequency of filters are user programmable.

Table 4.3 Comparison of conventional and computer-based systems.

Item	Conventional system	Computer-based system	Comments
Input transducer	1	1–*n*	May be different
Transducer supply voltage	Fixed	Variable	Programmable
Scanner/mux	None	1	Programmable
Amplifier	1	1–*n*	Programmable
Signal conditioner	—	1–*n*	Programmable
Modifier	1	1–*n*	Programmable
Filtering	1	1–*n*	Programmable
Sample and hold	—	1–*n*	Programmable
A/D and D/A conversion	—	1–*n*	8–16 bits
Storage, hard and soft	—	+	>1 Gb
Interfacing	—	+	All types
Output transducer	1	1–*n*	All types
Multiplier	—	1	1
Calibration	—	*n*	All channels
Communication	—	+	Serial, parallel
Autoranging	—	1–*n*	All channels
Autonulling	—	+	All channels
Rectifier	1	1–*n*	+
Calculations	—	Any	+
Number of measurements per unit time	1	1–*n*	Clock frequency dependent

A good example of a multichannel computer-based measuring system is the flight computer of an aircraft in which all of the above-mentioned features can be found.

4.2.11   Environmental conditions

Environmental conditions describe the specified circumstances to which the system is subjected during manufacturing. It also describes the environmental conditions under which a system may be used to remain within the specified regions: aspects such as shock and vibration resistances, humidity, laboratory or missile launch conditions and so on. We will see that the reliability of a system is also determined by the environmental conditions. A comprehensive discussion can be found in chapter 5.

Before we start our discussion of how to improve the specification of an electronic system, a review of types of error and their causes is given.

4.3   TYPES OF ERROR

Related to the specifications of a system are of course its limits, so we can say that every system has a limiting accuracy because of certain errors. Every stage will introduce some errors. These errors can be classified by their cause of occurrence. Three different types of error can be distinguished:

1   systematic errors
2   conditional errors
3   stochastic errors.

Obviously, all these errors will influence the accuracy. We will discuss these in some more detail.

4.3.1   Systematic errors

A systematic error is one that has a constant character and is system or process dependent. Good examples of systematic errors are

● the offset of an amplifier
● a non-linear behaviour of the process
● an undesired but yet required fixed scaling factor.

Such errors can be eliminated by an automatic nulling, or by a calibration procedure applying, for instance, a digital memory. Sometimes a ratio measurement, as discussed in chapter 2, can offer good results. Systematic errors are not very difficult to eliminate, if they are constant. Offset errors will be discussed in more detail in chapter 7.

4.3.2 Conditional errors

Conditional errors are those which are caused by external influences such as electro-magnetic interference (EMI) and electro-magnetic pulses (EMPs). Conditional errors are strongly dependent on the environmental and operational conditions the system is operating in. Measures to eliminate these types of error are as follows:

1 the choice of the right components which are less sensitive to the interfering source the system is subjected to
2 isolation, guarding, and stabilizing the system from the interfering source
3 compensation of the interfering source by an equal yet reverse signal, for instance applying a two-sensor configuration in one system can be considered as a compensation technique.

Generally the operational conditions have to be specified. Some types of conditional error and their precautions will be discussed in more detail in chapter 8 concerning guarding and shielding. In particular the new (European) EMC legislation procedures have to be outlined here.

4.3.3 Stochastic errors

Stochastic errors can be defined as caused by a stochastic process which basically has to be considered as a fundamental phenomenon in applied materials and components. Noise is one of the best examples of a stochastic process; others are the temperature of a material, the pressure in a gas volume and the current flow in a material. Some measures to reduce the noise influence are the following.

1 The temperature of the noise-sensitive stages can be drastically reduced.
2 A cross-correlation technique using the following expression can be applied:

$$R_{xy} = \lim_{T \to \infty} \frac{1}{T} \int_0^T x(t)y(t-\tau)\, \mathrm{d}t \tag{4.23}$$

where $x(t)$ is the received signal and $y(t-\tau)$ is a known signal shifted over a time τ. This technique is a so-called *posterior* technique, because it is performed after the signal detection; this is in contrast with a *prior* technique, which can be executed beforehand, as stated below.
3 The application of a microprocessor-based sampling technique in some circumstances can reduce noise behaviour considerably. This method can be called a *prior* technique because it is performed before every measurement is realized. An example will be described in section 4.5.
4 The frequency can be shifted to a higher frequency band where less $1/f$ noise is present.
5 Another sophisticated noise-reduction technique might be the application of optical recording techniques. Nowadays, this application is available because optically writing and reading devices are available. The storage capacity is enormous. For instance, an 8 in video long-play record can store over 4000 images, which is the equivalent of several gigabits of information.

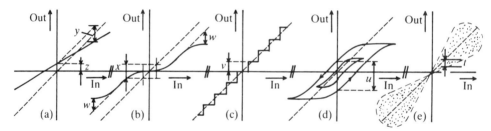

Figure 4.11   (a) Zero error z with scaling error y; (b) dead zone area x with saturation area w; (c) quantization error v; (d) hysteresis u; (e) stochastic error t.

Finally, in figure 4.11 five different representations of errors are shown, which can be assigned to one of the above-mentioned categories. In the following sections we shall discuss some widespread measures for improving electronic systems.

4.4   MEASURES FOR IMPROVING ELECTRONIC SYSTEMS

In this section we will discuss some measures for improving electronic systems. The errors that occur are common to all instruments, but which special precautions must be taken to reduce these errors depends on the signal-to-noise ratio and other required specifications. This is application dependent. Even costs aspects can play a role here, because every precaution will raise the costs of a system.

4.4.1   The feedback principle

In chapter 3 we have already discussed the principle of feedback. This is again illustrated in figure 4.12. Assume that a linear active element has a transfer function $H_1(j\omega)$ and a time constant τ_1. In the feedback loop an attenuator K is implemented. The input signal S_i is compared with the output signal $V_f = KS_o$, thus supplying a fraction of the output signal to the input of the amplifier. In the following discussion we further write $H_1(j\omega) = H_1$. The resulting signal is called the error signal S_f and is given by

$$S_f = S_i - KS_o. \tag{4.24}$$

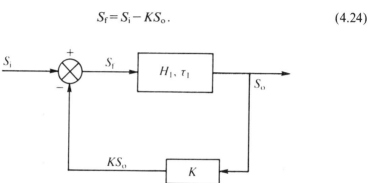

Figure 4.12   Principle of a feedback loop.

It arrives at the amplifier's input; then the resulting output signal S_o is given by

$$S_o = H_1(S_i - KS_o) \tag{4.25}$$

and the transfer function H_t becomes

$$H_t = \frac{S_o}{S_i} = \frac{H_1}{1 + H_1 K} \tag{4.26}$$

or

$$H_t = \frac{S_o}{S_i} = \frac{1}{K} \frac{1}{1 + 1/H_1 K}. \tag{4.27}$$

From this expression some important conclusions can be drawn.

1 If H_1 is very large, then H_t will approach $1/K$ and if H_1 is large enough H_t will become completely independent of the parameters of H_1 itself.
2 The relative sensitivity for changes in H_1 is defined as

$$S_{H1} = \frac{\partial H_t / \partial H_1}{H_t / H_1} = \frac{1/(1 + H_1 K)^2}{1/(1 + H_1 K)} \tag{4.28}$$

or

$$S_{H1} = \frac{1}{(1 + H_1 K)}. \tag{4.29}$$

It can be concluded from the expression for the sensitivity that a considerable decrease for changes in H_1 can be realized. The consequence of this feedback principle is that a very accurate transfer function can be realized which is completely determined by K only. Often this K is composed of passive components such as resistors, and can be manufactured very accurately.
3 The third conclusion which may be drawn will become clear when in (4.26) a frequency-dependent first-order transfer function is substituted for H_1. Assume that the transfer function of H_1 is given by

$$H_1 = \frac{A_0}{1 + j\omega\tau_1} \tag{4.30}$$

where A_0 is the open-loop gain of the amplifier. Then it can be found that the Bode plot for the amplitude versus frequency of the total transfer function H_t compared with the transfer function H_1 will appear to have an increased bandwidth. This effect is illustrated in figure 4.13.

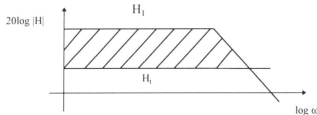

Figure 4.13 Bode plot versus frequency for open-loop gain H_1 and for the closed-loop feedback-controlled loop H_t.

We can observe from this discussion that frequency bandwidth and gain can be interchanged with respect to each other. Feedback reduces gain but improves bandwidth and vice versa.

*   *   *

Example 4.13
Suppose that a feedback loop is composed of an amplifier and a resistive attenuator consisting of two resistors connected in parallel to the output. The fraction of the output signal which is supplied to the input is 10^{-3}. The amplifier has a gain factor of 10^6 ; then, applying (4.26) we find for the total transfer function H_t

$$H_t = \frac{1}{10^{-3}} \frac{1}{1 + 1/(10^6 \times 10^{-3})} = 999.$$

Now assume that the gain factor of the amplifier decreases by 50% owing to thermal effects and deterioration; then the gain factor becomes 5×10^5 and the transfer function H_t is given by

$$H_t = \frac{1}{10^{-3}} \frac{1}{1 + 1/(5 \times 10^5 \times 10^{-3})} = 998$$

which means a reduction of only 0.2%. The transfer function with feedback applied has become dependent on external elements, and thus is almost completely independent of the amplifier's characteristics, if the open-loop gain factor is high.

*   *   *

The above-discussed type of correction applying feedback to improve the accuracy is only true for *multiplicative* errors and is not valid for *additive* errors. Additive errors require other precautions. This will be discussed in the chapter about transducers in

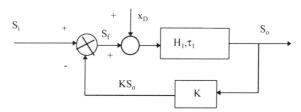

Figure 4.14 A feedback loop with an *additive* disturbance x_D applied at the input of the amplifier.

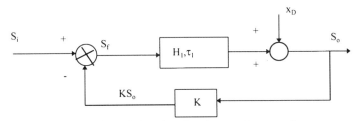

Figure 4.15 Interfering signal acting on the output.

more detail. In figures 4.14 and 4.15 it is shown that feedback gives *no* improvement for additive errors.

In figure 4.14 an additive disturbance x_D is acting on the input of an active element with the same transfer function H_1. By superposition the total output signal S_o is found to be

$$S_o = S_i \frac{H_1}{1 + H_1 K} + x_D \frac{H_1}{1 + H_1 K} \tag{4.31}$$

or

$$S_o = \frac{1}{K} \frac{1}{1 + 1/H_1 K} (S_i + x_D) \tag{4.32}$$

and the signal-to-noise ratio is not improved, but remains the same.

Secondly, in figure 4.15 an interfering signal x_D is applied at the output of the active element. Then considering the effect of x_D only and supposing $S_i = 0$, we find

$$x_D - H_1 K S_o = S_o \tag{4.33}$$

or

$$S_o = \frac{x_D}{1 + H_1 K} \approx \frac{x_D}{H_1 K}. \tag{4.34}$$

This is true as long as $H_1 K \gg 1$ and, compared with the original signal-to-noise ratio, the reduction will appear to have a maximum value of $1/H_1$, which is again very small.

4.4.2 Feedforward coupling

Feedforward coupling is another possible measure to improve the accuracy in a linear amplifier stage. The principle of operation is illustrated in figure 4.16 in which two amplifiers with transfer functions H_1 and H_2 and time constants τ_1 and τ_2 are shown. At both amplifiers a feedback coupling is applied and these are supposed to

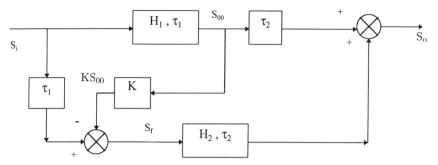

Figure 4.16 Principle of operation of feedforward coupling.

be equal with a transfer ratio of $1/K$, but in reality they will differ slightly. Suppose that the two transfer functions are

$$H_1 = \frac{1}{K}(1 - \delta) \tag{4.35}$$

and

$$H_2 = \frac{1}{K}(1 - \theta). \tag{4.36}$$

From figure 4.16 going from right to left we can derive the following equations:

$$S_o = S_f H_2 + S_{00} \tag{4.37}$$

$$S_f = S_i - K S_{00} \tag{4.38}$$

$$S_{00} = S_i H_1. \tag{4.39}$$

Substituting (4.38) and (4.39) in (4.37), we find

$$S_o = (S_i - K S_{00}) H_2 + S_i H_1. \tag{4.40}$$

Again substituting (4.39) in (4.40) and rearranging, we have

$$S_o = S_i (H_1 + H_2 - K H_1 H_2). \tag{4.41}$$

Then the total transfer function, obtained by substituting (4.35) and (4.36) in (4.41), is

$$H_t = \frac{S_o}{S_i} = \frac{1}{K}(1 - \delta + 1 - \theta - 1 + \delta\theta) \tag{4.42}$$

or

$$H_t = \frac{1}{K}(1 - \delta\theta). \tag{4.43}$$

If both fractions δ and θ are in the order of 1% then the total the error is 10^{-4}, but it is not very difficult to achieve an accuracy of 10^{-6}. Of course, there are some restrictions to be mentioned:

1 the accuracy of the subtraction or comparison and additive errors at the input of the second amplifier
2 the accuracy of the function of K
3 in practical circumstances the accuracy of the addition at the output will form the most crucial restriction
4 the high-frequency accuracy is restricted by the differences in the time delay, τ, which will result in addition at the wrong moments; these time delay differences can be compensated for by the depicted time delays τ_1 and τ_2 in both parallel loops.

Typical applications for this principle can be found in the field of gain improvements in very-high-frequency circuits and controlled processes with dead time zones.

It is interesting to note here that it can be proven that, for the application of feedback to systems of second order and higher order, apart from a frequency range where the accuracy improves, a frequency range also exists where the accuracy deteriorates. This mechanism is expressed in the so-called *Westcott* theorem which says that integrating the logarithm of the absolute value of the frequency-dependent factor of the transfer function over the complete frequency range equals zero, or

$$\int_0^\infty \log\left|\frac{1}{1+H(\omega)K(\omega)}\right| d\omega = 0. \tag{4.44}$$

The system will preferably be designed with the total signal range maintained in the negative part of $\log|1/(1+HK)|$. This is illustrated in figure 4.17.

Finally, we want to make some general remarks concerning feedback.

1 Total accuracy in the subtraction of the input signal and the feedback signal is required; any fault in this part will not be eliminated.

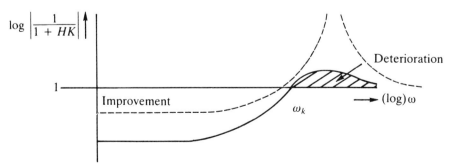

Figure 4.17 Illustration of the Westcott theorem for a second- or higher-order transfer function.

2 In an electronic system stable feedback can easily be realized, whereas this is not as easy in non-electrical systems. The stability criterion for electrical systems involves that the loop gain $|KH_1|$ should remain always less than 1 when the argument $\phi = \arg(KH_1)$ equals π rad.
3 Sometimes an undesired feedback is realized especially in the case of high-frequency behaviour. This can occur via a capacitive coupling between input and output and via the supply connections. We will return to this again in chapter 8.

4.4.3 The influence of feedback on input and output impedance

In this section we shall discuss the influence of the applied feedback configuration on input and output impedances. The principle of feedback involves coupling a part of the output signal back into the input again. The way the coupling is realized at the input as well as at the output determines the final impedance which will be seen at the input and output.

The signal at the output can be sampled in two ways, namely a series connection and a shunt or parallel connection. In the same way, at the input the feedback signal can be connected in series or in shunt with the signal generator. The two possibilities for input and output will result in four different possible configurations, with their specific characteristics. Some authors prefer to speak of current and voltage feedback

Table 4.4 Survey of four possibilities of applying feedback.

Series input	Series output
Series input	Shunt output
Shunt input	Series output
Shunt input	Shunt output

respectively for a series and a shunt connection. This is summarized in table 4.4, whereas figure 4.18 illustrates all four possibilities, realized with passive impedances only.

A general treatment requires a calculation of the input voltage V_{it} which arrives at the input terminals AB of the amplifier in all four cases. This voltage, V_{it}, can be considered as the superposition of the fraction of the output voltage V_o and the fraction of the signal source V_g, or

$$V_{it} = k V_g + \beta V_o \qquad (4.45)$$

where k describes the fraction of V_g arriving at the input terminals AB of the amplifier and β describes the fraction of the output voltage arriving at the input terminals AB. It will be clear that β is composed of two fractions: one fraction originates from the output connection and one fraction is developed in the input connection, each time dependent on the applied configuration. These fractions are often denoted by δ for the output and τ for the input; hence $\beta = \tau \delta$. We will provide τ and δ with indices 's' for series and 'p' for a shunt configuration. Further, we will assume that Z_t represents the impedance seen at the input nodes AB or CB if we look to the output. Summarizing this, we obtain table 4.5. The derivation of these expressions will be omitted.

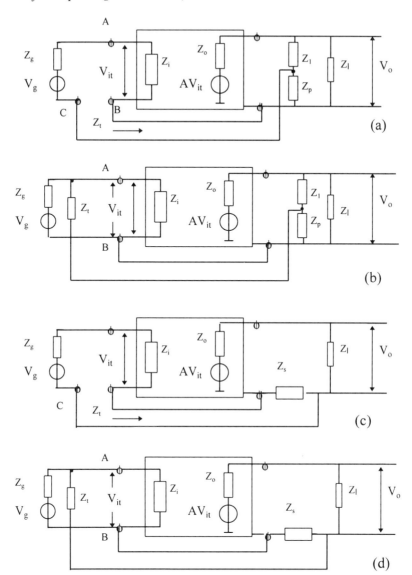

Figure 4.18 The four basic configurations applying feedback.

For V_{it} the expression

$$V_{it} = V_o / A \qquad (4.46)$$

always holds, in which A is the gain of the amplifier. If we substitute this in equation (4.45), the result is

$$\frac{V_o}{A} = kV_g + \beta V_o. \qquad (4.47)$$

Table 4.5 Fractions of source signal, V_g, and output signal, V_o, which will arrive at the input terminals AB of the amplifier producing $V_{it} = kV_s + \beta V_o$.

k (fraction of V_g)	τ, fraction of output due to input configuration	δ, fraction of output due to output configuration	$\beta = \tau\delta$, total fraction of output arriving at input AB
$k_s = \dfrac{Z_i}{Z_i + Z_g + Z_t}$	$\tau = \dfrac{Z_i}{Z_g + Z_t + Z_i}$	$\delta_s = Z_s/Z_i$	$\tau_s \delta_s$
$k_s = \dfrac{Z_i}{Z_i + Z_g + Z_t}$	$\tau = \dfrac{Z_i}{Z_g + Z_t + Z_l}$	$\delta_p = Z_p/(Z_p + Z_1)$	$\tau_s \delta_p$
$k_p = \dfrac{Z_i \| (Z_2 + Z_t)}{Z_g + Z_i \| (Z_2 + Z_t)}$	$\tau = \dfrac{Z_i \| Z_g}{Z_i \| Z_g + Z_2 + Z_t}$	$\delta_s = Z_s/Z_l$	$\tau_p \delta_s$
$k_p = \dfrac{Z_i \| (Z_2 + Z_t)}{Z_g + Z_i \| (Z_2 + Z_t)}$	$\tau = \dfrac{Z_i \| Z_g}{Z_i \| Z_g + Z_2 + Z_t}$	$\delta_p = Z_p/(Z_p + Z_1)$	$\tau_p \delta_p$

Rearrangment of (4.47) gives again the already well known expression for the amplifier with feedback except that k expresses the fraction of V_g which arrives at the input; hence

$$V_o = V_g \frac{kA}{1 - \beta A}. \tag{4.48}$$

In practice it is often difficult to recognize which part of β is τ and which is δ, because they can be completely interwoven, and often the distinction is relevant. The value of this approach is to obtain an insight into the fact that every feedback loop can be rearranged into one of these four basic configurations.

The input and output impedance
It will be clear that the applied configuration will determine the input and output impedance and that again four different cases can be distinguished. We will not provide the total derivation but will confine ourselves to summarizing the results in table 4.6. As can be seen from this table, a series connection will increase the impedance at the input and at the output. In its turn a shunt connection will decrease the input and the output impedance. Note that a factor $1 - A\beta$ is involved each time. For clarity the following expressions are used in the table:

$$Z_n = \frac{Z_o + Z_1}{Z_1} \tag{4.49}$$

in which Z_o is the output impedance of the amplifier without feedback, and

$$Z' = Z_n \frac{Z_s}{Z_o} \tag{4.50}$$

where Z'_i is the new input impedance and Z'_o is the new output impedance obtained with feedback. The factor β is as defined previously. In most cases, approximations

Table 4.6 The input and output impedance of the feedback amplifier in four basic configurations.

$\beta = \tau\delta$, total fraction of output arriving at input AB	Z_i' at nodes AC	Z_o'
$\tau_s\delta_s$	$(1 - A\delta_s)Z_i$	$(1 - A\tau_s Z')Z_o$
$\tau_s\delta_p$	$(1 - A\delta_p)Z_i$	$\dfrac{Z_o}{1 - AZ_n\tau_s\delta_p}$
$\tau_p\delta_s$	$\dfrac{(Z_2 + Z_t)Z_i}{Z_2 + Z_t - A\delta_p Z_t}$	$(1 - A\tau_p Z')Z_o$
$\tau_p\delta_p$	$\dfrac{(Z_2 + Z_t)Z_i}{Z_2 + Z_t - A\delta_p Z_i}$	$\dfrac{Z_o}{1 - AZ_n T_p\delta_p}$

can be made which simplify these expressions drastically, but it is sometimes worthwhile to have a general layout available. The above-outlined description is not exclusive; sometimes it is more convenient to express all quantities in admittances.

4.5 A MICROPROCESSOR-BASED MEASURING SYSTEM

In this section we shall introduce some basic concepts of microprocessor-based systems. We have seen already in chapter 2 some *structural* aspects of automated systems such as building a virtual instrument and possible I/O interfacing including DAQ cards. Here we will discuss the basic concepts of a microprocessor which often functions as the heart of an automated system and hence can be understood as one of a collection of system specifications. Nowadays, no instrument exists without a built-in microprocessor. The advantages are evident. A large increase in amplitude ranges, automatic nulling, calibration, built-in test procedures, increased accuracy and high-speed measurements are possible. As well as these aspects, automatic measurements provide long-term measurement capabilities. In other words, we can measure easily and continuously over time periods of 24 h and more, without manual monitoring. A microprocessor is found in numerous applications: missile launch equipment, flight computers and so on. Furthermore, reduced size and increased reliability are obtained. An application-specific software program forms an inseparable part of a specific system.

4.5.1 Microprocessor fundamentals
The most general layout of every digital computer shows three basic functional units.

1 an input/output unit (I/O)
2 a memory unit, which can be divided into an instruction memory and a data memory for temporary storage of the results of calculations or measurements
3 a central processing unit (CPU)
4 bus systems, consisting of three groups of wires connecting the three different units in parallel, i.e. (a) the *address bus*, (b) the *data bus* and (c) the *control bus*.

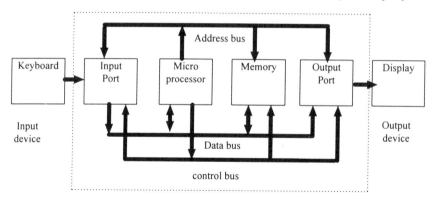

Figure 4.19   Basic architecture of microprocessor system.

The architecture is given in figure 4.19. We will discuss the different parts in some detail.

A general architecture for a microprocessor-based measuring data acquisition system is shown in figure 4.20. In this system the processor forms the core of the system. Software controls the measuring protocol. All data are stored on hard disk and floppy disks for back-up purposes. Also optical registration is already available. In particular, for long-term and multichannel measurements this is an ideal configuration. For instance, in research for new sensors stability and drift measurements are

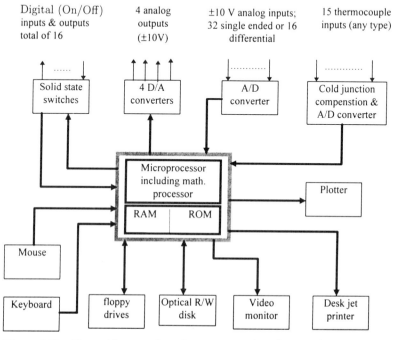

Figure 4.20   General layout of a microprocessor-based measuring set-up.

very time consuming and usually last for days before conclusions can be drawn. We will see examples of this later.

4.5.1.1 *An input/output unit*

An input port connects the processor to the keyboard or some other input device, for instance a scanner. The output port, to which the processor can send data, connects the processor to an output device, for instance a display. For the I/O ports several peripherals are available which can perform, in a user-programmable way, the input and output functions. Such a device is called a *peripheral* because it is not a part of the microprocessor itself. A good example is the Intel 8155. The principle of this I/O device is depicted in figure 4.21.

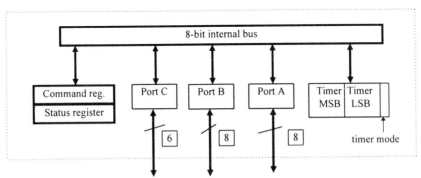

Figure 4.21   Layout of Intel 8255 I/O device.

We shall discuss this peripheral device in some more detail. It contains the following functional units:

1   256-byte static RAM
2   two 8-bit I/O ports (PA and PB) and one 6-bit port (PC)
3   a 14-bit counter
4   a command register (write only) to initialize the different modes of the ports mentioned
5   a status register (read only).

The command register must be programmed. It contains 8 bits and follows the next sequence of mode commands. The command register is depicted in figure 4.22. Four bits (0–3) are required to define the mode of ports A, B and C. The two bits PC_1 and PC_2 define how port C is used, just as input, as output or as a control port. When C is defined as a control port then three bits of C are assigned to A and the other three bits of C to B. This is given by ALT3 and ALT4; see figure 4.22. Two bits (4 and 5) are used to enable or disable the interrupt of port C when it is used as a control port. The last two bits (6 and 7) are used for the timer. To set the command register an OUT instruction is used; see below. The status register in figure 4.21 is directly connected to the command register and it is a read-only register. For further details reference is made to the manufacturer's data books. This device offers very flexible I/O facilities, but it is just one example of many possible I/O devices.

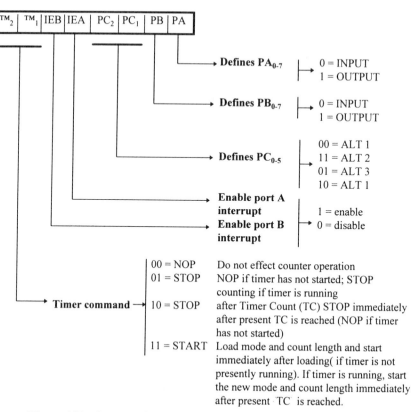

Figure 4.22   Layout of command register of Intel 8255 I/O device.

However, because of a large increase in integration density, microprocessors are now offered with on-chip I/O capabilities too.

Three-state buffers
It is important to know that all devices in a microprocessor exchange information over the same set of wires. To avoid any bus conflict it is necessary to use three-state buffers. A three-state buffer or driver has three possible states controlled by the *output enable* line. The output enable has two possible states 0 or 1, which determine whether the output buffer is floating or not. When the output is not floating then the output can be 0 or 1. These characteristics are depicted in figure 4.23.

4.5.1.2   *A memory unit*
No computer can function properly without a memory device used for temporary or permanent data storage of (intermediate) results and instructions. In general, a memory unit can be divided into an instruction memory and a data memory for temporary storage of the results of calculations or the measurement results. Usually,

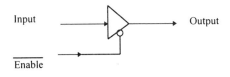

Enable	Input	Output
0	0	0
0	1	1
1	0	floating
1	1	floating

0 = low 1 = high

Figure 4.23 Three-state driver with truth table.

microprocessor applications the instructions are invariable and therefore they are stored in a permanent, non-erasable memory, a so-called read-only memory (ROM). Data are frequently modified and therefore stored in an allocated address of a memory location of which the contents can be changed any time. This type of read/write (R/W) memory is usually called a random access memory (RAM). Note that the *meaning* of the word *RAM* does not describe the fact that you can read and write in this memory.

Two types of RAM can be distinguished and it is useful to know these differences. In particular we have the *static* and the *dynamic* RAMs. Static RAMs use a flip–flop for each memory element. Each flip–flop can be set to store a 1 or reset to store a 0. Inside the RAM an address decoding circuit selects the particular flip–flop. The state of the flip–flop does not change unless new data are stored or the power to the RAM is switched off.

Dynamic RAMs use an on-chip capacitor for each storage element. A stored charge on the capacitor is a 1; no charge indicates a 0. To avoid leakage of the charge off the capacitor it is necessary to *refresh* these RAMs every 2 ms. Refreshing consists of reading and rewriting the same data back in the read location of the RAM. Dynamic RAM technology is simpler than that of a static RAM and therefore more dense dynamic RAMs can be constructed. Advantages of dynamic RAMS are less power consumption and they are less expensive. Usually, they are 1 bit wide, so eight 1 Mbit RAMs are required to construct a memory block of 1 Mbyte. Nowadays, a chip size of 256 Mbit MDRAM is obtainable and, before the year 2000, 1 Gbit dynamic RAMs are expected.

Read-only memories provide a tool for permanently storing data, programs and instructions. There are four different types of ROM: (1) mask-programmed ROMs, programmed by the manufacturer using a mask; (2) programmable read-only memories (PROMs), which can be programmed once electrically by the user; (3) the erasable programmable read-only memory (EPROM), which can be programmed electrically and erased by UV light; (4) electrically alterable read-only memory (EAROM, sometimes also called EEROMs), which can be erased electrically while in the circuit. For other possibilities reference is made to section 3.5.4. This is summarized in table 4.7.

Memories can be organized in a kind of *hierarchy* determined by their function. The lowest level is a *register* of which the accumulator is well known. Next is a *register file*; then we find a kind of *cache memory* which can speed up the execution of instructions considerably. Finally, we find a *work memory*, a fixed ROM memory, and a *disk R/W memory* which can store all the information. Disk memories can be constructed as magnetic or optical devices. This is depicted in figure 4.24.

Table 4.7 Survey of semiconductor memory devices (see also table 3.3).

Random access memory (RAM or R/W)	Read-only memory (ROM)
Static (flip–flop)	Mask programmed (ROM)
Dynamic (capacitive)	User programmable (PROM)
	Erasable PROM (EPROM)
	Electrically alterable (EAROM)

The memory locations are addressed by the central processing unit (CPU) through the address lines. The principle of an example of a RAM is given in figure 4.25. The 16 address lines A(0) to A(15) give an addressing capacity of $2^{16} = 65\ 536$, usually referred to as 64k RAM. To obtain a width of 1 byte, eight register files are connected in parallel to the corresponding data lines of the data bus.

An example of a possible memory map is given in figure 4.26 in which the RAMs and the ROMs are combined in one map. The addresses are given in hex(adecimal) form; each address (location, loc) requires an assignment of two bytes to obtain the 64k possible address locations. At least seven different modes of addressing memory locations are known. We will mention these modes without going into further detail; we distinguish

- implied addressing
- immediate addressing
- direct addressing
- direct indexed
- register indirect indexed
- relative addressing
- paging.

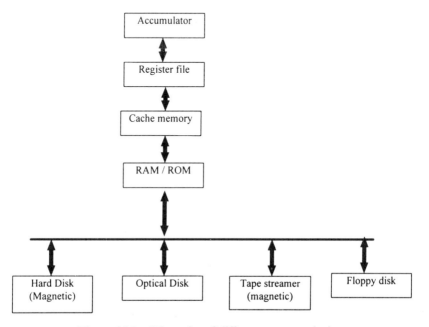

Figure 4.24   Hierarchy of different memory devices.

addressbus
A(0)
A(15)
D(0)
D(1)
D(2)
D(3)
D(4)
D(5)
D(6)
D(7)
Databus

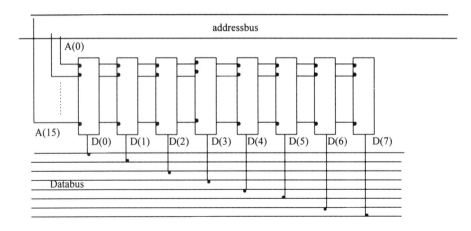

8 x 64 x 1 bit RAM

Figure 4.25 Principle of an $8 \times 64 \times 1$ bit RAM.

Other addressing modes are also known, depending on the manufacturer's definition and the type of microprocessor involved. Each memory address location can contain data, instructions or other addresses. Because of space limitations further reference is made to the literature.

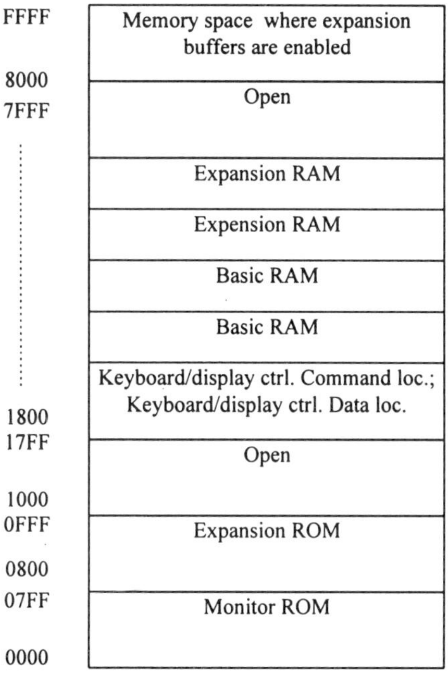

FFFF
Memory space where expansion buffers are enabled
8000
7FFF
Open
Expansion RAM
Expension RAM
Basic RAM
Basic RAM
Keyboard/display ctrl. Command loc.; Keyboard/display ctrl. Data loc.
1800
17FF
Open
1000
0FFF
Expansion ROM
0800
07FF
Monitor ROM
0000

Figure 4.26 Example of a memory map; the addresses are given in hex form on the left-hand side of the memory map.

4.5.1.3 *The central processing unit*
The central processing unit (CPU) is the actual core of the microprocessor system. In this device all the real actions take place. It can perform different kinds of operation such as

- data transfer operations
- arithmetic operations
- logic operations
- branch operations
- stack, I/O and machine control group operations.

To define these operations a number of bytes are required; this number ranges from one to six. A specific instruction set is defined by the manufacturer for each type of microprocessor. In general the task of the CPU is to generate the address for the next instruction, to interpret this instruction and to execute it. In the CPU depicted in figure 4.27, we can distinguish the following parts:

1 a program counter (PC)
2 an instruction decoder
3 a control and timing section
4 an arithmetic logic unit (ALU)
5 a number of general-purpose registers and one or more data paths.

The program counter (PC) contains the address of the next instruction. For most instructions the PC is automatically incremented to address the instruction. The instruction, once fetched from the memory, is usually stored in the instruction register (IR), which is connected to the instruction decoder. This decoder is typically an internal ROM which translates the machine code instructions for the arithmetic and logic unit (ALU), the registers and the data paths to be used by intervention of the control and the synchronization part of the processor. This part can function only if provided with clock pulses at the appropriate timing intervals. The manipulation of data is accomplished by the ALU which is capable of performing additions, subtractions and logic operations. After manipulation of the data according to the instruction, the destination of the result can be either register, a data memory location or an I/O port.

Microprocessors can be classified as

- one-chip microprocessors (the microprocessor contains ROM, RAM, I/O capability and an internal clock oscillator)
- general-purpose microprocessors (this type is made up of one or more chips)
- bit-slice microprocessors (the microprocessor ALU is divided into subsections)
- microcontrollers which perform a specific task with a limited number of available instructions.

Most recent microprocessors contain over 3 million transistors and operate on a 32 bit wide instruction length. Basically they still act as serial machines in which the instructions are handled in a fixed sequence. As a contrast, there exists also a type

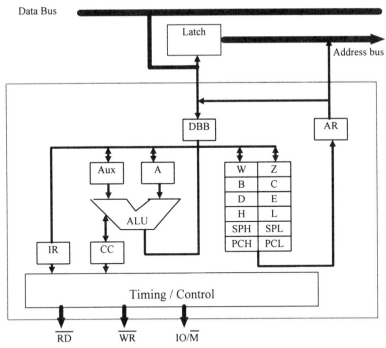

Figure 4.27 Architecture of a microprocessor.

of parallel machine known as transputers. They can be connected in parallel directly and have powerful on-chip I/O capabilities. Also, on-chip A/D and D/A conversion and I/O are possible which then provides a complete analogue–digital system. We will come to that again in chapter 6.

4.5.1.4 *The software*

Basically what can be understood by a microprocessor is only bits, zeros and ones. Therefore we have to provide the microprocessor and the memory locations with zeros and ones only. Each microprocessor has its own specific instruction set specified by the manufacturer. This instruction set is known as *machine code*. This machine code can be written in hex form, so a more convenient description is obtained. However, when large groups of instructions must be handled, this is still unsatisfactory. A more powerful approach is the use of symbolic names for each type of instruction. These symbolic names for each type of instruction are called *mnemonics* and a whole program written with the help of these mnemonics is called a description in *assembler*.

Format of instructions

All instructions have a certain predetermined format specified by the manufacturer. This means that in each instruction, whether it consists of one or more bytes, we find a part defined as the *opcode* and a part which is called the *operand*. The opcode

Figure 4.28   Three different instruction lengths.

tells us 'what' must be done, and the other part in the instruction tells us 'on what' it must be executed: data or an address. In figure 4.28 three examples are given.

To obtain an idea of how this works out we can specify it in more detail. An example of a simple transport instruction as the content of register B is moved (copied) to register A is written as shown in figure 4.29. The symbol D stands for destination and S stands for source. Programs can be written in a six-column format, as address, data, label, opcode plus operand and comments.

MOV A, B

Figure 4.29   A one-byte transport instruction.

* * *

Example 4.14
We give the following example, which demonstrates the addition of 10 numbers; the result is written in register A (accumulator). For reasons of simplicity we use the instruction set of the Intel 8085. See table 4.8.

Table 4.8   Example for summing ten numbers in mnemonics.

	Memory address (hex)	Opcode and operand (hex)	Labels	Mnemonic (Opcode)	Mnemonic (Operand)	Description and comments
1	0000	210001		LXI	H,100H	;initialize pointer
2	0003	060A		MVI	B,0AH	;initialize counter
3	0005	3E00		MVI	A,0	;initialize sum
4	0007	86	L	ADD	M	;sum←sum + number
5	0008	23		INX	H	;increment pointer
6	0009	05		DCR	B	;decrement counter
7	000A	C20700		JNZ	L	;all numbers used?
8	000D	76		HLT		;stop program execution

* * *

Finally, the start address of the program must be given; this is omitted here. For the explanation of the different opcodes reference is made to the Intel data books. Special programs are available to translate the assembler (mnemonics) into machine code. Today it is preferred to write programs in Pascal or C and then to use a special program to convert this into the basic machine code. Every machine has its own instruction set. Hence no generality can be claimed. This is just meant to provide an idea of how things work. Because of space limitations this discussion cannot be exhaustive. We conclude with an example to demonstrate the power of the use of a microprocessor.

$$* \qquad * \qquad *$$

Example 4.15
To demonstrate the capabilities of a microprocessor-based measuring system, a high-speed spectral density measurement set-up can deliver an output signal where a normal standard measuring system would only produce noise. Often spectral measurements of plasmas produce a very weak signal which is completely immersed in noise. A plasma is an ionized gas which can conduct an electrical current. For instance, the retrieved signal from the process will have the form

$$A(t) = S(t) + R(t) \tag{4.51}$$

in which $S(t)$ is the desired signal and $R(t)$ is the undesired noise contribution. If it is known that $S(t)$ shows a periodic behaviour with a constant time interval T then

$$S(t) = S(t + mT) \tag{4.52}$$

with m ranging from 0 to M, where M denotes the total number of measurements. The noise signal will show a completely random stochastic behaviour and will approach an *average* of zero if t goes to infinity. This is realized when the number of measurements M increases to a large value. The signal's average, $\langle A(t) \rangle$, can be calculated via a microprocessor:

$$\langle A(t) \rangle = \frac{1}{M} \sum_{m=0}^{M-1} [R(t + mT) + S(t)]. \tag{4.53}$$

As can be seen from expression (4.53) the noise term will diminish as m increases. The applied principle is illustrated in figure 4.30 in which two microprocessors are installed; one is only used for data collection and the other is used for numerical calculations. Both microprocessors have access to the same memory.

This principle is sometimes called a *dual-ported memory* because the two synchro-nized microprocessors have access to the same memory and the required tasks are strictly separated between the two processors. An example of a measuring result is given in figure 4.31. In the lower part of the graph, the completely noise-immersed signal is depicted, measured with a conventional system. The upper part of the graph shows the result obtained by the dual-ported memory system, and this may be considered as an impressive result. To obtain this result the required number of summations is less than 25.

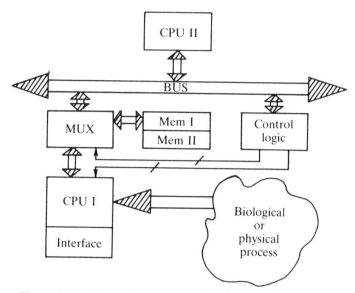

Figure 4.30 A two-microprocessor-based measuring system.

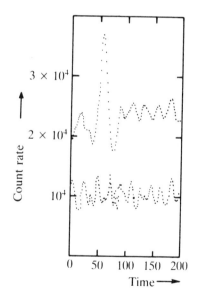

Figure 4.31 Comparison of the results of a conventional and a microprocessor-based measuring system for a completely noise-drowned signal.

PROBLEMS

4.1 To improve the accuracy in a linear amplifier the feedforward technique can be applied. What are the main characteristics of this technique and why is no stability problem involved?

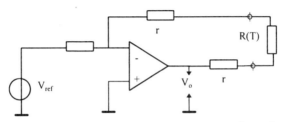

Figure 4.32 A temperature measurement configuration.

4.2 An amplifier has a gain setting between 10 and 1000 times. The equivalent noise voltage over the whole frequency bandwidth is 15 μV and the maximum allowed voltage range for input and output is 15 V. Determine the dynamic range for this amplifier with a required inaccuracy of 0.5%.

4.3 The inaccuracy of a DC voltage measuring system with a 4-digit display is ±0.05% fsd plus ±0.1%±$\frac{1}{2}$ digit of the indicated value. The full-scale deflection is 1.999 mV. Determine the absolute and the fractional error in an indicated value of 1.000 V.

4.4 In figure 4.32 a measuring configuration for temperatures is shown. In this circuitry $R(T)$ represents a calibrated platinum resistor of which the transfer function can be described with

$$R(T) = R_o(1 + aT + bT^2)$$

where $R_o = 100\ \Omega$, $a = 4.9 \times 10^{-3}\ \mathrm{K}^{-1}$, $b = -5.9 \times 10^{-7}\ \mathrm{K}^{-2}$ and T is the observed temperature in degrees Celsius. The resistance of the conductors is r and V_{ref} represents a reference voltage source of 6.4 V. R_1 is 1.0 kΩ and the amplifier is supposed to be ideal.

(a) Determine the maximum allowed resistance of the conductors for an inaccuracy smaller than 0.1 °C owing to r.

(b) The current through $R(T)$ will raise the temperature of the sensor by 10 °C per Watt dissipated in $R(T)$. Determine the fault signal in degrees Celsius due to the self-heating effect of $R(T)$ at $T = 0$ °C.

(c) Determine the non-linearity as a percentage value with respect to the case where b is supposed to be zero over a temperature range from −50 to +100 °C.

(d) If $b = 0$ prove that the sensitivity of this configuration is 2.5 mV K^{-1}.

(e) If $b = 0$ determine the maximum allowed offset voltage V_{off} of the amplifier when at $T = 0$ °C the fault signal is smaller than 0.1 °C owing to V_{off}.

(f) Answer the same question as in (e) for the maximum allowed bias current I_{bias} of the amplifier.

(g) Determine the absolute error in the reference voltage source V_{ref}, for an inaccuracy of ±0.1 °C at $T = 0$ °C.

(h) Answer the same question as in (g) for an inaccuracy in R_1.

4.5 Derive an expression in β for the CMRR and the discrimination factor F for the attenuator configuration connected to an amplifier as drawn in figure 4.33. Introduce β_1 and β_2 for the corresponding fractions arriving at the input terminals of the amplifier.

4.6 For the differential amplifier circuit depicted in figure 4.34 derive an expression for the CMRR. If the fractional error of the resistors is $\delta = 1\%$ and the gain A is 100, what CMRR can be guaranteed? What important conclusions can be drawn from this example?

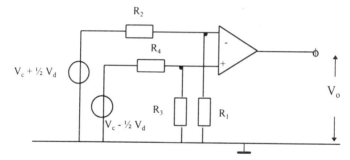

Figure 4.33   Circuit for problem 4.5.

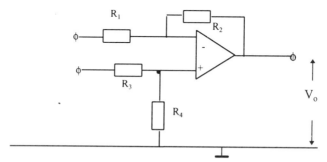

Figure 4.34   Circuit for problem 4.6.

4.7   A strain gauge has a sensitivity of $10\,\text{mV}\,\text{mm}^{-1}$ and a temperature sensitivity of $10\,\mu\text{V}\,\text{K}^{-1}$; then a variation of for instance 0.1 mV can be caused either by a temperature variation or by a mechanical load variation. Determine the parasitic sensitivity.

BIBLIOGRAPHY

[1] Crecraft D I, Gorham D A and Sparkes J J 1993 *Electronics* (London: Open University, Chapman & Hall)
[2] John Fluke Manufacturing 1979 *Fluke Documentation* 8050A
[3] Barney G 1988 *Intelligent Instrumentation, Microprocessor Applications in Measurement and Control*, 2nd edn (London: Prentice-Hall)
[4] Regtien P P L 1978 *Modern Electronic Measurement Systems* (Delft: DUP)
[5] Hewlett-Packard 1995 *Test and Instrument Catalog*
[6] Intel 1978 *Microprocessor Handbook* 8085, SDK-85
[7] Sudar T (ed) 1981 *Digital Instrument Course* part 3 (Eindhoven: Philips)

Five

RELIABILITY

5.1 INTRODUCTION

The formal study of reliability dates from the early 1930s and hence can be considered as a rather young science. The need for reliability predictions about systems and equipment has grown with the need to control society and with the still increasing complexity of applied systems. The concept of reliability first appeared in equipment for military purposes, where survival conditions involved reliability of equipment and systems play a most important role. In the early 1970s the importance of reliability became accepted worldwide. The concepts of reliability have become well established in a great variety of disciplines, for instance in transport, medicine, banking, the aircraft industry and even in software development and tools. Because of the nature of electronics the first major applications were found in this discipline. Today a tremendous change in approach to reliability engineering for VLSI and ULSI circuits is being carried through. How can the reliability of a chip with over 10^8 transistors be tested? The classical approach will not suffice here. We will mention in section 5.13 some aspects in relation to this.

Reliability is based on the quality of the different components of which every system is composed. In its turn, quality of components is based on the applied quality-control procedures in the manufacturing process. This strong relationship between reliability and quality will be discussed in more detail in a later section.

In the manufacturing process there exists a strong relationship between human behaviour and the effectiveness of reliability. In an organization a consensus between all departments is a basic requirement to make reliability a success. This has been proven over and over again. Nowadays, effective production protocols exist which must be followed and signed by all partners involved for every subsequent production step. Also, as will be shown, the ISO 9000 system is concerned not only with quality but also with reliability.

It will be shown that two types of reliability can be distinguished: *technical reliability* and an *economical reliability*. The maxim 'pay now, save later' has been demonstrated dramatically many times by the damage and loss of reputation caused when this rule is neglected. Regarding this, one other important question must be kept in mind: 'what are the consequences if a system fails?'

Ultimately, we want to be able to make predictions about systems, and about their lifetime and reliability under known conditions expressed in numbers. The treatment will have an introductory character but an elementary discussion of all relevant reliability parameters and aspects will be presented.

Because it is in the very nature of this discipline to *predict* specifications of components and systems of samples of a population, instantaneously and in the long run, a knowledge of probability theory is a basic requirement..

5.2 RELIABILITY CONCEPTS

We begin this section by presenting the required definitions and concepts of reliability about which a lot of confusion often arises.

5.2.1 Definitions
1 The *quality* of a product is defined as the instantaneous sample of technical specifications with which that product complies. There exists another definition which says 'it is the totality of features and characteristics of a product or service able to meet its specifications when it is required by the user'. It is sometimes defined as user's satisfaction. Note the word 'instantaneous' because here the specifications are determined under standardized conditions and at a certain moment.
2 The *lifecycle* or *life length* of a system is defined as the total time between 'birth' and 'death' of that system under the condition, if applicable, of an *a priori* fixed specified maintenance scheme. We will return to the phrase an *a priori* specified maintenance scheme.
3 The *reliability* of a system is defined as the probability that a system will still perform successfully the functions associated with its predetermined specifications after a certain operational period. Note in this statement the words 'probability' and 'after a certain period'. It will be clear from this statement that time is an inherent parameter in reliability. This is the main difference from quality, but if we look for a relationship between quality and reliability, the following statement also holds.
4 *Reliability* is quality over time, which can be expressed as

$$R(t) = Q(t)/t \tag{5.1}$$

where $R(t)$ denotes the reliability as a function of time. From this expression it will be clear that quality is also a time-dependent quantity. The quality of a system after two years will differ from the quality of a brand-new system. Also, it will be obvious that reliability decreases in time.

A few examples can further stress the importance of the above statements. Suppose the calculated economic lifetime of a large mainframe computer is 10 years. The maintenance scheme is fixed at twice a year, and after 10 years the computer will be considered as *economically* worn out because maintenance costs are exceeding the accepted operational cost level. From a technical point of view the computer can still perform its tasks but it is no longer acceptable with respect to the operational cost level. In the aircraft industry it will be obvious that we want to know the reliability of every system involved in an aircraft. The same philosophy is applicable as in the example of the mainframe computer, but neglecting maintenance for reasons of cost savings may result in disasters. Here again the principle of 'pay now and save later' is very well applicable.

5.2.2 Stochastic variables, probability distributions, functions and reliability

As stated already, probability is required to describe the reliability process. Two main reasons can be mentioned why this is true: first we want to predict the behaviour of a system over a certain period of time; second the systems have become too complex to make absolute statements possible, so in every reliability statement an amount of uncertainty will always be implied. However, as will be seen, the task of the reliability engineer is to minimize this uncertainty.

It is beyond the scope of this textbook to present a comprehensive discussion of probability theory, but for convenience the relevant parameters of this theory are summarized. In probability theory two types of stochastic variable can be distinguished, i.e., discrete and continuous variables, of which the parameters are defined in table 5.1.

Table 5.1 Parameter review of stochastic variables.

Discrete stochastic variables	Continuous stochastic variables
Probability function of x is	$f(x)$ probability density function
$p(x) = P(x_i)$ for $i = 1, 2, 3, \ldots$ $\quad = 0$ elsewhere	$f(x)\Delta x$ is called the probability element for the event Δx
To every possible event a number is assigned Distribution function of x $p(x \leq k) = \sum_i P(x = x_i) = F(x)$ for $x_i \leq k$	$F(x) = P(x \leq k) = \int_{-\infty}^{k} f(t)\, \mathrm{d}t$
The probability of the occurrence of a maximum of k events, sometimes called *cumulative distribution*	
Expectation of x $\mu = E(x) = \sum_i x_i P(x = x_i)$	$\mu = E(x) = \int_{-\infty}^{\infty} xf(x)\, \mathrm{d}x$
Often called *expectation of the first moment*	
Variance of x $\mathrm{Var}(x) = \sigma_x^2 = \sum_i (x_i - \mu)^2 P(x)$	$\mathrm{Var}(x) = \sigma_x^2 = \int_{-\infty}^{\infty} (x - \mu)^2\, \mathrm{d}\tau$
The square root of the variance is called the *standard deviation* (std)	

The result of throwing a dice is an example of a discrete variable; noise is an example of a continuous variable. It has been proven that the normal life length of a system can be considered as a continuous variable. The lifecycle of most systems can be described by a limited number of probability density functions: the *exponential* distribution is often the most likely one and, in addition, it is the most widely used in electronic systems. We shall illustrate most concepts of reliability with the exponential distribution function instead of a more general treatment.

Suppose the probability that a system will fail is p, where p is small, for instance 0.01; then the probability that the system will not fail is $1 - p = q$ by definition. Suppose further that np is large, where n is the number of systems to be considered, and x is the event that a system will fail; then the probability that one or more systems will fail may be calculated with the Poisson distribution:

$$f(x) = \frac{\exp(-\mu)\mu^x}{x!} = \frac{\exp(-np)(np)^x}{x!} \tag{5.2}$$

where $\mu = np$ is the mean of the Poisson distribution. Now suppose the event of failing or not failing is a function of time and that the system fails with an average of λ times during its lifetime; then the average number of systems that will fail in a given time t is λt, with the condition that the total number of working systems remains constant. Then the probability that x systems will fail in a given time t can again be described by the Poisson distribution. Basically, this distribution is a function of λ and t and becomes

$$f(x; \lambda, t) = \frac{\exp(-\lambda t)(\lambda t)^x}{x!}. \tag{5.3}$$

Finally, if the probability is calculated that no system will fail then $x = 0$. Substituting this value in equation (5.3) yields

$$f(0) = \exp(-\lambda t). \tag{5.4}$$

The failure rate λ can also be characterized as the cumulative per cent failure per unit of time.

There is another approach possible for the derivation of (5.4). Assume that N is the total number of systems still working after t and ΔN is the number of failing systems in Δt. Then the total number of failing systems in Δt is given by

$$\Delta N = -\lambda N \, \Delta t. \tag{5.5}$$

Taking the limit for $\Delta t \to 0$ and integration of (5.5) also gives (5.4).

The two expressions (5.3) and (5.4) are depicted in figure 5.1, from which the relationship between the two expressions can be understood. It can be seen that for $x = 0$ the graph $f(0)$, still as a function of t, lies in front of the figure along the t-axis. The function (5.4) can be interpreted as the probability of survival for the considered system and for that reason it is called the *survival function* or *reliability* of that system, often denoted by $R(t)$. Thus $R(t)$ is the probability that the system does not fail during the interval from 0 to t or is still functioning at the time t:

$$R(t) = f(0) = \exp(-\lambda t). \tag{5.6}$$

It can be seen from this graph that at $t = 0$ the probability that no system will fail is 1, as is correct of course, because all systems will work at $t = 0$. If t goes to infinity it will be clear that all systems will have failed. So the graph of the expression for $R(t)$ shows the probability that no system will fail in the interval from 0 to t, or still functions for $t \geq t_1$.

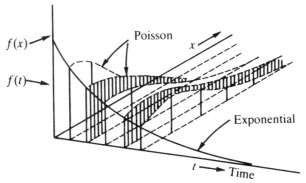

Figure 5.1 The relationship between the survival function and the Poisson function. (Reprinted by permission of John Wiley & Sons Inc.)

From (5.5) it can also be retrieved that the failure rate λ equals $\Delta N/tN$. Because life testing is usually performed at elevated temperatures, the failure rate is corrected with an acceleration factor, A_F, to obtain the applicable failure rate at the environmental operating temperature. We will come to that again later on.

An alternative representation for the survival function may be expressed with the aid of the *cumulative distribution function* of t and is defined as

$$R(t) = 1 - F(t) = \int_{t}^{\infty} f(t)\,dt \qquad \text{for } F(t) < 1 \tag{5.7}$$

and where

$$F(t) = \int_{0}^{t} f(t)\,dt \tag{5.8}$$

and $f(t)$ is again the considered probability density function. Note that the limits in the integral expression complement each other. This is illustrated in figure 5.2. $F(t)$ is called the *cumulative failure distribution function*, or *failure probability* or the *unreliability function*. For the exponential distribution it can easily be calculated that $F(t)$ is $1 - \exp(-\lambda t)$.

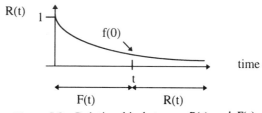

Figure 5.2 Relationship between $R(t)$ and $F(t)$.

5.2.3 Reliability parameters

In the application of probability distributions to reliability assessment studies of systems more basic concepts are needed, so here are some others. If the derivative of expression (5.7) is determined and we change sign in order to obtain a positive expression for $f(t)$ this yields

$$-\frac{dR(t)}{dt} = f(t) = \begin{cases} \lambda e^{-\lambda t} \\ 0. \end{cases} \tag{5.9}$$

For negative values of t this expression makes no sense so $f(t) = 0$ by definition for $t < 0$. The expression (5.9) is called the *failure probability density function*, often denoted by $f(t)$ only, and can be interpreted as the degree (slope) to which the reliability decreases at time t.

For λ the following definition holds: λ is the *average* number of failures per unit of time and per system and is called the *failure rate*. The dimension of λ is per unit of time as can be seen from expression (5.9). The applied unit of time is also often given as per 1000 h or per 10^6 h. For integrated circuits, the failure rate is usually specified in a fixed specified time period of 10^9 h. This failure rate is then called *failures in time* (*FIT*), which represents the number of failures in 10^9 operational hours.

*   *   *

Example 5.1

1 If for a certain repairable system it is known that in 1000 h 14 failures occur, then the failure rate λ equals $14/1000 = 0.014$ or 1.4% per hour. The statements 'a system fails λ times' or 'λ systems fail' in a given time t are interchangeable. No statement is made about when these failures occur or how these failures are distributed in time t. The failure rate λ can obtained by simple measurements.

*   *   *

2 A sample of 2500 ICs is operational during 1000 h. One integrated circuit has failed. The failure rate in FIT then is

$$\text{FIT} = \frac{\Delta N}{Nt} \times 10^9 = \frac{10^9}{2500 \times 1000} = 400.$$

Hence, this value equals $\lambda \times 10^9$ and is just a dimensionless number.

*   *   *

Another important parameter of reliability assessment studies is the mean of $R(t)$: this yields for the exponential distribution for $0 < t < \infty$

$$E(t) = \int_0^\infty R(t)\, dt = \int_0^\infty \exp(-\lambda t)\, dt = \frac{1}{\lambda}. \tag{5.10}$$

It is easily seen here that the result for the mean will be $1/\lambda$ with the dimension of time. How can this relationship be interpreted? The mean denotes an expectation, and hence the calculated value for the mean of $R(t)$ describes the mean time that a system will function or not fail. This expression therefore is called the *mean time between failure* or MTBF and describes exactly what it says but, as you will see later on, one distinction must still be added. MTBF can also be considered as the average time between two failures.

*   *   *

Example 5.2
The reciprocal value of the failure rate λ is equal to the MTBF, so for instance suppose that the failure rate λ of a component is $0.1 \times 10^{-4} \, h^{-1}$; then the MTBF is $1/\lambda = 1/(0.1 \times 10^{-4}) = 10^{-5} \, h$.

*   *   *

So far the failure rate, λ, has been considered as a constant but sometimes the failure rate may be considered as a function of time. This is especially the case when a system is brand new or almost worn out. In those circumstances we have to write $\lambda(t)$ instead of λ only. If this is true, λ is called the *failure rate density*, *hazard rate* or *force of mortality* and is denoted by $h(t)$ or sometimes by $z(t)$. The hazard rate is defined as the ratio of the failure probability density function $f(t)$ and the reliability $R(t)$ thus

$$h(t) = \lambda(t) = \frac{f(t)}{R(t)} = -\frac{dR(t)/dt}{R(t)}. \tag{5.11}$$

Solution of this first-order differential equation for $R(t)$ yields

$$R(t) = \exp\left[-\int_0^t h(t)\, dt\right] \tag{5.12}$$

which is another expression for $R(t)$ as a function of the hazard rate. If we apply both expressions (5.11) and (5.12) to the exponential failure probability density function $f(t) = \lambda \exp(-\lambda t)$ it is easily seen that the result for the hazard rate is λ, and hence it is a constant, and $R(t) = \exp(-\lambda t)$.

For an exponential failure distribution function the results obtained so far are illustrated in three different graphs in figure 5.3.

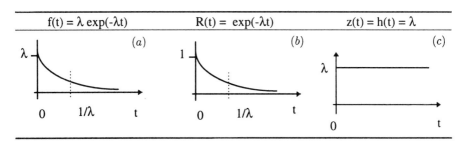

Figure 5.3 Graphical representation of three different parameters for the exponential failure distribution function: (*a*) the failure probability density function, $f(t)$, (*b*) the reliability function $R(t)$; (*c*) the failure rate λ.

In the following section a separate discussion is devoted to the MTBF, because a more detailed evaluation is required where repairable and non-repairable systems are considered. This is needed because the mathematical description of a repairable system differs from that of a non-repairable system.

5.2.4 The mean time between failure

In equation (5.10) a first definition for the mean time between failure is given, but as stated already some distinctions have to be made. In general, two classes of systems can be distinguished, which differ considerably in treatment and in principle of design. The two classes of systems are (a) the repairable and (b) the non-repairable systems. A few examples will illustrate this distinction. A launched satellite is an example of a non-repairable system in most cases and its design is characterized by the diagnosed and required longevity under extremely severe operating conditions. However, the Hubble telescope is also an example of a launched satellite which has been repaired in space. A car is an example of a repairable system and its design is largely based on possible replacement of components which are worn out or malfunction. All integrated circuits can be considered as non-repairable. This distinction can be found in all kinds of technical discipline. Some authors prefer to speak of maintainable and non-maintainable systems which is in principle the same. However, with regard to costs aspects of repairable systems, maintenance can be omitted.

The MTBF is defined as the mean over all possible time intervals in which the system has run out from its specifications when t goes to infinity. The MTBF is expressed as a time and is sometimes denoted by m, and hence by definition

$$\text{MTBF} = m = \int_{0}^{t \to \infty} t f(t) \, dt = \int_{0}^{t \to \infty} R(t) \, dt \qquad (5.13)$$

where $f(t)$ is the considered failure probability function. As will be clear from the definition, the MTBF is always related to a repairable system. For a non-repairable system the lifetime can be measured by the *mean time to failure* (MTTF), which then exactly represents the span of life. Sometimes a comparison is made between repairable and non-repairable systems, and then the time to the first failure of the repairable system can be measured and is expressed as a number called the mean time to first failure (MTTFF). We will discuss the related aspects of MTBF later.

It has been verified experimentally that the failure rate distribution as a function of time often shows a so-called 'bath tub' curve as is shown in figure 5.4. In this graph three different regions can be distinguished as will be explained. In region I, λ is a function of t and will decrease with time, because in this region, the so-called 'burn-in' period, a lot of failures can occur. This is often the case when a system is a totally new design and just manufactured. This region is also called the 'infant mortality' or 'early-failure' region. After a certain time t_1, λ appears to be a constant and is no longer a function of time, and region II commences.

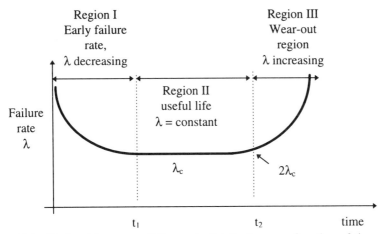

Figure 5.4 'Bath tub' curve or failure rate distribution as a function of time.

In region II, λ is a constant for a long period of time and this is therefore called the 'useful life' period, or normal operating region. Finally, region III, the so-called 'wear-out' period, starting at time t_2, is the region where λ now is an increasing function of time. The wear-out region is defined as beginning when the failure rate is twice the failure rate of the useful life region. In this respect an interesting analogy can be recognized. In living systems, the same bath tub curve can be recognized, but with large differences of shapes due to large differences in living circumstances and work loads. We shall conclude this section with another simple example, which stresses the concept of failure rate.

*   *   *

Example 5.3
Suppose a system has a failure rate of 1.4% per hour of operating time. What is the average number of failures that could have occurred after 500 operating hours? From the data it can be concluded that the failure rate per hour equals 0.014 and this equals the number of failures per hour and per system. Then with 500 operating hours it may be expected that $t\lambda = 500 \times 0.014 = 7$ failures have occurred.

*   *   *

5.3   FAMILIES OF LIFETIME DISTRIBUTIONS

Since lifetime is defined as a continuous variable, there are a number of probability density functions $f(t)$, corresponding survival functions $R(t)$ and instantaneous failure rate functions $\lambda(t)$ which have been considered as well suited to the description of possible lifetime distributions. We shall briefly describe some of them, but will not go into further detail. The following distribution functions are applicable depending on the circumstances to be considered.

1 *The exponential failure distribution.* As stated already the exponential failure distribution is one of the most important failure laws, most widely used, very suitable for electronic systems and components and applicable for normal operating conditions.

2 *The normal failure distribution.* The normal failure distribution is one of the best known and represents an adequate model for systems and components in which the failure is caused by some wear-out phenomena, so in region III as is illustrated in figure 5.4.

3 *The Rayleigh failure distribution.* The Rayleigh failure distribution is applicable when increased stress and accelerated wear-out are involved.

4 *The gamma failure distribution.* The gamma failure distribution (Γ) is a type of distribution function used when burn-in, wear-out and redundancy are involved. The concept of redundancy will be discussed later on.

5 *The Weibull failure distribution.* The Weibull distribution is applicable when wear-out, aging and overloading are involved simultaneously. The Weibull law basically contains all the above-mentioned distributions. This distribution function is also applicable for failure analysis of VLSI circuits.

In table 5.2 a summary is given for the reliability concepts discussed so far, with the results for the exponential failure distribution in the last column. Derivations of expressions for the reliability concepts with other failure distribution laws than the

Table 5.2 Survey of reliability parameters.

Parameter	Mathematical expression or definition	Result for exponential distribution
Survival function or reliability	$R(t) = 1 - F(t) = \int_t^\infty f(t)\,\mathrm{d}t$ $= \exp\left[-\int_0^t h(t)\,\mathrm{d}t \right]$	$\mathrm{e}^{-\lambda t}$
Failure rate	$\lambda = 1/\mathrm{MBTF}$	λ
Failure density function	$f(t) = -\dfrac{\mathrm{d}R(t)}{\mathrm{d}t}$	$\lambda\,\mathrm{e}^{-\lambda t}$
Cumulative distribution or unreliability	$F(t) = \int_0^t f(t)\,\mathrm{d}t = 1 - R(t)$	$1 - \mathrm{e}^{-\lambda t}$
Hazard rate	$\lambda(t) = h(t) = z(t)$ $= -\dfrac{\mathrm{d}R(t)/\mathrm{d}t}{R(t)} = \dfrac{f(t)}{R(t)}$	λ
MTBF = MTTFF	$1/\lambda = \int_0^\infty R(t)\,\mathrm{d}t = \int_0^\infty tf(t)\,\mathrm{d}t$	$1/\lambda$

expressions for the reliability concepts with other failure distribution laws than the exponential failure law tend to become complicated rapidly and are not very practical for an introductory chapter on reliability. Before continuing we consider two more examples.

*   *   *

Example 5.4
The useful lifetime of a component, with a failure rate of 0.0001 h^{-1}, is 1000 h. If it may be assumed that the failure distribution is exponential, calculate the MTBF and the reliability of that component for 10, 100 and 1000 h.

From the data given it may be concluded immediately that the MTBF is $1/10^{-4} =$ 10 000 hours. The numerical value for the reliability or survival function $R(t)$ can be found by substituting the corresponding values for t in the exponential function $\exp(-\lambda t)$. This yields

$$R(10) = \exp(-10^{-4} \times 10) = \exp(-0.001) = 0.999$$
$$R(100) = \exp(-10^{-4} \times 100) = \exp(-0.01) = 0.990$$
$$R(1000) = \exp(-10^{-4} \times 1000) = \exp(-0.1) = 0.905.$$

As can be seen from this example, the reliability decreases when time increases as may be expected.

*   *   *

Example 5.5
A circuit consists of 20 resistors with a failure rate of 10×10^{-9} h^{-1} per component and 30 soldered connections with a failure rate of 100×10^{-9} h^{-1} per connection. If it may be assumed that the distribution is exponential, calculate the total failure rate and the MTBF.

Here we first have to assume that the system is no longer operational when one of the components fails, including the soldering connections. The total failure rate can be found by simply summing all components with their failure rates, and thus

$$\lambda_{tot} = 20(10 \times 10^{-9}) + 30(100 \times 10^{-9}) = 3200 \times 10^{-9} \text{ h}^{-1}.$$

Then the MTBF is

$$1/\lambda_{tot} = 1/(3200 \times 10^{-9}) = 312\ 500 \text{ h} = 35.67 \text{ years.}$$

Here the strong influence of the soldered connections may be recognized.

*   *   *

5.4 RELIABILITY OF PARTS, COMPONENTS, EQUIPMENT AND SYSTEMS

In this section we shall present a few more definitions which are frequently used in the wrong context. We have used the concept of component and system several times without making any distinction between them; to avoid further confusion here are the correct definitions.

A *part* is the smallest building block of which a system is composed. Examples are resistors, transistors, integrated circuits (ICs), capacitors and so on. Parts are characterized by the fact that they are non-repairable and cannot perform a function on their own.

A *component* is a small structure composed of several parts which can perform a certain task, such as a power supply, a keyboard or a display. A component might be repairable, but a completely moulded power supply is an example of a component which is non-repairable.

Equipment is composed of several components and can perform a function or task as part of a larger system. In general, equipment is repairable. Examples are a temperature control, a fuel control system or a video terminal.

A *system* is defined as a structure of different pieces of equipment such as a large mainframe computer, a personal computer, an aircraft or a telephone station, which can perform large specific tasks.

It will be clear that the reliability of parts is the cornerstone of reliability of systems. In addition to the reliability of parts, a lot of other factors are involved which finally determine the overall reliability of the system. We will discuss several of these factors in the following sections such as

1   the required MTBF
2   the operational conditions
3   the purpose of use
4   the demanded reliability
5   the applied configuration
6   the maintenance protocol
7   the required availability.

From the viewpoint of system reliability, the reliability engineer is concerned with the identification and selection of an appropriate set of measures of reliability effectiveness and with defining the suitable reliability objectives. Very often the reliability engineer wants to approach the derived optimum lifecycle as a function of cost or vice versa as closely as possible. An example of such a lifecycle versus cost function is depicted in figure 5.5. Here, reliability is the chosen independent variable and cost the dependent variable. Curve (a) represents the manufacturing costs plus the cost of preventive maintenance of the system under development. Curve (b) represents

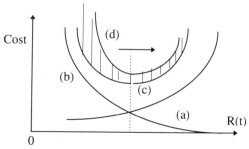

Figure 5.5   Cost versus reliability for a given system (courtesy Delft University Press).

(a) and (b). Finally, curve (d) gives an impression of the total to which is added the loss of production rate during non-estimated repair time of the system.

A poorly designed system means a cheap system with a poor reliability and with very high maintenance costs. A highly sophisticated designed system can provide a high reliability with very modest maintenance cost, but an extremely high sales price as a consequence. Somewhere there is an optimum which should be approached as closely as possible by the manufacturer.

It is noticed here that this approach is no longer valid for VLSI circuits. Here it appears to work in the reverse direction. For VLSI circuits no repair action is possible and all effort must be put in to make these circuits as reliable as possible. Built-in reliability and built-in self-testing features are a prerequisite for complicated ICs with many millions of transistors. The study of failure mechanisms has become extremely important, because exhaustive functionality testing in the old-fashioned way is no longer possible owing to the number of possibilities.

5.4.1 The ISO 9000 system

Here, we must stress again the relationship between quality $Q(t)$ and reliability. We have seen already that quality is the instantaneous sample of specification at a certain time; see also (5.1). A good quality of all parts in a system does not guarantee a good reliability; it may just be an essential condition, but not sufficient.

Nowadays, many manufacturers are confronted with the International Standards Organization (ISO) 9000 system. When a manufacturer complies with the requirements of certification for production, usually to be obtained via a long-term certification programme, the manufacturer can guarantee minimum quality for his or her products and services. For instance, suppliers are often only allowed to deliver parts and components to the car industry when they manufacture in conformity with the ISO 9000 system and protocols. In particular, for the manufacturer the following ISO 9000 quality systems may be applicable:

- ISO 9001, quality management system for procurement in the design, the development, the production, the installation and the after-sales service of products
- ISO 9002, quality management system for the procurement of the production and installation
- ISO 9003, quality management system for procurement in the final product approval and testing.

Basically, seven different ISO 9000 systems exist, ranging from 9000 to 9004/2; however, the previously mentioned are the most important ones for design and manufacturing purposes. How do they relate to reliability requirements? It will be obvious that minimum guaranteed product quality can guarantee minimum reliability specifications of the separate parts produced. However, overall system concepts, architecture of the applied configuration and environmental conditions will determine the final reliability achieved. We will conclude with a diagram of a production process which outlines the whole philosophy of the ISO 9000 systems, including

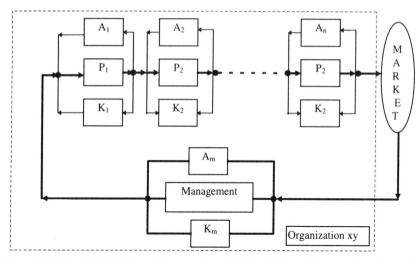

Figure 5.6 The ISO 9000 model as a closed feedback loop for a production and/or service organization.

the vital role of the management (figure 5.6). The (monitoring) audit step, A_i, in every production phase P_i gives the required feedback for each production step involved. K_i represents the predetermined control, calibration and references to be accomplished for each phase. Note that in this closed feedback loop every phase is a closed feedback loop in itself. Also, the management must be the subject of standard audit procedures and preset reference targets. If in this loop one of the elements is failing, the loop (organization) will always crash in due time. A complete discussion of the ISO 9000 system requires a separate chapter and is beyond the scope of this book. Also, legal regulations can require minimum guaranteed quality. We will discuss safety aspects, priorities and environmental conditions regarding reliability and costs in a later section.

5.5 AVAILABILITY; REPAIRABLE AND NON-REPAIRABLE SYSTEMS

In this section we shall go into some more detail concerning repairable and non-repairable systems. The idea of repairable and non-repairable systems implies that the system can be maintained, and so is synonymous with the concept of maintainable and non-maintainable systems, although a repair action differs from a maintenance action. The problem of developing mathematical tools for the reliability of systems that can be maintained while being in use can be approached by applying Markovian techniques. Markovian techniques are especially useful in describing the stochastic behaviour of systems under a great variety of failure and repair conditions. This means that Markovian techniques are able to take account of repair actions, which of course require a finite period of time to restore a system to an operational state. In short, the probability of every possible transfer state has to be defined and calculated. For instance, given that the system is working normally in a certain state, the question 'what is the probability that the system will achieve another state?' has to be answered. Therefore forward and backward differential equations in the time domain have to be developed, whereas in non-maintained systems only forward transitions have to be considered. This approach results in an additional figure of merit to describe the reliability of maintainable systems, the so-called *availability*, often denoted by $A(t)$. In general, availability is also a function of time.

The availability of the considered maintainable system is defined as the probability that the system is in an operational state at any time t given that the system was fully operational at the time $t = 0$. An alternative definition, more easily understood, is based on the ratio of the time that the system is operational, the so-called 'uptime', to the total amount of time that the system is needed. If the time for maintenance and repair is called the 'downtime', we can write

$$\text{availability} = \frac{\text{uptime}}{\text{uptime} + \text{downtime}}. \tag{5.14}$$

The downtime is defined as the product of the number of maintenance and repair actions which require a shutdown of the system and the average time for each such action. The average time for each action is called the *mean time to repair* (MTTR), and this concept results in the following expression for the availability:

$$\text{availability} = \frac{\text{MTBF}}{\text{MTBF} + \text{MTTR}}. \tag{5.15}$$

If the MTBF and thus the reliability increase, the MTTR can also increase; if the MTBF decreases, the average time for a repair action must also decrease in order to maintain the same availability. In a great majority of cases, involving many different types of component, equipment and system, here again an exponential distribution of the repair time can be justified. As stated already, to be able to make comparisons possible between maintained and non-maintained systems, the concept of *mean time to first failure* (MTTFF) can be introduced for maintained systems. We have to introduce here another definition to make calculations more accessible. The *repair rate* is defined as the frequency per unit of time with which repair actions are effected and is denoted by μ, often expressed in numbers per hour.

You will not be surprised that the reciprocal value of μ is the MTTR which is in complete analogy with the failure rate λ and the MTBF.

* * *

Example 5.6
Suppose the repair rate of a system is $\mu = 0.1 \text{ h}^{-1}$, then the MTTR is $1/\mu = 10 \text{ h}$.

* * *

If the distribution is exponential and the MTBF $= 1/\lambda$ and the MTTR $= 1/\mu$, then substitution of these values in the expression for the availability (5.15) results in

$$A(t) = \frac{1/\lambda}{1/\lambda + 1/\mu} = \frac{\mu}{\mu + \lambda}. \tag{5.16}$$

$$* \qquad * \qquad *$$

Example 5.7
If the MTTR of a computer system equals 5 h and the MTBF $= 5000$ h, then by substitution in expression (5.16) the availability $A(t)$ is given by

$$A(t) = \frac{1/5}{1/5 + 1/5000} = 0.9990$$

which is also a dimensionless number of course. It will be shown that, in general, the availability is also a function of time.

$$* \qquad * \qquad *$$

If in expression (5.16) $\mu = 0$ is substituted it may be expected that the expression reduces to an expression for non-maintained systems, but this is wrong and the result appears to be zero. As stated already availability is, in general, a function of time and the expression (5.16) is called the stationary or *steady-state availability* when t goes to infinity. The general expression for $A(t)$ is called the *momentary availability* and results in a more complicated expression requiring Markovian techniques for derivation. For instance, here is an example for the expression of $A(t)$ for a single system with single repair actions:

$$A(t) = \frac{\mu}{\lambda + \mu} + \frac{\lambda}{\lambda + \mu} \exp[-(\mu + \lambda)t]. \tag{5.17}$$

If now $\mu = 0$ is substituted we obtain the well known expression (5.6) for the reliability $R(t) = \exp(-\lambda t)$ for a single system, and it may be concluded that the availability is a generalization of the concept of reliability. The exponential term is called the *transient term* and as t goes to infinity it can be seen from (5.17) that this term vanishes and the steady-state availability results.

To summarize, the following availability concepts are used.

1 The *momentary availability* $A(t)$ is defined as the probability that the system is in an operational state at any time t given that the system was operational at the time $t = 0$, which equals the previously given definition. The momentary availability is well suited to describe systems which must be available any time.

2 The *steady-state availability* A_{SS} is defined as the mean of $A(t)$ if $t \to \infty$:

$$A_{SS} = \lim_{t \to \infty} \int_0^t A(t)\, dt. \tag{5.18}$$

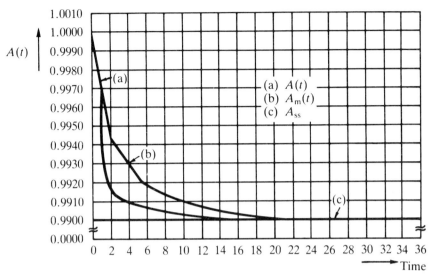

Figure 5.7   The availability $A(t)$ as a function of time.

3   The *mission availability* $A_m(t)$ is defined as the mean for a certain time interval t_1-t_2:

$$A_m(t) = \frac{1}{t_2-t_1} \int_{t_1}^{t_2} A(t) \, dt. \qquad (5.19)$$

In figure 5.7 the availability as a function of time is depicted from which it can be seen that, for $t \to \infty$, the three kinds of availability tend to coincide with the same steady-state availability A_{ss}.

*   *   *

Example 5.8
For a single subsystem having the parameter MTTR $= 1/\mu = 100$ h, the proportion spent in an operational state is 0.9901 and that spent in a failed state is 0.0099. These figures indicate that over a period of for instance 10 000 h the system is operable for 9901 h and occupies a failed state for 99 h.

*   *   *

5.6   BOOLEAN ALGEBRA AND RELIABILITY

This section considers the application of *Boolean* algebra to find the reliability of a system which is composed of multiple components or subsystems. In most circumstances we have to deal with such composite systems and other sophisticated tools are required to calculate reliability. In many applications failing or not failing can be characterized by two states, i.e. the system works or it does not work. This statement asks for a Boolean approach, that is to say we can apply Boolean algebra to calculate reliability of composite systems.

Suppose a system denoted by C consists of two subsystems A and B and assume that A, B and C denote the states of the respective systems. Then the following statement

$$C = A + B \tag{5.20}$$

can be read as 'the system C works if either the system A or B or both systems are in an operational state', which represents an OR function.

Consider now a situation in which the system C works when and only when both A and B are in an operational state, which represents an AND function:

$$C = AB. \tag{5.21}$$

Finally, if we define a working state with '1' and a non-working state with '0' it is easily seen that all basic rules of Boolean algebra can be applied to composite systems for reliability calculations. For convenience, here are the basic rules of Boolean algebra without any derivation:

$$
\begin{aligned}
A + 0 &= A & A0 &= 0 \\
A + 1 &= 1 & A1 &= A \\
A + A &= A & AA &= A
\end{aligned}
\tag{5.22}
$$

$$
\begin{aligned}
A + B &= B + A & A + \bar{A} &= 1 \\
(A + B) + C &= A + (B + C) & A\bar{A} &= 0 \\
AB + AC &= A(B + C) & A\bar{A} &= A \\
(AB)C &= A(BC).
\end{aligned}
\tag{5.23}
$$

Finally, De Morgan's theorem is

$$\overline{AB} = \bar{A} + \bar{B} \quad \text{and} \quad \overline{A + B} = \bar{A}\bar{B}. \tag{5.24}$$

We will demonstrate the use of Boolean algebra in reliability with an example.

*   *   *

Example 5.9
Suppose we want to compare the reliability of the two systems depicted in figure 5.8; then we have only to determine the Boolean expression for the working state of each of the system configurations.

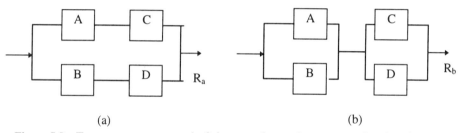

(a)   (b)

Figure 5.8 Two systems composed of the same four subsystems performing the same function, but with different reliability.

For a series-connected system, consisting of two or more subsystems, it will be clear that the system will function if all subsystems are functioning; on the other hand, for a parallel-connected system, the system will function if one of the subsystems is functioning. If it is further assumed that all subsystems are independent, and that the states are A, B, C and D, then for the state working, W, of the upper branch in figure 5.8(a) the following expression holds:

$$W_1 = AC.$$

For the lower branch

$$W_2 = BD$$

and hence for the complete system

$$W_{\text{syst a}} = W_1 + W_2 = AC + BD \tag{5.25}$$

and in figure 5.8(b) the Boolean expression becomes

$$W_{\text{syst b}} = (A + B)(C + D)$$
$$= AC + AD + BC + BD$$
$$= W_{\text{syst a}} + AD + BC. \tag{5.26}$$

By comparison of (5.25) and (5.26) the conclusion will be justified that the configuration of figure 5.8(b) will represent a larger reliability than that of figure 5.8(a). This result can easily be understood by inspection immediately of figure 5.8 because the system (b) still functions if either A and D or B and C are failing.

$$* \qquad * \qquad *$$

So far only the Boolean functions of both systems for the working state have been determined. If it is required to calculate the reliability, then we have to transfer the Boolean expression into a probability function as in the following example, where it is assumed again that all the subsystems are independent.

$$* \qquad * \qquad *$$

Example 5.10
Suppose the reliability of every subsystem in figure 5.8 is 0.1. Then the reliability for system (a) can be calculated as follows:

$$P(W_{\text{syst a}}) = P(AC + DB)$$
$$= P(AC) + P(DB) - P(ABCD)$$
$$= P(A)P(C) + (D)(B) - P(A)P(B)P(C)P(D)$$

and by substitution of $P(A) = P(B) = P(C) = P(D) = 0.1$ the reliability is

$$P(W_{\text{syst a}}) = 0.1 \times 0.1 + 0.1 \times 0.1 - (0.1)^4 = 0.0199.$$

For system (b) we find for the reliability function

$$P(W_{\text{syst b}}) = P(A+B)P(C+D)$$
$$= [P(A)+P(B)-P(AB)][P(C)+P(D)-P(CD)]$$

and by substitution

$$P(W_{\text{syst b}}) = (0.2-0.01)(0.2-0.01) = 0.0361.$$

The reliability of the configuration of figure 5.8(b) is 1.8 times larger than that of figure 5.6(a). So, a well chosen *configuration* can improve reliability considerably.

*   *   *

5.7 THE RELIABILITY AND MTBF OF SYSTEMS IN SERIES

Here we consider a system composed of components and subsystems connected in series and forming one complete system. At first no repair action will be taken into account, and hence only the failure rates of the components and subsystems are involved. The system fails when any of the subsystems attains a failed state. We will start with an example.

*   *   *

Example 5.11
Of an X_R missile type, applied to launch telecommunication satellites, it is known that 7 out of 10 launches are successful. Of the communication satellite X_S it is known that after launching 95 out of 100 are functioning in accordance with their predetermined specified characteristics. The applied launching system X_L fails 25 out of 250 launching operations, so we can conclude that the reliability of the launching is

$$R(X_L) = (250-25)/250 = 0.90.$$

Where the reliability of the whole system is a combination of all items involved, supposing that every system is independent we find for the total reliability

$$R(X_R)R(X_S)R(X_L) = (0.7)(0.95)(0.90) = 0.5985.$$

*   *   *

It is not very difficult to apply the same philosophy to a type of vital measuring system monitoring a critical temperature in a nuclear reactor or in an aircraft engine. This concept is illustrated in figure 5.9 and can be generalized. Suppose that a system

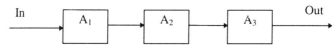

Figure 5.9   Series system consisting of three subsystems performing one function.

is composed of n subsystems. Then for the reliability of the complete system the following multiplying rule can be applied:

$$R_{(system)} = \prod_{i=1}^{n} R_i \qquad (5.27)$$

where R_i is the reliability of the subsystem i. If the reliability of every independent subsystem may be denoted by the exponential time function $\exp(-\lambda_i t)$, which is often the case, then expression (5.27) can be rewritten as

$$R(t) = \prod_{i=1}^{n} R_i(t) = \prod_{i=1}^{n} \exp(-\lambda_i t) = \exp(-\lambda_s t) \qquad (5.28)$$

with

$$\lambda_s = \sum_{i=1}^{n} \lambda_i = \lambda_1 + \lambda_2 + \ldots + \lambda_n. \qquad (5.29)$$

If all n subsystems are identical, i.e. $\lambda_1 = \lambda_2 = \ldots = \lambda_n = \lambda$, then

$$R(t) = \exp(-n\lambda t). \qquad (5.30)$$

The MTBF of a series system can be calculated as follows:

$$\mathrm{MTBF} = \int_{0}^{\infty} \exp\left(-\sum_{i=1}^{n} \lambda_i t\right) dt = \frac{1}{\sum_{i=1}^{n} \lambda_i} \qquad (5.31)$$

where again it is assumed that the systems follow an exponential distribution and are independent; if the systems are not independent a *conditional* reliability must be included which complicates calculations considerably. For n identical subsystems the MTBF is given by

$$\mathrm{MTBF} = 1/n\lambda. \qquad (5.32)$$

Figure 5.10 is a graphical representation of the reliability and the MTBF of a single system and a system composed of two subsystems in series with a failure rate of $\lambda = 0.25 \, \mathrm{h}^{-1}$ for each system. Note that the reliability and MTBF of a single system are *higher* than for a series system. As we have already seen, talking about the MTBF of a series system involves taking repair and maintenance actions into account, but this will be discussed later. The expressions rapidly increase in complexity because the repair rate μ is involved.

5.8 THE RELIABILITY AND MTBF OF SYSTEMS IN PARALLEL

Systems are often connected in parallel to improve reliability. In particular, when a vital system must be monitored and controlled, there is a demand for more than one system to be installed. For example, in an aircraft, as well as a lot of other doubly installed equipment such as the two flight computers as a basic prerequisite, sometimes the fly-by-wire system is installed fourfold. Very often power supplies are installed twice, and in power stations all vital equipment is installed more than once.

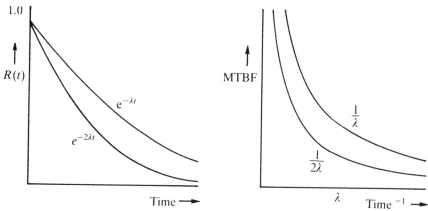

Figure 5.10 Comparison of the reliability and the MTBF of a single system and a series system for $n=2$ with exponential distribution and failure rate of $0.25 \, \text{h}^{-1}$.

The same idea can be found in telecommunications stations, computers, critical memory systems and a large variety of more sophisticated measuring instruments.

Every piece of equipment connected in parallel is able to execute the complete task on its own independently of the other pieces of equipment. The background of this philosophy can be described as *redundancy*, so if a system retains redundancy, more than one identical system is installed and, as we will see, reliability is increased considerably. In parallel systems we have to be more specific because the four following possibilities frequently occur, each with their own characteristics.

1   n systems are connected in parallel of which all n systems are active. This is called an *active on-line standby system*.
2   n systems are connected in parallel of which one system is active only. This is called a *cold on-line standby system*, or sometimes a *passive on-line system*. It is assumed here that, if one system fails, a new system is switched on immediately and the failed system is switched off. It is further assumed that switching off occurs perfectly.
3   n systems are connected in parallel, but to be operational only k out of n systems have to be active with $k < n$, sometimes called a *parallel k out of n system*.
4   Mixed configurations exist in which series and parallel connected systems are combined.

We first discuss the reliability and MTBF of these systems without considering repair and maintenance actions.

5.8.1   An active on-line standby system

In this mode of operation n systems are provided operating simultaneously, performing one function only and with equal failure rates. It can be seen that the system is still in an operational state when $n-1$ systems have reached a failed state since

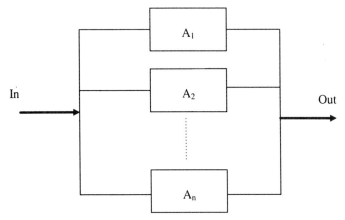

Figure 5.11 Configuration of an active on-line standby system composed of n identical systems.

any of the systems can execute the correct function. This is illustrated in figure 5.11.

In this system $n-1$ systems are redundant and the system still works when one system works; hence we can describe the reliability with the Boolean expression for an OR function of n systems:

$$R_{\text{syst}} = P(A_{\text{syst}}) = P(A_1 + A_2 + \ldots + A_n). \tag{5.33}$$

This expression can be rewritten with the basic rules of probability for independent systems, which is really a cumbersome affair. It is much easier to convert the problem, calculating the probability the system will fail and then taking the complement. The probability the system will fail is given by

$$F(A_{\text{syst}}) = \prod_{i=1}^{n} [1 - P(A_i)]. \tag{5.34}$$

Then the probability that the system will not fail equals

$$R(A_{\text{syst}}) = 1 - F(A_{\text{syst}}) = 1 - \prod_{i=1}^{n} [1 - P(A_i)]. \tag{5.35}$$

If again an exponential distribution is assumed then by substitution for n identical systems expression (5.35) yields

$$R(A_{\text{syst}}) = 1 - \prod_{i=1}^{n} [1 - \exp(-\lambda_i t)]. \tag{5.36}$$

Suppose that $n=2$ and the systems are identical with a failure rate of λ; then expression (5.36) yields

$$R(t, 2) = 2 \exp(-\lambda t) - \exp(-2\lambda t) \tag{5.37}$$

You can try to find for yourself an expression when the failure rates are unequal. To find an expression for the MTBF we have to apply the basic definition for the MTBF:

$$\text{MTBF} = \int_0^\infty R_S \, dt = \int_0^\infty \left[1 - \prod_{i=1}^n (1 - \exp(-\lambda_i t)) \right] dt. \tag{5.38}$$

To carry out a general calculation requires quite a lot of effort, and therefore we shall confine the calculation a little and perform the calculation for two systems with unequal failure rates.

$$\text{MTBF}(t, 2) = \int_0^\infty \{1 - [1 - \exp(-\lambda_1 t)][1 - \exp(-\lambda_2 t)]\} \, dt$$

$$= \int_0^\infty \{\exp(-\lambda_1 t) + \exp(-\lambda_2 t) - \exp[-(\lambda_1 + \lambda_2)t]\} \, dt. \tag{3.39}$$

Solving this multiple integral gives for the MTBF

$$\text{MTBF}(t, 2) = \frac{1}{\lambda_1} + \frac{1}{\lambda_2} - \frac{1}{\lambda_1 + \lambda_2}. \tag{5.40}$$

If the failure rates are identical then the MTBF becomes

$$\text{MTBF}(t, 2) = \frac{3}{2\lambda} \tag{5.41}$$

and in the same fashion for a triply identical system it can be found that the MTBF is given by

$$\text{MTBF}(t, 3) = \frac{11}{6\lambda}. \tag{5.42}$$

A graphical representation of these results can clarify the consequences for the reliability and MTBF of a parallel active on-line system. As can be seen from figure 5.12 the reliability and the MTBF will increase considerably in comparison with a single system. As stated already we have neglected any repair action, so in the following example the MTTF is calculated.

* * *

Example 5.12
Suppose that the failure rate of a non-repairable system equals $0.0015 \, \text{h}^{-1}$; then the MTTF of a single system is simply $1/0.0015 = 667 \, \text{h}$. Then for an active on-line

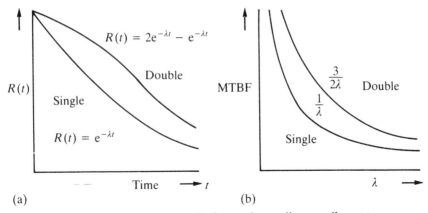

Figure 5.12 The reliability and MTBF of an active on-line standby system composed of two identical systems compared with a single system.

standby system consisting of two such systems the MTTF is found by applying (5.41):

$$\text{MTTF} = 3/2 \times 1/0.0015 = 1000 \text{ h.}$$

*   *   *

It can be shown that for an *n*-fold active on-line standby system when no repair action is involved, the MTTF is given by

$$\text{MTTF} = \sum_{i=1}^{n} \frac{1}{i\lambda} = \frac{1}{\lambda} + \frac{1}{2\lambda} + \ldots + \frac{1}{n\lambda}. \tag{5.43}$$

As can be noted, it makes less sense to connect many systems in parallel, because the slope of the MTTF decreases with an increasing number of systems.

5.8.2 A cold on-line standby system

In this configuration only one system is on standby; the other systems will be switched on when a system fails. No repair actions are performed and maintenance is executed when all systems have failed: this is called a non-maintained cold on-line standby system. The basic configuration is depicted in figure 5.13 for which the following assumptions have to be made.

1 The system fails when all subsystems have failed.
2 The failure probabilities of all systems are independent.
3 Cold systems not in operation cannot fail.
4 The exponential distribution is applicable for all systems.
5 The reliability of switching off and on of a system is 100%.

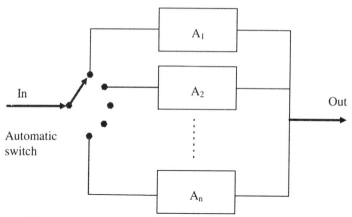

Figure 5.13   A parallel cold on-line standby system configuration of n identical systems.

So it can be said the system is still working when no system has failed, or if one system has failed or two systems etc up to $n-1$ systems. This refers to a Poisson distribution which can be described as follows:

$$R(t) = P(x \leq n-1) = \sum_{k=0}^{n-1} \frac{(\lambda t)^k \exp(-\lambda t)}{k!}$$

$$= \exp(-\lambda t)\left[1 + \lambda t + \frac{(\lambda t)^2}{2!} + \ldots + \frac{(\lambda t)^{n-1}}{(n-1)!}\right]. \qquad (5.44)$$

It is now easy to derive an expression for a cold on-line standby system composed of two identical systems. Substituting $n=2$ in expression (5.44) gives

$$R(t) = \exp(-\lambda t) + \lambda t \exp(-\lambda t) \qquad (5.45)$$

Again, applying the basic definition for the MTTF, it can be found that for the two-system configuration that

$$\text{MTTF} = \int_0^\infty R(t)\, dt = \int_0^\infty [\exp(-\lambda t) + \lambda t \exp(-\lambda t)]\, dt \qquad (5.46)$$

and by partial integration for the second term the result is

$$\text{MTTF} = 1/\lambda + 1/\lambda = 2/\lambda. \qquad (5.47)$$

If the same calculation is executed for n systems then we find

$$\text{MTTF} = n/\lambda. \qquad (5.48)$$

Note that here again repair actions are omitted so it makes more sense to talk about MTTF instead of MTBF of which the definition implies that repair actions can be performed.

In the following example we will compare both types of parallel configuration. It will be seen that a parallel cold on-line system delivers a slight advantage above a parallel active on-line system. However, it must be borne in mind that for the cold on-line system it is assumed that no switching problems exist which in practice is not always the case.

* * *

Example 5.13

A two-system parallel configuration is considered of which the active and passive configurations must be compared for reliability aspects. The failure rate of each system involved is 0.01 h^{-1}, an exponential distribution may be assumed and if a system fails ideal switching occurs. The mission time is 10 h.

We first calculate the reliability for an active on-line standby system. This is obtained by applying (5.36).

$$R(t, 2) = 1 - [1 - \exp(-0.01 \times 10)]^2 = 0.9909.$$

Application of (5.41) gives

$$\text{MTTF}(t, 2) = 3/2\lambda = 3/(2 \times 0.01) = 150 \ h.$$

For the cold on-line standby system the reliability is, from equation (5.45),

$$R(t, 2) = \exp(-0.01 \times 10) + (0.01 \times 10) \exp(-0.01 \times 10) = 0.9953$$

and on applying (5.47) the MTTF becomes

$$\text{MTTF}(t, 2) = 2/0.01 = 200 \ h$$

which is an improvement of 25% compared with the active on-line standby system.

* * *

5.8.3 A parallel *k* out of *n* system

Here again, a distinction must be made between active and cold on-line standby systems. For the *active* on-line standby configuration composed of *n* identical systems of which *k* out of *n* systems are active, a binomial distribution is applicable. Then it can be shown that the reliability, when an exponential distribution is assumed, is given by

$$R(n, k, p) = \sum_{i=k}^{n} \binom{n}{i} [\exp(-\lambda t)][1 - \exp(-\lambda t)]^{n-i} \qquad (5.49a)$$

and for the MTTF

$$\text{MTTF}(n, k, p) = \frac{1}{\lambda} \sum_{i=0}^{n-k} \frac{1}{n-i}. \qquad (5.49b)$$

If $k = 1$ then a cold on-line standby configuration is present, as discussed already in section 5.8.2, because only one system must be operational. If $k = n$ then by definition a *series* system is implied, because all systems must be active.

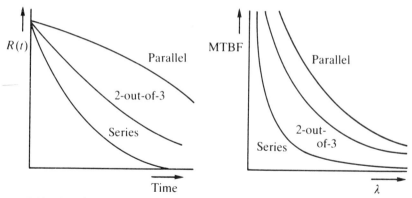

Figure 5.14 Graphs of the reliability and MTTF versus time for a series system, an active on-line standby system and a cold standby on-line k out of n system.

For a *cold* on-line standby system composed of n identical systems of which k out of n must be operational, the reliability for an exponential distribution is given by

$$R(n, k, p_{\text{cold}}) = \sum_{i=0}^{n-k} [\exp(-\lambda t)]^k \frac{(k\lambda t)^i}{i!} \qquad (5.50a)$$

and for the MTTF it can be shown that

$$\text{MTTF}(n, k, p_{\text{cold}}) = \frac{n-k+1}{k\lambda}. \qquad (5.50b)$$

From the above it can be seen that reliability increases as n increases. As claimed already, all systems are considered to be independent of each other. For comparison, figure 5.14 illustrates the reliability and MTTF behaviour versus time for a series system, an active on-line standby parallel system and a k out of n system.

5.9 METHODS TO DETERMINE THE RELIABILITY OF SYSTEMS

There are numerous methods available to determine the reliability of systems, some more powerful than others. We shall discuss five methods with illustration by examples. Four of these methods have one thing in common, that is how to simplify a complex system to a subset of known systems to make reliability calculations possible. The fifth method is called Markov techniques and is the most powerful one.

5.9.1 The network reduction method
This method implies a straightforward reduction of the system to simpler systems until just one single system remains.

*   *   *

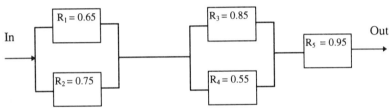

Figure 5.15 A mixed parallel–series system.

Example 5.14

We will determine the reliability of the mixed parallel–series system depicted in figure 5.15. The failure distribution is assumed to be exponential and the parallel subsystems are of the active standby type. The two parallel systems can be substituted by a

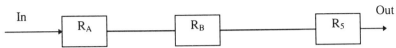

Figure 5.16 Reduction of the mixed parallel–series system of figure 5.15 to a series system only.

single system as depicted in figure 5.16; hence the result is a series system of three subsystems. Finally, one system will remain when this is replaced again as illustrated in figure 5.17.

The calculation can be carried out by first applying equation (5.36) for both parallel subsystems:

$$R_{syst} = 1 - \prod_{i=1}^{2} [1 - \exp(-\lambda_i t)]. \tag{5.51}$$

Then

$$R_A = 1 - (1 - 0.65)(1 - 0.75) = 1 - 0.0875 = 0.9125$$

and

$$R_B = 1 - (1 - 0.85)(1 - 0.55) = 1 - 0.067\,50 = 0.9325.$$

Then the system reliability can be found by multiplication:

$$R_{syst} = R_A R_B R_5 = (0.9125)(0.9325)(0.95) = 0.8084. \tag{5.52}$$

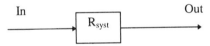

Figure 5.17 Completely reduced system of figure 5.15 representing the original system with the same function.

In the same way the MTTF can be calculated. This results in the following expression for the failure rate:

$$\lambda_{syst} = \lambda_A + \lambda_B + \lambda_5. \tag{5.53}$$

If λ is a constant and identical for all subsystems then the MTTF equals

$$\lambda_{syst} = 2\lambda/3 + 2\lambda/3 + \beta = 7\lambda/3. \tag{5.54}$$

The reciprocal value of (5.54) is the MTTF:

$$\text{MTTF}_{syst} = 1/\lambda_{syst} = 3/7\lambda. \tag{5.55}$$

Note that for parallel systems the failure rate decreases and for series systems it increases. If in this example five different failure rates were involved all expressions would be considerably more complicated.

*   *   *

5.9.2 The path-tracing technique
This method considers the different ways that a system can function satisfactorily. The systems are supposed to be independent. The method will be illustrated by the following example. The method is a rather simple one and is well suited for systems that are not too complicated.

*   *   *

Example 5.15
Applying Boolean algebra and probability theory to the mixed independent parallel–series system depicted in figure 5.18, we can write the following expression for the reliability, summing both paths AC and BC for a successful operational state:

$$R_{syst} = P(AC + BC) = P(AC) + P(BC) - P(ABC)$$
$$= P(A)P(C) + P(B)P(C) - P(A)P(B)P(C). \tag{5.56}$$

If the values for the reliability of the independent systems are known, the total reliability of the system can be found by substitution. For instance, if $P(A) = 0.95$, $P(B) = 0.85$ and $P(C) = 0.80$ then $R_{syst} = 0.794$.

*   *   *

5.9.3 The decomposition technique
The decomposition technique is well suited to the determination of the reliability of more complex systems. A good example of such a more complex system is a bridge

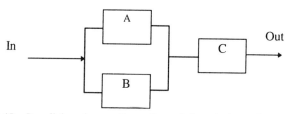

Figure 5.18 Parallel–series configuration of three independent systems.

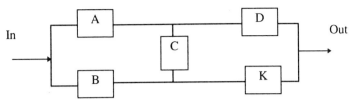

Figure 5.19 Bridge configuration of five independent systems.

configuration of systems as depicted in figure 5.19. The aim of this method is to reduce the system to a configuration of less complicated subsystems to make reliability calculations possible with reduced effort .

Here we have to follow a quite different strategy, but suppose first that we try to apply the path-tracing technique. Then the following expression can be found for the reliability of the complete system, because four different paths can be found for successful operation:

$$R_{syst} = P(AD + BK + ACK + BCD). \tag{5.57}$$

If you try to convert this expression, over 30 terms are found. The result can be obtained after a tremendous effort with a high chance of making mistakes. To solve this problem we use the following theorem, sometimes improperly called the *Bayes theorem*, but this is only part of it:

$$P(E) = \sum_i P(E/B_i)P(B_i) \tag{5.58}$$

with

$$\sum_i P(B_i) = 1. \tag{5.59}$$

This theorem says that if a certain event E depends on a number of mutually exclusive events B_i, then the probability that E occurs equals the product of the probability of event B_i and the conditional probability $P(E/B_i)$, summed over all possible events of B_i. This theorem will be applied to the bridge configuration by making a special choice of events. Here is how it works

*   *   *

Example 5.16
To solve this problem we apply the following procedure.

1 Remove a well chosen subsystem from the configuration under consideration. This removed system is sometimes called the 'key system'. In this example C will be removed.
2 Replace the removed system by another virtual subsystem V which can never fail, and calculate the conditional reliability.
3 Replace the removed system by another virtual system $\bar{V}$ which can never work, and calculate the conditional reliability.

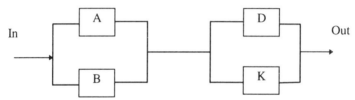

Figure 5.20 Bridge configuration of which the system C is replaced by a system V which never fails.

4   Find the reliability of the complete system E by applying equation (5.58) combined with the above statements 1, 2 and 3, then

$$R_{\text{syst}} = P(E) = P(E \text{ works}/V \text{ works always})P(V \text{ works always})$$
$$+ P(E \text{ works}/\bar{V} \text{ never works})P(\bar{V} \text{ never works}). \qquad (5.60)$$

You can see from this equation that for V and $\bar{V}$ two conditional probabilities for the given events are introduced in accordance with equation (5.58), and hence two new systems can be drawn, the reliability of which $R(E_1)$ and $R(E_2)$ can be calculated by standard equations. These two newly introduced systems are depicted in figures 5.20 and 5.21. For the system depicted in figure 5.20 the reliability is (with abbreviated notation)

$$R(E_1) = P(E/V) \qquad (5.61a)$$

and for the system depicted in figure 5.21 the reliability is

$$R(E_2) = P(E/\bar{V}). \qquad (5.61b)$$

Then the reliability of the total system E is

$$R(E) = R(E_1)P(V) + R(E_2)P(\bar{V}) \qquad (5.62)$$

or

$$R(E) = P(E/V)P(V) + P(E/\bar{V})P(\bar{V}). \qquad (5.63)$$

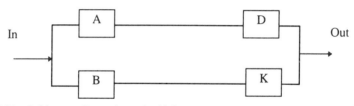

Figure 5.21 Bridge configuration of which system C is replaced by a system $\bar{V}$ which never works.

For the parallel–series system of figure 5.20 we can write

$$R(E_1) = \{1 - [1 - P(A)][1 - P(B)]\}\{1 - [1 - P(D)][1 - P(K)]\} \qquad (5.64a)$$

and for the parallel system of figure 5.21 we can write

$$R(E_2) = 1 - [1 - P(A)P(D)][1 - P(B)P(K)]. \qquad (5.64b)$$

Now one final step must be set for the virtual element V: if we equate the reliability of our key element C to the reliability of V then

$$P(V) = P(C) \qquad P(\bar{V}) = 1 - P(V) = 1 - (C) \qquad (5.65)$$

and the reliability of the bridge configuration becomes

$$R_{syst} = R(E_1)P(C) + R(E_2)[1 - P(C)]. \qquad (5.66)$$

All systems are assumed to be independent. As you can see, it is not that easy but it is the shortest way to find the reliability for a more complex system such as a bridge configuration. If all systems are identical and have the same reliability R you can verify that the reliability of the bridge configuration is

$$R_{syst} = 2R^2 + 2R^3 - 5R^4 + 2R^5. \qquad (5.67)$$

When the reliability is found the MTTF can be determined by applying the proper definition for the MTTF.

* * *

5.9.4 The minimal-cut-set technique

The minimal-cut-set technique is a method well suited to computer calculations. The method relies on the fact that in the system under consideration one searches for the minimum collection of elements that must be removed from the system so that the system fails. It can be found by determining the collection of (reduced) minimum cut sets for which the system will fail. It can also be called the critical path technique. If all possible minimum cut sets $A_1, A_2, \ldots, A_n$ are determined then the reliability is

$$R_{syst} = 1 - P\left(\bar{A}_1 + \bar{A}_2 + \ldots + \bar{A}_n\right) \qquad (5.68)$$

where $\bar{A}_1 + \bar{A}_2 + \ldots + \bar{A}_n$ is the event that the cut set will fail.

* * *

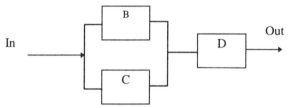

Figure 5.22   Minimum-cut-set method applied to a parallel–series system.

Example 5.17

We will demonstrate this method by the example illustrated in figure 5.22. By investigation of the system the number of cut sets A for which the system fails is five. They are

$$A_1 = BC \qquad A_2 = BD \qquad A_3 = CD \qquad A_4 = D \qquad A_5 = BCD$$

but these are not the minimum cut sets because A_4 is involved in A_5 and so are A_3 and A_2. The minimum cut sets are A_1 and A_4 only (verify that this result can be found by applying Boolean algebra), so the reliability of the system is

$$R_{\text{syst}} = 1 - P(\bar{A}_1 + \bar{A}_4) \tag{5.69}$$

$$R_{\text{syst}} = 1 - P(\overline{BC} + \bar{D})$$
$$= 1 - [P(\bar{B})P(\bar{C}) + P(\bar{D}) - P(\bar{B})P(\bar{C})P(\bar{D})] \tag{5.70}$$

in which it is again implicitly assumed that the systems are independent.

*   *   *

5.9.5   Markov techniques

Markov techniques are the most powerful methods available for reliability calculations. The application of this method makes it possible to account for dependent and independent systems, *time*-dependent failure and repair rates which are not necessarily equal. As well as these aspects, systems with more than two possible states can be taken into consideration. This makes it possible to take into account possible transitions between different states. We first have to introduce some necessary definitions before demonstrating this method with an example.

A Markov state model of order k and with m possible events for every stochastic variable x_i can be in m^k possible states. The number of symbols with which every state must be described is called the *order* of the Markov chain. Suppose there is a system with $m=2$ and the two stochastic variables $x_i=1$ or 0; then the number of possible states equals 4 and these can be denoted as $S_{0,0}$, $S_{0,1}$, $S_{1,0}$ and $S_{1,1}$. In table 5.3 this is illustrated with still another example.

If a Markov chain is in a certain state, a transition can occur from one state to another state. In particular, if a possible state is left, another state can occur, but the original state can be obtained too. Thus after a state is left m different events can occur; therefore the number of transitions is determined by the expression m^{k+1}.

Table 5.3 Examples of possible states related to the order of a Markov chain.

Description	S	S_m	S_m	$S_{m,m}$	$S_{m,m,m}$
Order k	0	1	1	2	3
Number of states S	1	2	3	4	8
States	S	S_0, S_1	S_0, S_1, S_2	$S_{0,0}, S_{0,1}, S_{1,0},$ $S_{1,1}$	$S_{0,0,0}, S_{0,0,1}, S_{0,1,0},$ $S_{0,1,1}, S_{1,0,0}, S_{1,0,1},$ $S_{1,1,0}, S_{1,1,1}$

If for instance if $k=1$ and $m=2$ with $x_i=0$ or 1, then the following four transitions can occur:

$$0 \rightarrow 0, \ 0 \rightarrow 1, \ 1 \rightarrow 0 \text{ and } 1 \rightarrow 1.$$

As can be seen from table 5.3 for $m=2$ and $k=3$ we find eight different states and 16 possible transitions. The next step is to assign a probability to every *transition*. This means that, given a certain state, we require the probability that a possible transition will occur. This describes precisely the definition of a conditional probability. The collection of transitions and states is often depicted in a so-called *state diagram*. An example of such a diagram is given in figure 5.23. As can be seen from this diagram, the indices of S denote the sequence numbers of the states and the value of the symbol used is indicated after the equals sign, so for instance $S_2=1$ is the second possible state with value 1. A transition $P(1/0)$ should be read as the probability the state 1 will occur given that the system is in state 0. For this state diagram the following two equations will hold:

$$P(0/0) + P(1/0) = 1 \tag{5.71}$$

$$P(0/1) + P(1/1) = 1. \tag{5.72}$$

You can check for yourself that this has to be true. In general for a Markov chain the statement will hold that given a certain state S_a the sum of all possible transition probabilities is 1, or

$$\sum_i P(S_i/S_a) = 1 \tag{5.73}$$

where S_i denotes another possible state to be reached.

Finally, to apply Markov techniques some other assumptions and definitions are required.

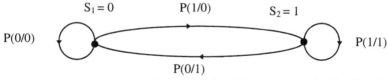

Figure 5.23 State diagram for a Markov chain with two states of order 1 and $m=2$.

1   Every state is mutually exclusive and states are independent of each other.
2   The sum of probabilities for every possible state is 1.
3   In the time interval Δt one event can occur.
4   Failure rates and repair rates are supposed to be constant.
5   The probability a system will fail in the time interval Δt equals $\Delta t\,\lambda$.
6   The probability a system will not fail in the time interval Δt equals $1 - \Delta t\,\lambda$.
7   The probability of repairing a system in the interval Δt is $\Delta t\,\mu$.
8   The probability a system will not be repaired in the time interval Δt equals $1 - \Delta t\,\mu$.
9   The state where all systems have failed will be denoted by $S_1 = 0$; this is called an absorbing state, because certainly $P(0/0) = 1$.
10 Where one of the systems is operational and the other systems are to be repaired or are waiting for repair actions, this is denoted by $S_2 = 1$.
11 Where all systems are working is a state often called the initial state and denoted by $S_3 = 2$.
12 An exponential distribution will be assumed.

*   *   *

Example 5.18

We will start with the most simple example possible. Consider a system which consists of one system only and cannot be repaired. Then $S_2 = 1$ cannot occur, and hence the state diagram is as is depicted in figure 5.24. This is a Markov chain of order 1 with two symbols, the number of states equals $m^k = 2$ and the number of transitions is $m^{k+1} = 4$, of which one transition cannot occur, because repair is excluded. This is

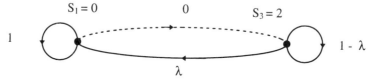

Figure 5.24   State diagram of a single system without repair actions.

depicted with the dashed line in figure 5.24. Of the two possible states S_1 and S_3 the probability has to be determined at the time $t + \Delta t$. Applying equation (5.58), we can write the following two equations for the two states to be considered:

$$S_1(t + \Delta t) = P(0)_{t + \Delta t} = P(0/0)P(0) + P(0/2)P(2) \tag{5.74a}$$

$$S_3(t + \Delta t) = P(2)_{t + \Delta t} = P(2/0)P(0) + P(2/2)P(2). \tag{5.74b}$$

It is important now to recognize now that if S_3 is calculated, the reliability of the system is found, because this is the probability the system will work at the time $t + \Delta t$

by definition of the above-given statements. From the state diagram the transitions with their (conditional) probabilities can be read and can be written as a matrix:

$$\begin{array}{cc} 2 & 0 \leftarrow \text{at time } t + \Delta t \end{array}$$

$$P = \begin{array}{c} 2 \\ 0 \end{array} \begin{bmatrix} 1 - \lambda & \lambda \\ 0 & 1 \end{bmatrix}. \tag{5.75}$$

$$\uparrow$$
$$t = \text{time } t$$

At the top of the matrix the states at time $t + \Delta t$, called the final states, are depicted, and to the left of the matrix the states at time t, called the initial states, are written. Within the matrix we write the transition probabilities for $t + \Delta t$. For instance, $P(2/2) = 1 - \lambda$ is the probability that state 2 will occur at $t + \Delta t$ given that the system is in state 2 at t, etc. By substitution of these values in equations (5.74a) and (5.74b) the following difference equations are found, in accordance with statements 5 and 6:

$$P(0)_{t + \Delta t} = 1 P(0)_t + \Delta t \, \lambda P(2)_t \tag{5.76a}$$

$$P(2)_{t + \Delta t} = 0 P(0)_t + (1 - \Delta t \, \lambda) P(2)_t. \tag{5.76b}$$

Rearrangement of both equations yields two differential equations:

$$\lim_{\Delta t \to 0} \frac{P(0)_{t + \Delta t} - P(0)_t}{\Delta t} = P'(0)_t = \lambda P(2)_t \tag{5.77a}$$

$$\lim_{\Delta t \to 0} \frac{P(2)_{t + \Delta t} - P(2)_t}{\Delta t} = P'(2)_t = -P(2)_t \tag{5.77b}$$

or

$$P'(0)_t - \lambda P(2)_t = 0 \tag{5.78a}$$

$$P'(2)_t + \lambda P(2)_t = 0. \tag{5.78b}$$

To solve these equations two boundary conditions must be known, namely at time $t = 0$ the system works so $P(2)_{t=0} = 1$ and at $t = 0$ the system cannot fail, thus $P(0)_{t=0} = 0$. Applying Laplace transforms gives

$$sP(0) - P(0)_0 = \lambda P(2) \tag{5.79a}$$

and

$$sP(2) - P(2)_0 = -\lambda P(2). \tag{5.79b}$$

Substituting the boundary conditions gives

$$sP(0) - 0 = \lambda P(2) \tag{5.80a}$$

$$sP(2) - 1 = -\lambda P(2). \tag{5.80b}$$

Solving (5.80*a*) and (5.80*b*) yields

$$P(2) = \frac{1}{s+\lambda} \tag{5.81}$$

and

$$P(0) = \frac{\lambda}{s(s+\lambda)}. \tag{5.82}$$

Inverse transformation delivers

$$S_3 = P(2) = \exp(-\lambda t) \tag{5.83}$$

and

$$S_1 = P(0) = 1 - \exp(-\lambda t). \tag{5.84}$$

As stated already, the reliability $R(t)$ is represented by S_3 *and is in accordance with equation* (5.6).

*   *   *

We will give now a more complicated example of an active on-line standby parallel system.

*   *   *

Example 5.19
The system to be considered is depicted in figure 5.25 and can be interpreted as a measurement and control system in an active on-line standby configuration monitoring and controlling a process. The failure rate and repair rate are supposed to be constant. Both systems are active and controlling each other's operational behaviour; when one of the two systems is operational the system works. Note that this is a very common configuration which can be met in many applications. The reliability can be described as

$$R(t) = 1 - P(0)_t. \tag{5.85}$$

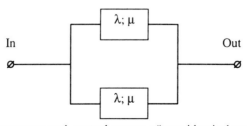

Figure 5.25 Measurement and control system of two identical active on-line standby systems.

The system can be interpreted as a Markov chain of order $k=1$ with three state symbols, namely

$S_1=0$ both systems have failed
$S_2=1$ one system is operational and the other system is in repair
$S_3=2$ both systems are active.

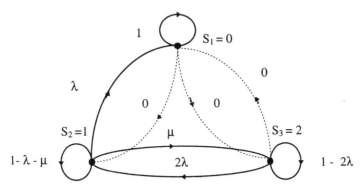

Figure 5.26   State diagram for example 5.19.

Hence the number of states, S, is three and the number of transitions is $m^{k+1}=9$. Some transitions cannot occur. The state diagram is depicted in figure 5.26 from which several important conclusions may be drawn.

1   The probability the system will come into state S_1 given that the system is in state S_1 is a sure event, thus $P(0/0)=1$. In other words, a failed system cannot become operational.
2   The probability the system will come into action again when both systems have failed is 0, and hence $P(2/0)=0$.
3   The probability the system will fail when both systems are active is 0, and hence $P(0/2)=0$.

The state diagram can be represented as a matrix:

$$
\begin{array}{c}
\quad\quad 2 \quad\quad\quad 1 \quad\quad\quad 0 \quad \leftarrow \text{state at } t=t+\Delta t \\
P = \begin{array}{c} 2 \\ 1 \\ 0 \end{array}
\begin{bmatrix}
1-2\lambda & 2\lambda & 0 \\
\mu & 1-\mu-\lambda & \lambda \\
0 & 0 & 1
\end{bmatrix}.
\end{array}
\tag{5.86}
$$

$$\uparrow$$
$$t=\text{time } t$$

Note that the sum of the conditional probabilities in every row of the matrix is 1; this is a check for the validity of the matrix. The appropriate equations belonging to this matrix are

$$P(0)_{t+\Delta t}=P(0/0)P(0)_t+P(0/1)P(1)_t+P(0/2)P(2)_t$$

$$P(1)_{t+\Delta t}=P(1/0)P(0)_t+P(1/1)P(1)_t+P(1/2)P(2)_t \tag{5.87}$$

$$P(2)_{t+\Delta t}=P(2/0)P(0)_t+P(2/1)P(1)_t+P(2/2)P(2)_t.$$

Substituting the appropriate values gives a set of difference equations

$$P(0)_{t+\Delta t} = 1 P(0)_t + \Delta t \; \lambda P(1)_t + 0 P(2)_t$$

$$P(1)_{t+\Delta t} = 0 P(0)_t + (1 - \Delta t \; \lambda - \Delta t \; \mu) P(1)_t + 2\lambda \; \Delta t \; P(2)_t \qquad (5.88)$$

$$P(2)_{t+\Delta t} = 0 P(0)_t + \Delta t \; \mu P(1)_t + (1 - \Delta t \; 2\lambda) P(2)_t.$$

By rearranging and taking the limit for $\Delta t \to 0$, the set of difference equations can be transformed into a set of differential equations:

$$P'(0)_t - \lambda P(1)_t = 0$$

$$P'(1)_t + (\lambda + \mu) P(1)_t - 2\lambda P(2)_t = 0 \qquad (5.89)$$

$$P'(2)_t - \mu P(1)_t + 2\lambda P(2)_t = 0.$$

Applying the Laplace transform of (5.89), by definition the following boundary conditions at the moment $t = 0$ are valid, namely $P(2)_{t=0} = 1$, because the system will work at that time and $P(1)_{t=0} = P(0)_{t=0} = 0$, for the reason both systems are operational at that time. We find

$$sP(0)_s - \lambda P(1)_s = 0 \qquad (5.90)$$

$$-(\mu + \lambda + s) P(1)_s + 2\lambda P(2)_s = 0 \qquad (5.91)$$

$$-\mu P(1)_s + (2\lambda + s) P(2)_s = 0. \qquad (5.92)$$

If we solve this set of equations for $P(1)_s$ only, then the other quantities are also known. Combining (5.91) and (5.92) gives for $P(1)_s$

$$P(1)_s = \frac{2\lambda}{2\lambda^2 + (3\lambda + \mu + s^2)}. \qquad (5.93)$$

This equation can be written as

$$P(1)_s = \frac{2\lambda}{(s - s_1)(s - s_2)} \qquad (5.94)$$

where s_1 and s_2 are the two values of

$$\tfrac{1}{2}[-(3\lambda + \mu) \pm (\lambda^2 + 6\mu\lambda + \mu^2)^{1/2}]. \qquad (5.95)$$

Fractional separation of (5.94) gives

$$P(1)_s = \frac{2\lambda}{s_1 - s_2} \left(\frac{1}{s - s_1} - \frac{1}{s - s_2} \right). \qquad (5.96)$$

Taking the inverse Laplace of (5.96) gives again $P(1)$ in the time domain:

$$P(1)_t = \frac{2\lambda}{s_1 - s_2} [\exp(s_1 t) - \exp(s_2 t)]. \qquad (5.97)$$

We are not yet finished because we have to find first $P(0)_t$ before ultimately the reliability can be found. Rearranging the first equation of (5.89) yields

$$P'(0)_t = \lambda P(1)_t \tag{5.98}$$

and then $P(0)_t$ becomes

$$P(0)_t = \int_0^T P'(0)\,\mathrm{d}t = \int_0^T \lambda P(1)\,\mathrm{d}t \tag{5.99}$$

or after integration

$$P(0)_t = 1 + \frac{s_2 \exp(s_1 t) - s_1 \exp(s_2 t)}{s_1 - s_2} \tag{5.100}$$

and finally by applying equation (5.84) with substitution of (5.100) the reliability is given by

$$R(t) = 1 - P(0)_t \tag{5.101}$$

or

$$R(t) = 1 - P(0)_t = \frac{s_1 \exp(s_2 t) - s_2 \exp(s_1 t)}{s_1 - s_2} \tag{5.102}$$

where s_1 and s_2 are as given in equation (5.95).

*   *   *

This method requires some effort. However, once the Markov state diagram is determined for a given system, the reliability can be found by a systematic application of the concepts given above. The elegance and the power of this method lie in the fact that we can take into account time- and state-dependent states and transitions.

5.10 DESIGN ASPECTS

The design process of a system is very complex because numerous aspects have to be taken into account before the right decisions can be made. Note also that *built-in reliability* must be realized from scratch and cannot be realized afterwards. Some basic rules are available and a very systematic approach is needed. The best way to go ahead is to apply a type of checklist and to investigate every part of it. To give you an impression of such a checklist, which can be used in a design procedure, here is an example.

1   Define the application and purpose of the system.
2   Define the required reliability with respect to the environmental conditions the system will be used in.
3   Use a type of index system to describe the effectiveness of the reliability.

4   Define the system and divide the system into pieces which are independent of each other.
5   Find for every subsystem the optimum configuration with respect to costs and reliability.
6   Determine the costs of manufacturing for every subsystem.
7   Define a complete system configuration and a maintenance philosophy relating to all other factors and conditions.
8   Assign failure rates and repair rates to every part of the system in conformance with the operational environmental conditions.
9   Predict the reliability of every subsystem and the whole system. Use for this part of the design procedure the latest published data and failure rate handbooks. The most reliable sources on data for components and systems are issued by the US Department of Defense (Washington, DC).
10 Compare the calculated results with the predicted values. Repeat the procedure where required to determine best possible strategy.

An important note can be made here. As we have seen in figure 5.5, costs versus reliability always play a vital role in the design procedure. However, every time a disaster has occurred, apparently basic human factors were involved. It is interesting to mention that some of these *human factors* are very often responsible for great disasters. Some factors are running out of time, hurry, arrogance, greed (fraud), commercial stress, negligence, wrong or failing communication procedures in the organization, etc. That this is true can easily be verified by analysing disasters such as a space shuttle failure and a nuclear power plant blow. Well known are also the causes of many aircraft incidents. No system is only technique; human factors are always involved and these are the most difficult to grasp. Therefore, despite all possible implemented precautions 100% reliability cannot exist. There are some other points of view to be considered here. It will be clear that in the design procedure the manufacturer will use different arguments from the consumer. Last but not least safety aspects can be important as is the case in the aircraft industry, in the medical field and when the environment is involved. These aspects can drastically increase the costs of design.

In general reliability must be designed into a system; it does not happen by chance. Of course, it is based on the reliability of each component separately, but using good engineering practice during all phases of system development is essential for reliable system operation.

It is interesting to mention here some of these design considerations used in the car industry. In the car industry the characteristics interacting with each other are,

Table 5.4   Examples of classification of some car systems.

System	Sensor application	Reliability priority	Accuracy
Ignition	Revolution	High	Middle
ABS brakes	Revolution of wheels	High	High
Instruments	Speed	Middle	Low
Emission	Lambda sensor	High	Low
Airbag	Acceleration	Extremely high	High

for instance, maintenance, fuel consumption, comfort, safety, required reliability of components and ergonomics and allowed emission levels. Priorities must be balanced with cost aspects, which results in clearly optimized choices for each component involved. In table 5.4 a classification is given with some examples.

These considerations determine to a large extent the cost aspects. It will be obvious that safety aspects also influence reliability requirements. It is hoped that the airbag system is never needed but reliability must be extremely high. Environmental conditions with their effects must be well known. Some of these conditions with their effects are summarized in table 5.5.

Table 5.5 Examples of some environmental conditions and their effects in car components.

Environmental load	Effects
Temperature and UV radiation	Aging, cracking
Temperature cycles, day and night, seasons	Fatigue cracks
Humidity	Corrosion, leaking
Dust and mud	Short circuit
Chemicals	Corrosion
Oil and fuel	Swelling of rubber

Finally, EMC regulations may require protective measurements against an electrical field strength of 200 V m^{-1} and a frequency range of 20 MHz–1 GHz. (See also chapter 8.)

It is beyond the scope of this textbook to present a detailed discussion of all types of design procedure. We will confine the treatment to some relevant aspects to stress their importance. Design procedures can be used at a systems level but also on the part level. In the integrated circuits industry dedicated design procedures are developed. In electronics systems special attention must be devoted to low-noise power distribution, signal fidelity with regard to spiking and reflections and common I/O devices where high contention currents can flow because of bus conflicts.

In general in the design stage, eight different quality classes can be distinguished ranging from highest military standard class through industrial class, to consumer class with the least severe requirements. To each class a so-called quality factor B_Q can be assigned, as summarized in table 5.6.

Concerning the environmental conditions eleven different environmental classes are distinguished. These classes range from benign laboratory conditions to severe environmental conditions during a missile launch with large shock and vibration levels. The environmental classes with their symbols are summarized in table 5.7.

An example of failure rates for integrated circuits is given in table 5.8. In this table the circuit complexity is given as the number of gates per circuit. In each column the failure rate can be read as a function of the environmental condition and circuit complexity in failure rates per 10^6 h. For instance, to find the failure rate of a system the following formula can be used:

$$\lambda_{BE} = \sum_{i=1}^{n} N_i (\lambda_G B_Q)_i B_L \qquad (5.103)$$

Table 5.6 Summary of quality classes with assigned quality factors.

Quality class	Description	B_Q
S	Procured in full accordance with MIL-M-38510, class S requirements	1
B	Procured in full accordance with MIL-M-38510, class B requirements	2
B-1	Procured to screening requirements of MIL-STD-883, method 5004, class B, and in accordance with the electrical requirements of MIL-M-38510 slash sheet or vendor or contractor electrical parameters. The device must be qualified to requirements of MIL-STD-883, method 5005, class B. No waivers are allowed	5
B-2	Procured to vendor's equivalent of screening requirements of MIL-STD-883, method 5004, class B, and in accordance with vendor's electrical parameters. Vendor waives certain requirements of MIL-STD-8X3, method 5004, class B	10
C	Procured in full accordance with MIL-M-38510, class C, requirements	16
C-1	Procured to screening requirements of MIL-STD-883, method 5004, class C, and in accordance with the electrical requirements of MIL-M-38510 slash sheet or vendor or contractor electrical specification. The device must be qualified to requirements of MIL-STD-883, method 5005, class C. No waivers are allowed	90
D	Commercial (or non-military standard) part, hermetically sealed, with no screening beyond the manufacturer's regular quality assurance practices	150
D-1	Commercial (or non-military standard) part, packaged or sealed with organic materials (e.g. epoxy, silicone or phenolic)	300

where λ_{BE} is the total failure rate in the applicable environment, λ_G is the specific failure rate of the ith component at the applicable environment (see for instance table 5.8), B_Q is a quality factor (see table 5.6), B_L is a learning factor which equals 10 when a change in the design is made, N_i is the number of components of type i and n is the number of different categories of components.

In addition to the application of redundancy, one of the most effective measurements to improve the reliability or to reduce the failure rate of a system is *derating*. This technique implies a reduction in workload for every part or component involved in the design. Successful application of this method requires a thorough knowledge of the components' failure rates. Its effect can be demonstrated by the example depicted in figure 5.27 in which the failure rate of a capacitor versus temperature is shown with normalized voltage load as parameter. Derating can easily deliver factors of improvement of 1000, as shown.

However, six other different environmental circumstances which influence the reliability can be distinguished, and these are summarized in table 5.9. If any of these conditions can occur, additional factors in the calculation of the failure rate have to be account for. See also table 5.5.

For electronic components, during test procedures, a life test is often performed at elevated temperatures. For this reason an acceleration factor, A_F has been introduced which is defined as the ratio of the time at the life test temperature, T_1, to obtain the mortality $\Delta N/N$ to the time at the temperature, T_2, to obtain the identical

Table 5.7 Summary of the 11 environmental classes with symbols used.

Environment	Symbol BE	Nominal environmental conditions	Number
Ground, benign	G_B	Nearly zero environmental stress with optimum engineering operation and maintenance	1
Space, flight	S_F	Earth orbital. Approaches ground, benign conditions without access for maintenance. Vehicle neither under powered flight nor in atmospheric re-entry	2
Ground, fixed	G_F	Conditions less than ideal to include installation in permanent racks with adequate cooling air, maintenance by military personnel and possible installation in unheated buildings	3
Ground, mobile	G_M	Conditions more severe than those for G_F, mostly for vibration and shock. Cooling air supply may also be more limited, and maintenance less uniform	4
Naval, sheltered	N_S	Surface ship conditions similar to G_F but subject to occasional high shock and vibration	5
Naval, unsheltered	N_U	Nominal surface shipborne conditions but with repetitive high levels of shock and vibration	6
Airborne, inhabited, transport	A_{IT}	Typical conditions in transport or bomber compartments occupied by aircrew without environmental extremes of pressure, temperature, shock and vibration, and installed on long-mission aircraft such as transports and bombers	7
Airborne, inhabited, fighter	A_{IF}	Same as A_{IT} but installed on high-performance aircraft such as fighters and interceptors	8
Airborne, uninhabited, transport	A_{UT}	Bomb bay, equipment bay, tail or wing installations where extreme pressure, vibration and temperature cycling may be aggravated by contamination from oil, hydraulic fluid and engine exhaust. Installed on long-mission aircraft such as transports and bombers	9
Airborne, uninhabited, fighter	A_{UF}	Same as A_{UT} but installed on high-performance aircraft such as fighters and interceptors	10
Missile launch	M_L	Severe conditions of noise, vibration, and other environments related to missile launch, and space vehicle boost into orbit, vehicle re-entry and landing by parachute. Conditions may also apply to installation near main rocket engines during launch operations	11

Table 5.8 Example of failure rates per 10^6 h for integrated circuits as a function of circuit complexity and environmental conditions.

Circuit complexity	G_H and S_F	G_F	A_{IT}	A_{IF}	N_S	G_M	A_{UT}	A_{UF}	N_U	M_L
1–20 gates	0.010	0.048	0.099	0.16	0.14	0.12	0.21	0.30	0.25	0.24
21–50 gates	0.048	0.19	0.31	0.40	0.43	0.34	0.73	0.86	0.92	0.52
51–100 gates	0.076	0.31	0.48	0.59	0.68	0.54	1.2	1.3	1.5	0.78
101–500 gates	0.19	0.82	1.2	1.4	1.7	1.3	3.1	3.4	3.9	1.7
501–1000 gates	0.32	1.4	2.0	2.3	2.8	2.1	5.1	5.5	6.4	2.6
1001–2000 gates	0.74	3.1	4.6	5.2	6.4	4.8	12.0	13.0	15.0	6.0
2001–3000 gates	2.0	8.4	13.0	14.0	17.0	13.0	33.0	35.0	41.0	16.0
3001–4000 gates	5.4	23.0	35.0	39.0	47.0	36.0	90.0	96.0	111.0	44.0
4001–5000 gates	15.0	62.0	94.0	105.0	128.0	97.0	241.0	258.0	299.0	121.0
ROM ≤bits	0.021	0.087	0.13	0.15	0.18	0.14	0.33	0.36	0.42	0.17
ROM 321–576 bits	0.031	0.13	0.19	0.22	0.27	0.20	0.49	0.53	0.62	0.26
ROM 577–1120 bits	0.048	0.20	0.31	0.35	0.42	0.32	0.78	0.84	0.98	0.41
ROM 1121–2240 bits	0.072	0.30	0.45	0.52	0.63	0.48	1.2	1.3	1.5	0.61
ROM 2241–5000 bits	0.11	0.46	0.70	0.80	0.96	0.74	1.8	1.9	2.2	0.94
ROM 5001–11 000 bits	0.17	0.70	1.1	1.2	1.5	1.1	2.7	2.9	3.4	1.5
ROM 11 001–17 000 bits	0.25	1.1	1.6	1.9	2.2	1.7	4.1	4.5	5.2	2.2
Linear, ≤32 transistors	0.011	0.052	0.12	0.20	0.16	0.15	0.22	0.35	0.27	0.33
Linear, 33–100 transistors	0.023	0.11	0.24	0.41	0.35	0.31	0.48	0.73	0.60	0.66

Header spanning columns G_F through M_L: Environmental conditions

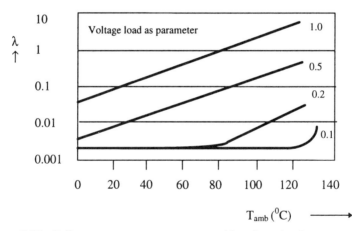

Figure 5.27 Failure rate versus temperature with voltage load as parameter.

Table 5.9 Environmental circumstances.

Thermal	Infrared radiation and large temperature gradients
Mechanical	Pressure, shocks and vibration
Pollution	Moisture, sand, dust, chemical contamination
Electrical	Electromagnetic interference (EMI), transients, lightning etc
Radiation	Electromagnetic pulses (EMPs) caused by nuclear blasts, neutrons, gamma radiation etc
Others	Human factors, maintenance and repair procedures

mortality. The failure rate rate is then expressed as

$$\lambda = \frac{\Delta N}{N} \frac{1}{A_F t} \qquad (5.104)$$

and the acceleration factor, A_F, is defined as

$$A_F = \exp\left[\frac{E_a}{k}\left(\frac{1}{T_1} - \frac{1}{T_2}\right)\right] \qquad (5.105)$$

where E_a is the activation energy of the specific failure mechanism, k is Boltzmann's constant (1.38×10^{-23} J K^{-1}), T_1 is the test temperature and T_2 is the operational temperature. For integrated circuits three major failure mechanisms are mentioned in table 5.10.

Table 5.10   Some failure mechanisms for integrated circuits.

Failure mechanism	Activation energy, E_a (eV)
Positive mobile ion contamination (PMIC)	1.0
Metal electromigration (EMG)	0.5
Gate oxide faults (GOX), with $E_a = 0.3$ eV	0.3
All others	0.7

A major important development in reliability is self-testing of components and systems as a result of built-in reliability procedures. When critical parameters are at the end of their lifetime, warning and/or automatic replacement of redundant systems is realized.

5.11   CAUSES OF FAILURE

A failure can be defined as the event where the system has failed or has run out of its specifications. The system has reached a repairable or non-repairable state. In general, failures can be catastrophic or non-catastrophic. A catastrophic failure is one in which the system has stopped functioning; in a non-catastrophic failure the system has stopped functioning within its specifications. Failures can be divided into three main technical categories and one non-technical category. We note that, when a system exceeds predetermined maintenance costs, it is said the system is running out of its economic life cycle, and it can be concluded that the system has also reached a failed mode.

1   *Design failures*
1   The design is inherently incapable of performing the actual mission or operation, because either the system's functional logic does not correspond to the reqirements or an unworkable combination of parts is implemented.
2   The environment is beyond the capability of the system because the environmental conditions were misestimated or the system was never qualified for that environment.

2 Manufacturing failures

The system equipment could have worked if it had been fabricated and maintained in accordance with the drawings and specifications determined in the design phase.

1 A faulty part was built into the hardware.
2 The hardware was damaged during manufacturing, test, reprocessing or repair, handling, installation or field service.

3 Random failures

1 Some part of the system has worn out owing to the action of chemical or physical processes inherent in the normal functioning of that part.
2 Some component or set of components shows accumulated environmental damage sufficient to cause malfunction.

4 Human factors

Human factors in the causes of failure are all those not falling in one of the categories mentioned above. Examples are as mentioned earlier:

1 miscommunication
2 hurry, greed, fraud, lack of time
3 failing organization structure (ISO 9000 system)
4 price versus market policies and acceptance
5 failing quality control department (ISO 9000 system)
6 system management and manufacturing philosophy.

Basically, in most cases where human factors are involved, when there is no conformity with natural laws, the system will stop functioning properly. Figure 5.28 is an example illustrating, by means of a sector digram, the division of different system failures in ground test circumstances of a satellite.

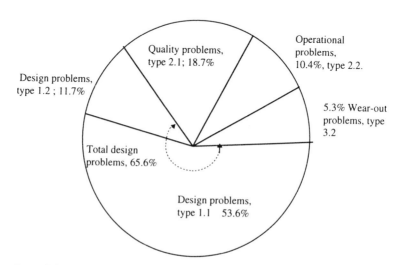

Figure 5.28 Division of different system failures of a satellite in ground test circumstances. (Courtesy McGraw Hill Publishing Co.)

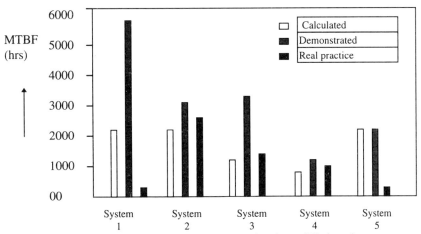

Figure 5.29 MTBFs of five different systems under three differing circumstances.

This chapter concludes with one of the most dramatic graphs to be shown relating to reliability. Figure 5.29 shows the large discrepancy which still exists between calculated values and measured values in laboratory circumstances and in real practical circumstances. A very important reason for such discrepancies is that tools developed so far are still not satisfactory, because the real world is more complex than we can understand. In most cases all tests performed to determine the reliability of a system are single *sequential* tests executed one after the other. Furthermore, failure mechanisms under test conditions are considered as independent. The real world does not act as a sequential test environment, but all environmental conditions influence the system in parallel and simultaneously. This makes things more complicated than we are able to understand and handle at this moment. However, reliability techniques have become a useful science and tools and are still in development.

For integrated circuits failure analysis has become a specialism and a great number of tools have been developed and are available. The scanning electron microscope is one of these major tools.

5.12 SOFTWARE

No attention has been paid to the reliability of software programs. This is a separate discipline; however, the same concepts as discussed above are used. For instance, in software redundancy also is built in to perform self-tests and to detect faults. Fault-tolerant systems are an absolute necessity where human safety aspects are involved. Examples are (1) the computer-aided traffic control system (COMTRAC), (2) COPRA, which is a reconfigurable multiprocessor system for space aircraft purposes and (3) electronic switching system (ESS), Bell Laboratories, for a telecommunications exchange.

To calculate reliability numerous programs are available with which reliability of a system can be determined under different operational conditions. With these software packages fault and critical path analysis can be performed. We will confine ourselves to mentioning a few packages without going into details. International Reliability Analysis System (IRAS) provides facilities for failure mode and electrical critical path analysis. CARAD is a program that can be used also for hybrid and temperature analysis. Also, a distinction must be made as to whether the reliability analysis is performed on 'chip' level or on system level.

5.13   FUTURE DEVELOPMENTS

We will make a few remarks about future developments in reliability engineering. The approach towards reliability engineering is changing considerably for VLSI devices. We will mention some aspects. Because VLSI devices are complete miniaturized systems we have to deal with very complicated systems in which many functions can be built in. For these devices important aspects in reliability are as follows.

1   Devices are operating at greater speed with higher current densities and larger electrical field strengths. Here metal interconnections, transistors and insulators are coming under severe stress.
2   Because of extremely reduced dimensions there is an increasing sensitivity to contamination.
3   When more than 10^8 transistors are integrated and a failure rate of 0.1% in one year is required, the chip MTTF must become larger than 10^{12} years which exceeds the lifetime of the universe.
4   We have to deal now with ULSI reliability and there is a shift from failure testing to the built-in reliability approach.
5   Electrostatic discharge (ESD) is also becoming an important factor in packaging design.
6   Electrical overstress is responsible for 50% of field failures.
7   For ULSI circuits ultraclean processing is an absolute prerequisite.

The interested reader is referrred to the literature. In particular reference 4 gives an excellent overview of the reliability problems in ULSI devices to be encountered in the years to come.

PROBLEMS

5.1   In a manufacturing process of integrated circuits, five of every 100 ICs satisfy the specifications. Calculate the probability that from a sample of 10 ICs just two of them pass the specifications.

5.2   Calculate the failure rate and MTBF of a circuit consisting of 20 resistors and 30 soldered connections when the failure rates are 10×10^{-9} h^{-1}/component and 1×10^{-9} h^{-1}/connection respectively.

5.3   A system with a useful lifespan of 1000 h has a failure rate of 0.0001 h^{-1}/system and is supposed to have an exponential distribution. Calculate the reliability for that system for 10, 100 and 1000 h of operation.

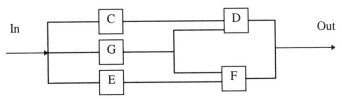

Figure 5.30 System configuration for problem 5.7.

5.4 The power supply unit of an aircraft consists of two identical systems with a failure rate of $350 \times 10^{-6} \, h^{-1}$/system. The system configuration is an active on-line standby system with exponential distribution. Determine the reliability and the MTBF for that power supply unit.

5.5 In a telecommunication satellite, three identical systems are installed in a cold on-line standby system configuration. If the MTTF equals 10 000 h determine the survival function.

5.6 A ground communication system consists of three identical systems in an active on-line standby configuration. If two of the three systems are operational, the communication system works satisfactorily. The failure rate of each system is $400 \times 10^{-6} \, h^{-1}$. Calculate the system reliability and the MTBF for a 24 h period.

5.7 Apply the decomposition technique to determine the reliability of the system depicted in figure 5.30. The system works when one of the traces CD, GD, GF or EF works error free.

5.8 Draw a possible Markov state diagram for a system of second order with $m = 2$ and $x_i = 0$ or 1. Every state has two possible transitions. Write down for every state the appropriate transition probabilities. Determine the number of transitions which are free to be chosen.

5.9 A system has two states with symbols 0 and 1 with $P(1/0) = p$ and $P(0/1) = q$. Calculate the probabilities $P(0)$ and $P(1)$. Hint: first draw the state diagram.

5.10 A control and measuring system has three states: $S_1 = 0$ is the state the system fails, $S_2 = 1$ is the state that the system works without control capacity, $S_3 = 2$ is the state that the system works normally. The appropriate state diagram is depicted in figure 5.31.

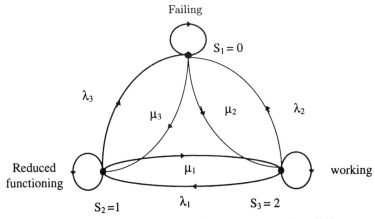

Figure 5.31 State diagram for the system of problem 5.10.

The boundary conditions are $P(2)_{t=0}=1$, $P(1)_{t=0}=P(0)_{t=0}=0$. Arrange the Laplace-transformed differential equations for this system.

5.11 Discuss the relationship between quality and reliability. Try to find the applicable regulations for quality assurance in your country.

BIBLIOGRAPHY

[1] 1981 *Reliability, IEEE Spectrum Compendium* (October)
[2] INMOS 1984 *Quality assurance and Reliability* 42-1027-000
[3] International Organization for Standardization 1991 *Quality Management and Quality Assurance* ISO 9000–9004/2 (Geneva)
[4] Chenming Hu (ed) 1993 *VLSI Reliability, Proc. IEEE Special issue* (May).
[5] Arsenault J E and Roberts J A 1980 *Reliability and Maintainability of Electronic Systems* (New York: Pitman)
[6] Dhillon B S 1983 *Reliability Engineering in Systems Design and Operation* (New York: Van Nostrand Reinhold)
[7] DOD 1965 *Reliability Stress and Failure Rate Data for Electronic Equipment* (Washington, DC: Department of Defense)
[8] Trent R L 1977 *Reliability/Availability Electronics Designers Handbook* 2nd edn (New York: McGraw-Hill) section 28

Six

TRANSDUCERS

6.1 INTRODUCTION

It is the author's belief that transducers belong to the most intriguing, challenging and complex fields of electronics today. This is because all types of discipline meet each other in the science, the research, the development and the manufacture of transducers. The study of sensors is multidisciplinary and requires a comprehensive knowledge of general physics, solid-state physics, electronics, technology and semi-conductor-manufacturing techniques.

With the help of sensors we are able to observe, to measure and to control our environment. Basically, sensors are energy converters, but, as we have seen already, energy *information* conversion is the objective of a sensor. The information available in one energy form must be converted into the same or another energy form, with exactly the same information content as the originating energy form. It is said the information is carried by a certain type of energy. In chapter 1 we have seen the great advantages in handling information in electrical form. However, as we will also see, the optical form of information is becoming of increasing importance. In addition, there is an increasing trend to use semiconductor technology in the manufacture of transducers. If this is successful, large cost and size reduction may be expected as a result. In all types of measuring system, transducers are found at the input and at the output. In several cases energy conversion will be found within the system also. This is illustrated in figure 6.1.

The identification unit contains a device which reacts to the desired physical quantity and converts it into an electrical quantity. The input transducer can be found here, normally called a sensor. In the modifier stage the electrical energy is converted into another shape as discussed extensively in section 2.2. In the presentation unit, two possibilities arise depending on the nature of the 'system' to be served. If a machine or process must be supplied with new information, the output transducer must be able to provide it. In that case, often a type of actuator may be required to execute the right action. In measurement systems to be perceived by humans, we need a type of display to which our senses react, or can take action. So if we state that the output transducer is an actuator we are always right, whether it activates a technical system or a living (biological) system.

In this chapter a detailed survey of possible energy conversions will be presented. However, because of the already-stressed great advantages of conversion into and/

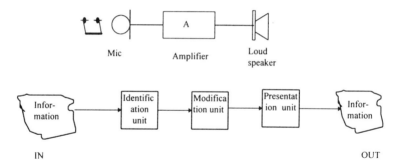

Figure 6.1 Functional block diagram of a measurement system.

or from the electrical energy domain, the main emphasis is laid on these conversions. In a separate section we shall discuss the most recent possibilities of semiconductor materials and technology for the manufacture of (three-dimensional) sensors. In the last ten years there has been tremendous growth from a technological point of view and we have to devote attention to that too. The development of sensors is influenced to a large extent by these new technologies, and smart sensors are common devices now.

We will discuss such developments as those in microsensors, smart sensors, micromachining, packaging of sensors and technologies. Also, optoelectronics and some basic interfacing circuits for sensors will be discussed. It can be concluded that we still are at the beginning of a tremendous area of possibilities which will result in completely integrated single-chip microsystems able to perform different kind of task.

To avoid any confusion we will start by presenting the required definitions and terminology used.

6.2 CLASSIFICATION OF TRANSDUCERS

Generally speaking the word 'transducer' can be thought to be derived from the Latin verb *traducere* which means *'to convert'*. So a *transducer* can be defined as a device capable of converting energy from one form into another. Transducers can be found both at the input as well as at the output stage of a measuring system. The word transducers can be considered as a collective noun. The input transducer is called the *sensor*, because it senses the desired physical quantity and converts it into another energy form. The output transducer is called the *actuator*, because it converts the energy into a form to which another independent system can react, whether it is a biological system or a technical system. So, for a biological system the actuator can be a numerical display or a loudspeaker to which the visual or aural senses react respectively. For a technical system the actuator could be a recorder or a laser, producing holes in a ceramic material. The results can be interpreted by humans.

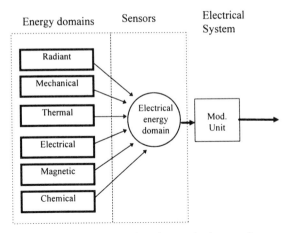

Figure 6.2 The six different energy domains at the input of a measuring system.

6.2.1 Types of energy form

We can distinguish six different energy domains: (1) radiant, (2) mechanical, (3) thermal, (4) electrical, (5) magnetic and (6) chemical. If certain information is already available in the electrical domain it can be claimed that it requires no energy conversion, but in general there is 'shape' conversion left and this is just the domain which belongs to the field of electronics and electrical science and engineering. A good example of such a sensor only sensitive to electrical energy is the probe of an oscilloscope, with which a good adaptation to the signal source is realized. In the modifier stage we meet other examples of shape converters, for instance the A/D and D/A converters. If only the identification unit of figure 6.1 is considered we can draw figure 6.2 in which the possible energy conversions at the input are depicted schematically. In the same way, the six different domain conversions at the output can be drawn. This is illustrated in figure 6.3, where compared with figure 6.2 the only difference is the reversed direction of the arrows.

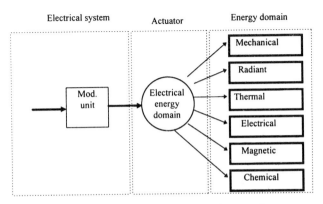

Figure 6.3 The six different energy conversions at the output stage.

6.2.2 Types of energy source

If the energy sources at the input, or the objects the actuators are acting on at the output, are considered, we can distinguish between *technical* systems and *biological* systems. Let us consider that in some more detail. It will be clear that technical systems can produce all six energy forms. Hence at the input side the six different types of energy source can always be recognized. For biological systems this is not so clear, but a more careful consideration will reveal the same six different types of energy form.

1   Radiant energy is produced by all biological systems. This is normally infrared radiation and can be detected although it is not visible, for instance, with thermographic cameras. It is interesting to note that the sensitivity of these cameras is incredibly high. For instance, if a person stands for 1 min by a wall, without touching it, and steps aside, the IR camera can easily detect where the person's position was before that wall. Radiant energy is generated by the system's proper temperature. It is well known that some kinds of fish produce radiant energy in the visible region between 400 and 800 nm.
2   Biological systems can also produce mechanical energy as a result of movements or the liquid pressure in the vessels.
3   Thermal energy is produced by all systems in which oxidation takes place.
4   Electrical energy for instance is produced by the heart muscle at a potential of several millivolts. The measurement of an electrocardiograph (ECG) is based on this. Some types of eel can produce over 700 V.
5   Magnetic energy is also produced by the human heart muscle. Also, magnetic brain activities can be monitored with the help of superconducting quantum interference devices, so-called SQUIDs.
6   In biological systems chemical energy is produced in all types of process and they can act as an energy source also.

At the output we can find the same six energy domains for both types of system. Again for technical systems this will be clear, but biological systems require more explanation.

1   Biological systems with their vision sense can react to radiant energy, as for instance is the case with the information displayed on a cathode-ray tube (CRT).
2   Biological systems are sensitive to mechanical forces and will react to them.
3   Thermal energy influences biological systems to a large extent and for instance determines the velocity of growth and movement.
4   Biological systems are sensitive to electrical energy and muscles for instance will react to electrical pulses or even produce electrical energy.
5   It is already known that magnetic fields can cause chemical changes in biological tissues and cell structures and so change the information content.
6   Numerous examples exist in which biological systems are sensitive to chemical actuators (substances). Examples are all types of medicine.

The above is summarized in figure 6.4 representing the same situation as figure 6.1 but supplied with the two types of system at input and output. We will see that the

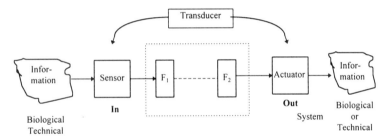

Figure 6.4 At input and output a biological or a technical system can act as source and as object for all six different energy domains.

possibility of measuring biological quantities gave rise to a huge effort to manufacture so-called biosensors.

6.2.3 Modulating and self-generating transducers

Another important characteristic related to transducers implies the distinction of transducers into two categories. This is because some transducers require an auxiliary energy source to become operational. Hence a distinction is made between *modulating* and *self-generating* transducers.

A *modulating* transducer is defined as a transducer which requires an auxiliary energy source to convert energy from one domain into another. A good example of a modulating transducer which is found at the input of a system is a strain gauge. This type of transducer requires an electrical energy source to become operational. The electrical current flowing in the strain gauge is modulated by a mechanical force which is converted into an electrical voltage change via a change of the resistive elements. As a consequence the input energy modulates the energy of the auxiliary energy source. At the output of a system a liquid-crystal display (LCD) is an example of a modulating transducer, because again an auxiliary energy source is required to perform the correct transduction. If no incident light is falling onto an LCD, the information cannot be detected. So the electrical modulating energy, in which the information is present, modulates the incoming auxiliary radiant energy and is presented as outcoming radiant energy.

A *self-generating* transducer is defined as a transducer which requires no auxiliary energy source to convert energy. Examples of a self-generating transducer which can be found at the input of a system are a solar cell or a thermocouple, and at the output we can find for instance a heating element converting electrical energy into thermal energy without the need of an auxiliary energy source.

We can illustrate these concepts in two simple diagrams. In figure 6.5(a) the diagram represents a self-generating transducer and can be considered as a two-port device with respect to information (signal) and energy. Figure 6.5(b) represents a modulating transducer and can be considered as a three-port device with respect to energy and as a two-port device with respect to information.

To avoid confusion at this point it is necessary to mention another classification for transducers. Sometimes the words 'active' and 'passive' are used in this respect. An active transducer is one which is capable of modulating energy. Normally the

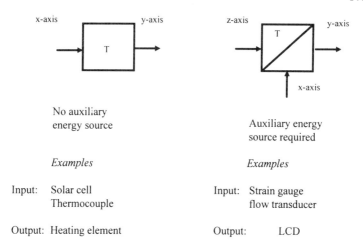

Figure 6.5 Symbolic representation for (a) self-generating transducers and (b) modulating transducers.

efficiency or power gain is much greater than unity. A transistor, a mechanically driven switch, a magnetic valve or an LCD are examples of active devices. A passive transducer is defined as a device able to convert energy without an auxiliary energy source. Examples are a solar cell, a thermocouple or a piezo element. The efficiency is always smaller than unity.

In the discipline of electronics, we meet the same types of definition. Components such as transistors, which are able to deliver gain, are called active components. This in contrast to resistors, coils and capacitors, etc which are not able to deliver gain. These devices are called passive components. For convenience the definitions are summarized in table 6.1. We will consider the first given definitions as more characteristic.

A convenient way of describing the different transduction possibilities is a three-dimensional energy space diagram as is illustrated in figure 6.6.

1   The x axis represents the input energy domain. It contains the required auxiliary energy only, if a modulating transducer is involved. In the case of a self-generating transducer it represents the energy and information bound to it.
2   The y axis represents the output energy domain in which the outcoming information content is available.
3   The z axis represents the modulating energy domain in which the incoming information is available if an auxiliary energy source is required.

Table 6.1 Alternative names for transducers.

Type of transducer	Efficiency	Alternative name	Examples
Modulating	$\gg 1$	Active transducer	Transistor, LCD, switch
Self-generating	< 1	Passive transducer	Resistor, coil, capacitor

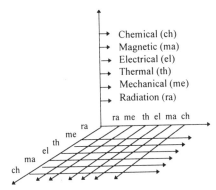

Figure 6.6 The x–y–z energy space and information diagram as basis for describing all possible transducers.

In the x–y plane or the so-called input–output plane all 36 possible self-generating transducers can be found. An interesting comment can be made on what is found on the main diagonal in the x–y plane and the space diagonal. In the transducers found on the main diagonal no energy conversion is involved; only an energy *shape* conversion is carried out. This type of transducer is normally called a modifier, because it only modifies the available information in the *same* energy domain. Examples of modifiers or shape transducers are the already-discussed A/D and D/A converters, all types of modulator, transformers, etc. So we can distinguish between the self-generating modifiers and the modulating modifiers.

In the mechanical energy domain, a gear set, with which the number of revolutions per minute can be changed, is a realization of a modulating modifier. Other examples are a transistor in the electrical energy domain and colour filters in the radiant energy domain. Note that here the words 'modifier' and 'modulating' can easily lead to confusion. On the basis of the three-dimensional energy space diagram it is not very difficult to see that the modulating and self-generating transducers can be described by a three-dimensional vector $[x, y, z]$ in which the z component represents the modulating energy domain. For this purpose and in analogy with the so-called Miller indices here Miller indices for *energy* can be applied. This convention results then in [input energy, output energy, modulating energy]. For *signal* or *information*, for modulating transducers this results in [00, information out, information in] and for self-generating transducers we find [information in, information out, 00]. For both kinds of transducer the energy and information content for the x, y and z axes are depicted in table 6.2. Table 6.3 lists several examples applying the use of Miller indices and this will be used later. In this table the following abbreviations are used:

Table 6.2. Survey of energy and information content of x, y and z axes.

	Modulating transducer			Self-generating transducer		
	x	y	z	x	y	z
Axis						
Energy	In	Out	In	In	Out	—
Information	—	Out	In	In	Out	—

Table 6.3 Applications of Miller indices for transducers.

Transducer	Miller index	Type description
Transistor	[el, el, el]	Modulating shape transducer
Thermocouple	[th, el, 00]	Self-generating input transducer
pH meter	[ch, el, 00]	Self-generating input transducer
LED display	[el, ra, 00]	Self-generating output transducer
LCD display	[ra, ra, el]	Modulating output transducer
Coil	[el, ma, 00]	Self-generating output transducer
Magnetoresistor	[el, el, ma]	Modulating input transducer
Photoconductor	[el, el, ra]	Modulating input transducer
Lateral photodetector	[ra, el, me]	Modulating input transducer

el, electrical; th, thermal; ch, chemical; ra, radiation; ma, magnetic; me, mechanical. Note that the type of energy in y functionally determines the kind of transducer with respect to input or output; hence if $y =$ el then an input transducer is involved and if $y \neq$ el then an output transducer is implied.

Referring to the LCD again, the input energy is radiation (x axis), the output energy is radiation (y axis) and the modulating energy is electrical (z axis); with respect to information, the z axis represents the incoming information in electrical form and the y axis the outgoing information in radiant energy.

In the x–y plane every energy domain contains one modifier, five self-generating input transducers and five self-generating output transducers. This is illustrated in more detail in figure 6.7 for electronic devices only. In the same figure, the main diagonal shows the shape transducers.

To illustrate the use of the above-outlined concepts we will give the example of a digital thermometer with LCD readout as depicted in figure 6.8. In a digital thermometer the thermal energy is converted into electrical energy by a thermocouple or a p–n junction; then the signal is modified by an amplifier stage and an A/D converter and finally the electrical signal is converted again by a modulating output transducer to produce radiant energy to be perceived by the eye (LCD).

IN \ OUT	Radiant	Mechanical	Thermal	Electrical	Magnetic	Chemical
Radiant	Filter			Photodiode		
Mechanical		Gear box		Piezo resistor		
Thermal	hot wire		Heat exchanger	Seebeck effect		
Electrical	LED	Piezo-electricity	Peltier effect	Transistor	Coil	Electrical plating
Magnetic				Magneto resistor	magnetic circuit	
Chemical				pH-meter		Chemical reaction

Figure 6.7 Matrix of the x–y plane showing a collection of electronic self-generating transducers with examples of 'shape' transducers on the main diagonal where no energy conversion is involved.

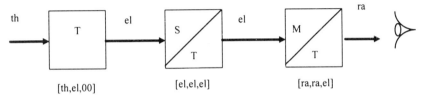

Figure 6.8 A digital thermometer with LCD display in a functional block diagram applying Miller indices.

Concerning the different possibilities of silicon and other compatible technologies for the production of transducers, we have to consider four classes of transducers. In each case an energy conversion is involved in which the electrical signal energy is implied for input or output. The four classes are summarized in table 6.4, in which for every class the electrical signal information is indicated and x, y and z represent the six types of energy. This classification holds for all energy domains. The numbers in parentheses denote the (theoretical) number of sensors or actuators. In the last column the abbreviations applied to the four classes are given. The five different energy conversions in semiconductors are discussed in more detail in the following sections. You can see here that six different energy domains imply five energy conversions only, because each time two energy domains are involved for one energy conversion. The corresponding energy domains to be discussed each give a detailed overview of basic important physical energy mechanisms and their related known physical effects.

So far we have presented a classification of transducers. In the next section we will introduce a formal description for transducers

6.3 STATE DESCRIPTIONS

The state of a transducer can be described by combining physical quantities of which the product represents an energy. This state description reveals some other common characteristics of transducers. A most important characteristic to be kept in mind is that no transducer is sensitive to one physical energy only. The ideal situation exists when the transducer can be isolated from all physical influences except the desired one. One information source should be measured at a time. How well this can be realized is determined to large extent by the accuracy of the transducer, which is often the weakest link in a measuring chain. In addition to this aspect, reliability aspects, as discussed in chapter 5, play a vital role and even safety aspects come into

Table 6.4 Review of the four different classes of transducers.

Type		Notation	Abbreviation
Modulating (30)	In	$[x, el, z]$	MI
Modulating (30)	Out	$[x, y, el]$	MO
Self-generating (5)	In	$[x, el, 00]$	SI
Self-generating (5)	Out	$[el, y, 00]$	SO

scope here, when life-saving systems are involved. Excellent examples can be found in intensive care departments where the personnel are completely dependent on the information delivered by the systems monitoring the patient. Another good example is the acceleration sensor in an airbag system of a car. It must operate extremely reliably and safely; however, one hopes it is never necessary. For a transducer, two state descriptions are available, namely

1   steady state description
2   dynamic state description.

We will discuss both of them.

6.3.1   The steady-state description

Practically every transducer is sensitive to every type of energy, so if we consider a small volume dV in which the transducer is placed, the energy content dW in this volume contains the summation of all possible energies, or

$$dW = \sum I_i e_i = \sigma \, dl + P \, d\rho/\rho + V \, dq + E \, dD + H \, dB + T \, dS + w_r \qquad (6.1)$$

where dW (J m^{-3}) is the total volume density of energy, I_i represents the intensive quantity, e_i represents an extensive quantity, σ (N m^{-3}) is a mechanical force per unit volume, dl (m) is a displacement, P (N m m^{-3}) is pressure, $d\rho$ (kg m^{-3}) is a volume density of mass, V (N m C^{-1}) is a voltage, dq (C m^{-3}) is volume density of charge, E (N C^{-1}) is the electrical field, dD (C m^{-2}) is charge per unit surface density, H (A m^{-1}) is the magnetic field strength, dB (V s m^{-2}) is the magnetic induction, T (K) is absolute temperature, dS (N m K^{-1} m^{-3}) is entropy per unit of volume and w_r (J m^{-3}) is radiation density per unit volume. Note that in this energy relation every product contains an intensive and an extensive quantity and that all products are expressed in energy by volume, as they should be. Between a specific intensive quantity I_i and an extensive quantity e_i the following relation exists:

$$I_i = \frac{dW}{de_i} \qquad (6.2)$$

with all other parameters kept constant. For instance, the intensive quantity temperature can be defined as $T = dW/dS$ and the intensive quantity voltage as $V = dW/dq$. The steady state of matter and, hence, of a transducer, can be expressed by an extensive quantity accounting for all possible intensive quantities. When the variations are small enough, a linear relation for all intensities may be assumed and when the superposition theorem is applicable we may write for a certain infinitesimal extensive quantity de_i

$$de_i = \sum_{m=1}^{n} s_{1,m} \, dI_m \qquad (6.3)$$

in which $s_{1,m}$ denotes the *sensitivity* 1 for an energy of type measurement. Note also the previously given definition for sensitivity in chapter 4 where (6.3) can be considered as a generalization of the concept of sensitivity where all sensitivities are assumed to be independent of each other. In the same way we can express the intensive quantity $\mathrm{d}I_1$ as the superposition of all types of extensive quantity, or

$$\mathrm{d}I_1 = \sum_{m=1}^{n} M_{1,m}\,\mathrm{d}e_m \tag{6.4}$$

in which $M_{1,m}$ denotes the *stiffness*, *rigidity* or *moduli* for the energy of type m. We will demonstrate these concepts with an example.

* * *

Example 6.1
Consider a self-generating pressure transducer composed of a piezo element. This pressure transducer is subjected to and sensitive to three types of intensive quantity, temperature T, a pressure P producing a force σ on the sensing elements and electrical field strength E. Note that in this case the pressure is converted into a mechanical stress resulting in an electrical field E. This is sometimes called a tandem effect. Usually, these types are constructed of resistive elements integrated on a semiconductor membrane. Then we have to deal with a modulating transducer and the change of the mechanical distortion is measured with the help of an auxiliary energy source by applying a current through it.

Now we want to describe the related extensive variations in length $\mathrm{d}l$ of the pressure elements, the change in entropy $\mathrm{d}S$ and the change in electrical flux $\mathrm{d}D$ as functions of the mentioned intensive quantities. We can write the following equations:

$$\mathrm{d}l = \left.\frac{\partial l}{\partial \sigma}\right|_{T,E} \mathrm{d}\sigma + \left.\frac{\partial l}{\partial T}\right|_{\sigma,\epsilon} \mathrm{d}T + \left.\frac{\partial l}{\partial E}\right|_{T,\sigma} \mathrm{d}E \tag{6.5}$$

$$\mathrm{d}S = \left.\frac{\partial S}{\partial \sigma}\right|_{T,E} \mathrm{d}\sigma + \left.\frac{\partial S}{\partial T}\right|_{\sigma,E} \mathrm{d}T + \left.\frac{\partial S}{\partial E}\right|_{T,\sigma} \mathrm{d}E \tag{6.6}$$

$$\mathrm{d}D = \left.\frac{\partial D}{\partial \sigma}\right|_{T,E} \mathrm{d}\sigma + \left.\frac{\partial D}{\partial T}\right|_{\sigma,E} \mathrm{d}T + \left.\frac{\partial D}{\partial E}\right|_{T,\sigma} \mathrm{d}E. \tag{6.7}$$

To obtain some insight these equations can be presented in a matrix depicting all nine sensitivities.

$$\begin{bmatrix} \mathrm{d}l \\ \mathrm{d}S \\ \mathrm{d}D \end{bmatrix} = \begin{bmatrix} \left.\dfrac{\partial l}{\partial \sigma}\right|_{T,E} & \left.\dfrac{\partial l}{\partial T}\right|_{\sigma,E} & \left.\dfrac{\partial l}{\partial E}\right|_{T,\sigma} \\ \left.\dfrac{\partial S}{\partial \sigma}\right|_{T,E} & \left.\dfrac{\partial S}{\partial T}\right|_{\sigma,E} & \left.\dfrac{\partial S}{\partial E}\right|_{T,\sigma} \\ \left.\dfrac{\partial D}{\partial \sigma}\right|_{T,E} & \left.\dfrac{\partial D}{\partial T}\right|_{\sigma,E} & \left.\dfrac{\partial D}{\partial E}\right|_{T,\sigma} \end{bmatrix} \begin{bmatrix} \mathrm{d}\sigma \\ \mathrm{d}T \\ \mathrm{d}E \end{bmatrix}. \tag{6.8}$$

On the main diagonal the shape conversions can be recognized, because no type of energy conversion is involved. These effects are sometimes called *direct* effects. All other effects imply energy conversions combining two forms of energy and are sometimes called *cross-effects*. For instance in this case

$$\frac{\partial l}{\partial \sigma}\bigg|_{T,E} \qquad\qquad (6.8a)$$

shows a direct effect or shape conversion due to a mechanical force and describes the sensitivity for mechanical stress and is called the *elasticity*; in the same way

$$\frac{\partial D}{\partial E}\bigg|_{T,\sigma} \qquad\qquad (6.8b)$$

is a direct effect also and is the definition of the *permittivity*. In the same way we read

$$\frac{\partial S}{\partial T}\bigg|_{\sigma,E}. \qquad\qquad (6.8c)$$

The following sensitivity describes a cross-effect and is called the specific *thermal expansion*:

$$\frac{\partial l}{\partial T}\bigg|_{\sigma,E}. \qquad\qquad (6.8d)$$

Of course we can ask for the expression for the intensive quantities temperature T, electric field E and mechanical force σ, and in this case the relations become as shown in (6.9). In this matrix again the direct effects are to be recognized on the main diagonal and the others are the cross-effects. Here all coefficients are called *moduli*. All coefficients can be given the appropriate name.

$$\begin{bmatrix} d\sigma \\ dT \\ dE \end{bmatrix} = \begin{bmatrix} \dfrac{\partial \sigma}{\partial l}\bigg|_{S,D} & \dfrac{\partial \sigma}{\partial S}\bigg|_{l,D} & \dfrac{\partial \sigma}{\partial D}\bigg|_{l,S} \\[2mm] \dfrac{\partial T}{\partial l}\bigg|_{S,D} & \dfrac{\partial T}{\partial S}\bigg|_{l,D} & \dfrac{\partial T}{\partial D}\bigg|_{l,S} \\[2mm] \dfrac{\partial E}{\partial l}\bigg|_{S,D} & \dfrac{\partial E}{\partial S}\bigg|_{l,D} & \dfrac{\partial E}{\partial D}\bigg|_{l,S} \end{bmatrix} \begin{bmatrix} dl \\ dS \\ dD \end{bmatrix}. \qquad (6.9)$$

In this example we have considered three parameters only, which is a simplification of course.

*　　*　　*

6.3.2 The dynamic-state description

If we consider dynamic effects then current densities J_i are the extensive quantities and the gradients G_i are the intensive quantities. The gradients can be considered as the driving forces and the current densities can be considered as current flows apart from the type of energy involved. The products of G_i and J_i represent a heat generation per unit time and volume, and hence appropriate units are joules per second per cubic metre or watts per cubic metre. So here power by volume is involved. For an arbitrary current density this can be written

$$J_i = \sum_{m=1} \sigma_{i,m} G_{i,m} \qquad (6.10)$$

in which $\sigma_{i,m}$ is called the *specific conductivity* for the respective process. For the gradient we can write

$$G_i = \sum_{m=1} \beta_{i,m} J_{i,m} \qquad (6.11)$$

in which $\beta_{i,m}$ is called the *specific resistivity*. For instance, a driving force for heat flow is a temperature gradient dT/dx and for an electric current a voltage gradient dV/dx. It should be kept in mind that all dynamic state equations ought to be represented as *vector* equations, and only for simplicity has this been omitted here. In the same way as before, cross-effects, combining two types of energy, and direct effects can be distinguished. In the following example as an application of these equations the current flow through a p–n junction is described.

$$* \qquad * \qquad *$$

Example 6.2
In a semiconductor the flow of charge carriers in a p–n junction consists of holes and electrons, and hence two equations are required to describe this transport process. For the flow of charge carriers two driving forces are responsible. These driving forces are a voltage gradient, or electric field, and a charge carrier concentration gradient. If electrons and holes flow in only one dimension (the x direction is considered), we find

$$J_n = q\mu_n n \, dV/dx + qD_n \, dn/dx \qquad (6.12)$$

$$J_p = q\mu_p p \, dV/dx + qD_p \, dp/dx \qquad (6.13)$$

where J_n (A m^{-2}) is the electron current density, J_p (A m^{-2}) is the hole current density, q (C) is the specific charge for electrons, μ_n (V^{-1} s^{-1} m^2) is the electron mobility in the semiconductor, μ_p (V^{-1} s^{-1} m^2) is the hole mobility in the semiconductor, n (m^{-3}) is the concentration of electrons as a function of position, p (m^{-3}) is the concentration of holes as a function of position, dV/dx is the voltage gradient in x direction, D_n (m^2 s^{-1}) is the diffusion coefficient for electrons, D_p (m^2 s^{-1}) is the diffusion coefficient for holes, dn/dx (m^{-4}) is the concentration gradient for electrons and dp/dx (m^{-4}) is the concentration gradient for holes. For instance, the products $q\mu_n n$ and qD_n represent the specific conductivities related to the two different driving forces.

In equations (6.10) and (6.11) you can easily recognize Ohm's law. We mention here two other well known cross-effects, namely the Peltier effect, producing a heat flow as a result of a voltage gradient, and the Seebeck effect, producing a voltage gradient from a temperature gradient.

6.4 TRANSDUCER PARAMETERS, DEFINITIONS AND TERMINOLOGY

In this section we will introduce frequently used sensor parameters, definitions and terminology. In order to obtain a quick overview the most recent technologies will be defined here. In the section on technology we will then discuss these terms in more detail.

6.4.1 Transducer parameters

In chapter 4 we have seen already a large number of specifications with which a system can be characterized. Here we will give some additional parameters for transducers although there is not always a clear conformity between different manufacturers concerning these parameters. It makes sense to present these also for comparison. It will be obvious that for a typical type of transducer a whole range of specifications are applicable devoted to a typical application. For instance, for a pressure transducer we can specify absolute pressure, differential pressure, barometric pressure, altimetric pressure, vacuum, reference pressure, maximum pressure reference temperature and so on. For application-dependent parameters the manufacturer's data sheets should always be consulted. See table 6.5.

6.4.2 Definitions and terminology

In recent decades an overwhelming number of techniques and technologies have become available. They are applied in the manufacture of new types of transducer. We will give here an overview to sort this out; however it cannot be exhaustive owing to space limitations. It can be said that this part is the most rapidly changing area in the field of transducers. New technologies and new materials are still being developed and they are leading to major new processing steps in the manufacture of (smart) transducers. We give frequently used terms and definitions belonging to the field of transducers, with their description.

Actuator
An actuator is a transducer at the output of a system able to convert the processed information into another, arbitrary, energy domain, without loss of information.

Biophotonics
Biophotonics is the science and engineering required to use photonic technology in medical or biotechnology products and procedures.

Biosensor
This term biosensor is rather confusing because it sometimes refers to sensors for the measurements of ion concentrations in living systems or in organic compounds. However, it can also refer to sensors made from polymers and other selective

Table 6.5 Specific transducer parameters.

Parameter	Expression or unit	Description
Settling time	s	The time between application of a step input signal and the indication of its magnitude within a rated accuracy
Response time	s	Same as settling time
Rise time	s	Usually the time required for a signal to change from 10% to 90% of its peak-to-peak value
Excitation	V or A	Power supply current or voltage needed for normal operation
Output voltage	V	Minimum and maximum limits of output voltage
Null voltage shift	V	Zero drift under no-signal condition
Sensitivity	dV/dx_i	The rate of change at the output and the change at the input
Hysteresis	Any	Permanent deviation from zero when input signal is set to zero
(Zero) offset voltage	mV	Output voltage obtained under (zero) reference input conditions
Temperature coefficient (TC)	ppm K^{-1}	The rate of change of reading as a function of temperature
Reproducibility	%	The closeness of agreement between measurements of the same quantity carried out with a stated change in conditions
Repeatability	%	The closeness of agreement between successive measurements carried out under the same conditions
Stability	$V_o K^{-1}$	Compare with drift
Maximum and minimum temperature	$T(K)$	Maximum and minimum temperature; operating span for specified accuracy

membrane material. Basically, specific enzymatic reactions at a sensitive interface can produce an electrical signal or a change of that signal. Proteins can form a specific selector. In various applications it is also possible to detect these changes optically. The measurement of glucose is a good example.

Mechatronics
Mechatronics is the science and engineering discipline which combines mechanical and electronic elements into a larger functional unit. Good examples are a photo-camera, a printer, a video recorder. The word *mechatronics* was first introduced in Japan in 1974.

Micromachining
Micromachining is the *technology* to fabricate very small mechanical structures in the micrometre or nanometre range. Micromachining technology can be classified into bulk and surface micromachining technology. Three different groups of micro-mechanical devices can be distinguished:

1 *static devices* such as fixed three-dimensional structures, for example nozzles, cavities, capillary canals, miniature connectors etc

2 *dynamic devices* are micromechanical structures, such as diaphragms, membranes, microbridges, cantilever beams, resonators and so on
3 *kinematic* micromechanical structures represent the family of gears in micromotors.

Micromechanics
Micromechanics is the design, the development and the production of extremely small mechanical devices ranging from nanometres to several millimetres in size.

Microsystem
A microsystem is a functional unit, sized in dimensions from several micrometers up to several millimetres, which can perform a specific task. A functional unit is an integrated single-chip circuit composed of a sensor, signal-processing unit and an actuator. These units are also called *smart* sensors

Optoelectronics
Optoelectronics is the science and engineering discipline combining optics or photonics and electronics in one device. For instance, an optocoupler is a device which connects or separates the optical and the electronic domains. Also, optically operating bus systems exist, where the bus system is operating completely optically and at the input and output the transducer interface is made between the optical and the electrical domains.

Photonics
Photonics is the science and engineering used to handle all information optically. In this case the information is bound to photons in the same way as the information is connected to electrons in the electrical domain. Examples are lasers as light sources, optical switching devices, splitters, glass fibres, optical amplifiers, etc. It is interesting to note that an optical computer already exists. A more sophisticated definition is the technology of generating and harnessing light and other forms of radiant energy whose quantum unit is the photon.

Processing system
A processing system is a functional unit which can process information. Usually, this is performed electronically, where the electrons are the information carrier. However, an increasing trend is present to process information optically. Basically, information can be processed in any energy domain. For instance, mechanically operating amplifiers are known as fluidics and they work pneumatically. The advantages of processing information electronically were discussed in chapter 1. Because of its increasing importance we will come back again to that in the section on the optical domain.

Sensor
A sensor can be defined as a transducer at the input of a system capable of transforming acquired information from an arbitrary energy domain into another without loss of information.

Smart sensor
A smart sensor or intelligent sensor is a single-chip functional unit combining sensing and processing functions. See also 'Microsystems'. A good example is a complete integrated accelerometer used in the automotive industry for airbag systems.

Transducer
This is a collective noun for sensors and/or actuators.

6.5 CHARACTERISTICS OF INTEGRATED CIRCUITS AND TRANSDUCERS

In this section we will present a brief historical overview of what is possible from a technological point of view in making integrated circuits and (smart) transducers. Also, what may be expected in the near future will be discussed. In order to provide a reference frame to see where we are, we start with a brief historical flash-back. This is performed best with the simple diagram illustrated in figure 6.9, where it is shown how the developments at component and system levels have merged into small intelligent systems. The microprocessor is taken here as the most striking

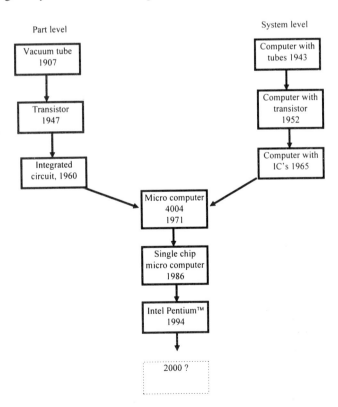

Figure 6.9 The development of electronics at part and system levels.

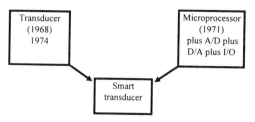

Figure 6.10   Development of the smart transducer.

reference vehicle, but many other systems developments can be taken as such. Predicted single-chip microprocessor systems will have a word length of 64 bits and a clock frequency of over 200 MHz. Also, a tremendous research effort is being put into optical telecommunications and computing and predictions covering more than a five year period are difficult to make.

The development of transducers started in the early 1970s. We can draw the same type of diagram for the development of transducers. This is illustrated in figure 6.10. This development has now resulted in completely integrated single-chip smart transducers combining all the required functions.

To provide an overview of what electronics implies, reference is made to figure 6.11. In this figure groups of families of electronics are depicted which now merge into complex systems. The combination of analogue with digital electronics on one chip results in so-called hybrids. Special design tools must be used when hybrids are developed.

In the early 1960s J S Kilby of Texas Instruments and J A Hoerni and R N Noyce of Fairchild invented planar silicon integrated technology. Nobody at that time was able to predict the powerful impact on the electronics industry and the huge social implications in daily life that this technology would have. The relatively simple planar technology is made possible by application of a photolithographic method and the use of oxide window masking. As has been revealed, these techniques have the important characteristics that passive and active components as well as the (aluminium) interconnections can be fabricated on one side of a silicon wafer. For some devices such as CCDs, this requires over 300 process steps of which most are very critical for the overall performance. It will be obvious also that here statistical process control (SPC) comes into scope. Note that the throughput time can cover a 50 day period. One mistake made at the beginning of the whole manufacturing process will

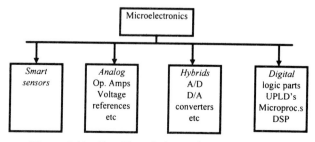

Figure 6.11   Families of electronics components.

Table 6.6 Specification of maximum number of allowed dust particles for Class 1, according to US Federal Standard 209D.

	Number of particles for the following particle size (μm)			
Class	0.1	0.2	0.3	0.5
1	35	7.5	3	1
10	350	75	30	10

be disastrous for the final result. One of the most critical aspects in the fabrication of semiconductors is contamination by dust. Because of the reduced size of ICs, clean rooms of 'class 1' are required now. This refers to the maximum contamination per cubic foot (28.4×10^{-3} m^3) as indicated in table 6.6. For comparison class 10 is also mentioned.

Today we can meet a great variety of applications of electronics in all types of market sector from the domestic to the car industries and from the watch industry to the toy industry. It is observed that the car industry is one of the most demanding. For instance, it is expected that the automotive sensor market will be more than $6 billion by the year 2000 (see also table 1.1). In the automotive industry over 20% of costs in a car are electronics. A very futuristic outlook of a possible society is presented by Alvin Toffler in his book *The Third Wave*, in which large emphasis is put on a communication society, all made possible by modern electronics and global networks. Meanwhile, IC technology has also influenced a great variety of measuring techniques. Nowadays in this process four groups of devices can be recognized:

1 integrated circuits, fabricated with standard IC technology, whose performance can improve transducer characteristics; these are often general-purpose ICs
2 custom-made ICs designed for one specific task, also made in standard IC technology, known as application-specific integrated circuits (ASICS): these are fabricated with analogue and digital functions on them
3 transducers fabricated in standard IC technology for one specific physical quantity
4 integrated circuits fabricated in combination with other compatible technologies resulting in a specific transducer.

In the early 1970s the great advantages of planar silicon technology raised the question of whether standard silicon technology could also be used for the manufacture of transducers. S Middelhoek and A F P van Putten of the Delft University of Technology made a successful first attempt to apply standard IC technology to the manufacture of transducers. This has resulted in a silicon anemometer or gas flow velocity meter based on the principle of the generation of a temperature gradient due to a gas flow in a silicon chip, whose average temperature is maintained well above ambient temperature. We will come back to that later.

In the next sections we first discuss some important advantages of semiconductor technology, which appear also to be of importance for the manufacture of transducers. Then the specific physical effects of silicon, which can be applied to transducers, will be discussed. It has been found that other semiconductor materials can be used for the manufacture of transducers, but the most promising with respect to costs are those which are compatible with silicon technology. So, let us first have a look at the great advantages of silicon technology.

6.5.1 Silicon technology
The main characteristics of silicon technology can be summarized in the following strongly related aspects:

1   size and linewidth
2   complexity
3   power consumption
4   reliability
5   electrons versus photons as information carrier
6   technologies
7   price versus costs.

We will discuss these in some more detail in the following sections.

6.5.2 Size and linewidth
Photolithographic techniques have made it possible to produce integrated circuits with very tiny details for which the micrometre is the normal measuring instrument. The ultimate resolution obtained by these techniques is determined by the applied wavelength of the source in the photolithographic process. It has been shown already that linewidths of less than 0.1 μm can be fabricated and the smallest transistor ever made cannot be detected with a normal light microscope with maximum magnification. This very-large-scale integration (VLSI) technique is made possible by applying an electron beam or an x-ray beam as the 'light' source in the photolithographic process. A complete electronic circuit requires a few square millmetres of silicon and apart from this effect the applied chip areas are still increasing. As an example a complete microprocessor requires a chip area of 6×6 mm^2. The wafer size on which the chips are fabricated has increased in time. Note the tremendous increase in costs with increasing wafer diameter. This is illustrated in figure 6.12 for digital circuits.

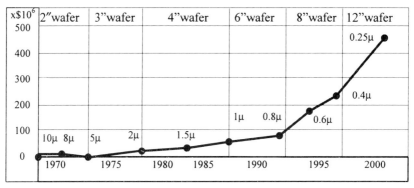

Figure 6.12 Investment for a production capacity of 1000 wafers per week for digital integrated circuits in relationship to wafer size and applied linewidth. (Source: Dataquest.)

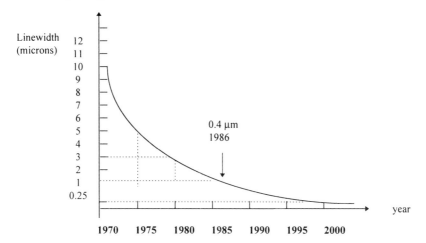

Figure 6.13 Linewidth achieved versus time of IC technology.

As also can be seen from figure 6.13, in manufacturing ICs over the last decades, a large reduction in linewidth has been achieved. Attempts are being made to reduce the linewidth further and research is going on to obtain a linewidth of 0.1 μm. The costs of introducing a reduction of linewidth and an increase in wafer size in production areas are enormous, because each processing step and all equipment must be adapted. In practice, this means building a new facility and training new personnel.

6.5.3 Complexity
The large number of components which can be produced on a single chip have made it possible to integrate complete electronic systems on a single chip. Today a memory chip contains not only the memory cells but all the required addressing and sensing circuitry such as decoders, registers, buffer control logic and so on. Today we are confronted with 32-bit microprocessors on a single chip of 10×10 mm^2, with all necessary peripheral electronics such as control circuitry decoders, bus interface, registers, RAM and ROM capacity, timing controller, floating-point unit etc all on the same chip. For instance, the Pentium microprocessor contains over 3.5 million transistors. For 16 Mb memory devices over 20 million transistors are integrated on a single chip.

For electronic measurement systems, transducing elements integrated with a large part of the modifier are realized. This results in an increased reliability and a large cost reduction. Examples are a complete integrated accelerometer or an IC temperature chip with digital read-out. To illustrate this increasing complexity an illustration of Moore's law is depicted in figure 6.14.

6.5.4 Power consumption
To illustrate the reduction in power consumption in electronic components and systems over the past decades a comparison between the power consumption of the first computer, the ENIAC (built in 1946), and a simple handheld programmable pocket calculator is most suitable. The ENIAC built with vacuum tubes had a

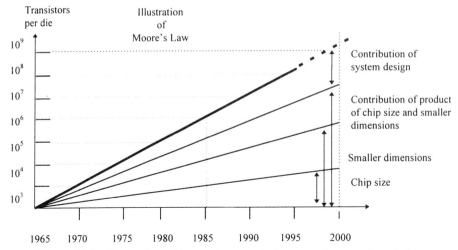

Figure 6.14   Illustration of increasing complexity according to Moore's law.

power consumption of about 150 kW, whereas the pocket calculator with much more computing power than the ENIAC consumes less than 1 W. In table 6.7 another comparison is made between the ENIAC and a more recently developed micro-computer which illustrates some striking characteristic developments.

The power consumption of logic gates and memory cells is often represented by a figure of merit called the *power–delay product*, which is the product of power consumption and gate delay time and is expressed in picoJoules (pJ). In table 6.8 the power–delay product of some technologies is summarized. Note that GaAs is also mentioned. This material offers advantages over silicon with respect to operating temperature and speed of charge carriers.

6.5.5   Reliability
In chapter 5 we have already discussed aspects of reliability in detail. Integrated circuit technology achieves a large increase in reliability compared with vacuum tube

Table 6.7   A comparison between the ENIAC computer and a 1994 microcomputer.

Specifications	ENIAC	Intel Pentium™ processor
Year	1946	1994
MTBF	2 h	>5 years
Power consumption	150 kW	<20 W
Clock frequency	100 kHz	>100 MHz
Power supply voltage	400 V	±5 V
Number of active components	18 000	>3 × 10^6
Mass	30 000 kg	0.1 kg
RAM	1 kbit	32 MB
Price	>$100,000	>$500

Table 6.8 Power–delay product of some technologies.

Technology	Power–delay product (pJ)
p-channel silicon gate	1
n-channel silicon gate	0.5
CMOS silicon gate	0.5
BICMOS	0.2
SOS CMOS	0.1
I^2L	0.01–1
Silicon (bipolar)	1
GaAs (bipolar)	0.5
Josephson gate	10^{-6}

technology. Integrated circuits can achieve a MTTF (non-repairable system) of 10^{11} h, whereas vacuum tubes were not better than 10^5 h; hence an improvement of 10^6 is obtained. Compared with discrete transistors the improvement is roughly a factor of 10^3. The most important parameters determining the reliability of integrated circuits are:

1 contamination and dust in clean rooms
2 wafer-handling procedures
3 accuracy of aligning the several masks (for some devices more than 20)
4 visual, mechanical and electrical test procedures
5 wire-bond testing
6 packaging of the chips.

Wire-bond (pull) testing is mentioned explicitly, because it can make up to 30–60% of manufacturing costs. For instance, some chips require 400 wire bonds or more and, for instance, to satisfy military standard specifications they should all be tested separately.

6.5.6 Electrons versus photons as information carrier

We will make here a comparison between electrons and photons as information carrier. As noted already, there is an increasing trend to handle information optically where the photons are the information carrier; there are definitely advantages also for the manufacture of transducers. In table 6.9 the main characteristics of both information carriers are summarized.

In the section on radiant energy a diagram is given in which the possible energy conversions 'in and to' the optical domain are outlined where we also will discuss the possible and potential applications. A huge world-wide research effort and investment are being made to obtain optical communication networks. In addition, the necessary optical devices are being developed simultaneously.

6.5.7 Available technologies for transducers

In this section we will give a brief overview of technologies available for making transducers. These technologies partly cover standard technologies for fabricating

Table 6.9 Comparison of electrons and photons as information carrier.

| Characteristic | Information carrier | | Remarks |
	Electrons (1)	Photons (2)	
Range of power	10^{-9}–10^9 W	pW–kW	Lasers (2)
Speed of propagation	$\leqslant 3 \times 10^8$ m s^{-1}	3×10^8 m s^{-1}	In vacuum (1), (2)
Acceleration	$>18 \times 10^{10}$ m s^{-2}	?	At $E = 1$ V m^{-1} (1)
Time domain	ps–h	?	
Transport technology	Copper/coax cable	Glass fibre	
Transport speed	Mb s^{-1}	>25 Gb s^{-1}	ATM technique (1)
Analogue and digital signal processing	Yes	Yes	
Amplification	$>10^8$	Yes	
Switches	Transistor	Optical transistor	
Modulation	AM, FM, PWM, etc	AM, FM, PWM, etc	
Intrinsic safety	<200 mW	Yes	Explosion (1)
EMC sensitivity	High	Much less	
Weight reduction	Large because of size reduction	Extremely large	
Bandwidth	$>10^9$ Hz	10^{15} Hz	

AM, amplitude modulation; FM, frequency modulation; PWM, pulse width modulation; ATM, asynchronous transport mode.

ICs, but newly developed technologies offer the possibility to make three-dimensional devices by applying different kinds of etching technique. The aim of this section is just to demonstrate that these new technologies can lead us to the design, development and production of completely new measuring systems, which could not be realized in earlier stages of technology.

Etching
Etching is the key process to achieve three-dimensional structures. It is also understood as a type of micromachining. Mechanical properties and the crystallography of the semiconductor material must be well known. *Wet etching* with EDP (ethylenediamine–pyrocatechol–water) and KOH are frequently used for sensors. However, also numerous other anisotropic wet etchants are used for silicon.

Bulk micromachining technology is based on single-crystal etching or etching of composite materials deposited or grown on a silicon substrate. Three-dimensional structures can be obtained. The crystallographic orientation determines the etch rate. For instance, the etching effect in the [111] crystal direction is 400 times slower than in other directions and because of this effect it can act as a stopper plane.

Surface micromachining is an etching technique for fabricating three-dimensional (3D) microstructures from multilayer stacked and patterned surface thin films. This can be characterized as isotropic etching of surface layers deposited on silicon with various techniques such as chemical vapour deposition (CVD), electron molecular beam (EMB) deposition and plasma-enhanced chemical vapour deposition. Often structured polysilicon material is deposited. The use of *sacrificial layers* in the fabrication process is characteristic of this technology. A sacrificial layer is a temporary layer used to obtain free moving parts. The layer is used as a kind of support layer and removed afterwards by an etching step. Gears and micromotors are made in this process.

Doping selective etching gives a higher degree of freedom to obtain 3D structures. Doped p and n regions have different etch rates when an Si wafer is immersed in KOH and connected to a rectifying current circuit. This is due to the formation of a Schottky-type junction at the semiconductor–electrolyte interface. It is possible to etch off p-doped material whereas n-doped material is retained.

LIGA

LIGA stands for Lithographie und Galvanischer Abformung (originates from German). 3D structures can also be made in this technology. LIGA combines deep lithographic, microelectroplating and micromoulding process steps. The highest precision with submicrometre resolution can be obtained with synchrotron radiation. However, where less resolution is required synchrotron radiation is not needed and the process can be performed anywhere. When sacrificial layers are needed then also LIGA technology can be used. The technique is then called SLIGA (*sacrificial LIGA technique*).

Laser micro-engineering

Laser micromachining can be applied to all kinds of bulk and thin film material and is an alternative microfabrication technique in an extremely wide range of applications. Materials used are silicon, glass, ceramics, polymers, metals, etc. Applied techniques are, for instance, thermal ablation, dry etching, deposition of material, anisotropic etching, cutting and microdrilling and laser trimming of ICs and thin- and thick-film hybrids.

Bonding
Here bonding is not meant as the technology of wire bonding to connect the bond pads on a chip with the interconnecting pads to the outside world in a type of package; here we mean the bonding or *lamination* of two similar or dissimilar materials without any adhesives. A glass–metal or glass–semiconductor bonding can be realized by applying an electric field between the two materials and a temperature of about 300–400 °C. This process is called anodic bonding. Another bonding technique requires the wafers to be immersed in boiling nitric acid and then brought together and heated to about 1000 °C. This process is called *silicon direct bonding*. For instance, this process can be used to obtain small hermetically sealed cavities or small chambers for pressure transducers.

Packaging
For transducers packaging or *encapsulation* is a major problem, because a transducer must always have a *window* with the outside world, which can discriminate strongly for the one physical parameter desired. Each transducer application requires its own adapted transducer package. Contamination, saturation and pollution play a vital role in the reliability of the transducer. Some transducers only have a MTTF of several hours owing to saturation problems. For instance, a pH sensor requires quite a different packaging from an accelerometer or a thermal flow sensor.

6.5.8 Price–performance ratio
Today, integrated circuits are made in fully automated production processes, 7 days per week and 24 h per day. Many different circuits can be simultaneously processed on one wafer. The yield can be over 5000 chips of size 10 mm^2 from one 10 in (250 mm diameter) wafer. The result is a drastic reduction in costs per component and often packaging and testing costs surpass the chip costs. Prices of integrated circuits have fallen to unbelievable depths and for comparison reasons this is often expressed in a price–performance ratio (price per gate). Figure 6.15 illustrates this development showing the development in price per gate for a circuit made in MOS technology. Sometimes another example is given to illustrate the dramatic developments realized in semiconductor integrated circuit technology. For this purpose a well-known comparison is made between some characteristics of a model T Ford of

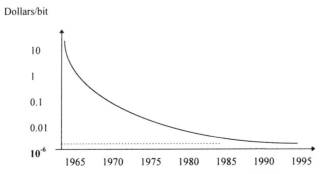

Figure 6.15 Price per gate development for a MOS circuit.

Table 6.10 Illustration of the dramatic semiconductor development by comparison of a model T Ford and a hypothetical 'Rolls-Royce' (Source: Moore).

	T Ford	'Rolls-Royce'
Year	1919	1979
Speed	60 km h^{-1}	160 000 km h^{-1}
Fuel economy	5 km l^{-1}	20 000 km l^{-1}
Life span	10 years	Centuries
Price	$1000	$0.10

1919 and a type of hypothetical Rolls-Royce if the car industry had made the same development and progress as the semiconductor industry. The comparison is made between the years 1919 and 1979 (table 6.10).

Considering these features of silicon technology, it will be clear that silicon or other silicon-compatible semiconductors can also offer great advantages for manufacturing transducers. However, for that purpose, the next step is to investigate all known present physical effects in silicon or other semiconductor materials which may be well suited for transducers.

6.5.9 Applications areas
The areas of applications for semiconductor transducers are numerous and expanding daily. The automotive market is one of the most demanding and over 20% of costs of fabrication are electronics. Another most expanding market is the telecommunications market, in particular for optoelectronic devices. Another rapidly expanding application area is medicine in which optical (photonic) devices play an increasing role in therapeutic treatments. For instance, a new journal *Biophotonics*, launched in 1994, is completely devoted to applications in this area. (See also chapter 1.)

6.6 TRANSDUCER EFFECTS IN SILICON AND OTHER COMPATIBLE MATERIALS

In this section we present known transducer effects in silicon and other compatible materials. The choice of materials and their related technologies is overwhelming and a survey can only obtained when a rough division is made.

6.6.1 Transducer effects in silicon
In table 6.11, a summary is presented of known physical effects in silicon. In this table the main emphasis is on silicon to be used for input and output transducers respectively for energy conversion into and from the electrical energy domain.

6.6.2 Transducer effects in compatible materials
The technology for making transducers is not limited to silicon only. Materials such as III–V and II–VI compound semiconductor materials are also used. Also, polymers

Table 6.11 Summary of known physical effects and examples of applications in silicon where electrons are the information carrier.

In/Out		Modulating			
Energy domain	Self-generating	Resistor, inductance, capacative	Diode	Transistor	Examples of smart transducers
Radiation	Volta effect, solar cell	Photoconductor	Photodiode	Phototransistor	Photo-IC, CCD
Mechanical	Not known	Piezoresistivity	Piezojunction	Piezotransistor	Accelerometer, Piezo IC
Thermal	Seebeck effect, thermocouple	$R=f(T)$	Reverse biased, $I_{rev}=f(T)$	Forward biased, $U_{BE}=f(T)$	Temperature IC
Electrical	Thermal energy, resistance	Electric field MOSFET	Electric Field FET	Dual-gate MOSFET	All types of IC
Magnetic	Maxwell diffused coil	Magnetoresistor	Magnetic diode	Hall effect	Hall IC
Chemical	Galvanic	Ion concentration	Not known	ISFET	Smart nose

and enzyme materials have to be considered, of which some are very complex. Sometimes complicated multilayered structures are used with a large variety of materials.

Table 6.12 depicts some compatible technologies, where compatible has to be understood as technologies which are about the same as applied for silicon technology. For instance, the vacuum deposition of a nickel layer on silicon requires the same technique and the same equipment as for the aluminium interconnection pattern on silicon wafers. The same holds for gallium–arsenic of which we have noted already that the material can perform the same functions with higher speed and at higher temperatures than can be realized in silicon. The technology to make ICs is different and more complicated because oxide masking on GaAs is not possible. Furthermore GaAs possesses an additional characteristic related to direct bandgap materials. Also, recent developments are found in GeSi material, which provides another extreme acceleration in processing speed of electronics.

6.6.3 Transducer effects in optical semiconductors

Because of its increasing importance, we will present an equivalent diagram to that for silicon. Here photons are the information carriers. It will be obvious that this table does not represent an exhaustive overview; however, it gives us an idea of what

Table 6.12 Some silicon-compatible technologies for input/output transducers.

Energy domain	(Chemical) vacuum or sputtering deposition	Remarks
Radiant	CdS/ZnS	On top of Si substrates
Mechanical	CdS/piezo-FET	On top of Si substrates
Thermal $R(T)$	Metal layer $R(t)$	On top of Si substrates
Electrical	Deposition of R_s	Transistors, ICs
Magnetic	Deposition of Ni layer	Recording ICs
Chemical	Aluminium oxide layers	

is possible and what will become possible in the near future. Hence, not all required energy conversions are present in silicon and sometimes another material is more appropriate for a specific application. This is the reason to look for other compatible materials. GaAs technology comes into scope here and, for the manufacture of LEDs, GaAs is the preferred material. It is noticed that although the emission effect

Table 6.13 Some physical effects and examples where photons are the information carrier.

Energy domain	In/Out Self-generating	Modulating Resistor, inductance, capacitive	Diode	Transistor	Others	Smart transducers
Radiation	Fibres, filters, splitters	Photoconductor		Switches, multiplier, amplifiers	Fibres, QW laser	Photo-IC, CCD, switches
Mechanical	IR pressure cavity, SLM	PSD			Fibres	
Thermal	Thermal radiation	IR detectors				
Electrical	LED, CCD, solar cells		Lasers, LEDs, photodiode	Thyristors	Thyristors	Video telephone communications ICs, diode arrays
Magnetic	Spin excitation	Kerr effect			Optical, recording, storage and R/W	
Chemical	Oxidation, photosynthesis				SAW, fluorescent dyes	

SLM, spatial light modulator; PSD, position-sensitive device; LED, light-emitting diode; CCD, charge coupled device; SAW, surface acoustic wave; QW, quantum well.

in silicon is present the efficiency is very poor. In table 6.13 the same type of information is given as in table 6.11.

As can be seen from these tables GaAs is applied in LEDs but also complete microprocessors, and very fast multipliers are manufactured in GaAs and commercially available. Numerous research centers are investigating GaAs and its optical potential for making ICs. Meanwhile, GaAs technology has become a specialism with dedicated design tools and fabrication facilities.

In the following sections we will discuss the different transducers in each energy domain. In each section basic physics is presented. An overview of known semiconductor transducers and their related mechanisms is presented. Because of space limitations, it will be obvious that in this respect to obtain completeness is impossible. We will present the following sections:

- the optical energy domain
- the mechanical energy domain
- the thermal energy domain
- the magnetic energy domain
- the chemical energy domain.

In each section, for all transducers the electrical in–out conversion is taken as a reference.

6.7 THE OPTICAL ENERGY DOMAIN

In this section we will discuss the energy conversions for transducers in the optical domain. We will start with a brief outline of terms and some basic physics relevant for understanding optical transducers. As discussed already, increasing credit must be given to the optical domain, in particular to the communications applications. We will see how the interaction of photons with semiconductor materials works out. As will be shown, the p–n junction is an elementary structure in photonics and plays a basic role in many optical devices.

6.7.1 Physics

The very first thing to do is to understand what the main basic mechanisms are in optoelectronic devices, so let us first look to the energy of a photon. This can be expressed as

$$E = h\nu = hc/\lambda \tag{6.14}$$

where h is Planck's constant (6.6326×10^{-34} W s^2), ν (Hz) is the frequency, λ (m) is the wavelength and c is the speed of light (2.998×10^8 m s^{-1}). Hence the photon energy is a function of frequency or wavelength. The frequency at which optoelectronic devices operate covers a wide range and runs from 10^{23} to 10^{12} Hz or in wavelength from 10^{-15} to 10^{-3} m. Note that the product of frequency and wavelength

Table 6.14 Review of frequency and wavelength of the electromagnetic spectrum for the optical domain.

Name	Frequency (Hz)	Wavelength (m)	⟨Energy⟩ (eV)
Cosmic rays	$>10^{23}$–10^{22}	$<10^{-15}$–10^{-14}	4.1×10^8
γ rays	10^{21}–10^{18}	10^{-13}–10^{-10}	4.1×10^5
X-rays	10^{21}–10^{16}	10^{-13}–10^{-8}	4.1×10^3
UV	10^{17}–10^{15}	10^{-9}–10^{-7}	4.1×10
Visible	10^{15}–0.5×10^{14}	10^{-7}–2×10^{-7}	4.1
IR	0.5×10^{14}–0.5×10^{11}	2×10^{-6}–2×10^{-3}	4.1×10^{-3}
Microwave	10^{12}–10^9	5×10^{-3}–10	4.1×10^{-4}

IR, infrared; UV, ultraviolet.

is always the light speed, c. We can summarize this in table 6.14. In the last column the related average energy per frequency range is given in electrovolts. It will be clear that in optical devices an interaction exists between radiation and solid state and different mechanisms can occur which can be described in a macroscopic way, such as

- refraction
- transmission
- absorption
- interference
- diffraction.

Refraction is governed by Snell's law and describes the relationship between incident radiation and transmitted radiation at the interface of two media:

$$n_i \sin \varphi_i = n_t \sin \varphi_t \qquad (6.15)$$

where n_i is the index of refraction of medium i, n_t the index of refraction of medium t, φ_i the angle of incident radiation with respect to the normal and φ_t the angle of transmitted radiation with respect to the normal. Under the condition of perpendicular incident radiation, the fraction of reflected radiation I_r at the surface of the interface of two different materials is found:

$$I_r = I_0 \frac{(n_t - n_i)^2}{(n_t + n_i)^2} \qquad (6.16)$$

where I_0 is the intensity of the incident radiation and n_i and n_t are the indices of refraction of the two media.

Absorption of photons results in attenuation of the incident radiation in a medium during the process of energy conversion. The cause is the interaction of the photons with the underlying atoms of the radiated material. The absorption can be described with

$$I(x) = I_s \exp(-\alpha x) \qquad (6.17)$$

where $I(x)$ is the intensity of the radiation in the material at depth x, I_s is the intensity of the radiation at the surface and α is the absorption coefficient determined by the medium and the wavelength of the incident radiation. Particle physics radiation from the nucleus can generate Coulomb interaction. The result can be ionization of the material, electron excitation or dispersion.

Interference occurs when two waves from different sources with equal frequencies but different in phase coincide. The result is a wave with the same frequency but of which the amplitude and phase are determined by the amplitudes and differences in phases of the original waves. When the phase difference is a multiple of 2π the amplitude is the sum of both amplitudes. If the phase difference is an odd multiple of π the result is the difference of both amplitudes and total annihilation can occur. *Diffraction* occurs when light waves are broken into separate light bands as in a prism.

Polarization is the characteristic of a fixed orientation in a certain direction of a physical parameter. We can distinguish electric, magnetic and light polarization. Polarized light is the propagation of a transverse wave in a fixed plane. Light can be polarized in a linear, a circular and an elliptic form. No polarization means that all planes of propagation are present equally.

In the next section we will introduce all relevant units used in the optical domain. To most of us those units are not very familiar and so it may be a useful overview.

6.7.2 Applied units in the optical domain

In table 6.15 we present an overview of units used in the optical domain. We make a distinction between *radiometric* and *photometric* units. Radiometry is the science

Table 6.15 Quantities and units in the radiant domain for radiometric and photometric parameters.

Radiometry			Photometry			Expression
Name	Symbol	Unit	Name	Symbol	Unit	
Radiant energy	Q, Q_e	J (energy)	Luminous energy	Q, Q_e	lm s	
Flux of radiation	Φ, Φ_v	W (power)	Luminous flux	Φ, Φ_v	lm	$\Phi = dQ/dt$
Irradiance	E, E_e	W m^{-2} (input power per unit area)	Illuminance	E, E_v	lm m^{-2}	$E = d\Phi/dA$
Radiant excitance	M	W m^{-2} (power per unit area)	Luminous intensity	M	lm m^{-2}	$M = d\Phi/dA$
Radiant intensity	I, I_e	W sr^{-1} (power per unit solid angle)	Luminous intensity	I, I_v	cd	$I = d\Phi/d\Omega$
Radiance	L, L_e	W m^{-2} sr^{-1} (power per unit area and per solid angle)	Luminance	L, L_v	cd m^{-2}	$L = dI/dA \cos\theta$

of electromagnetic radiation. Photometry is the science of electromagnetic radiation in the visible region to which the eye reacts.

A solid angle is the *space angle*. The unit is the steradian with symbol sr. This is the solid angle which, when the top coincides with the centre of a sphere, produces an area on that sphere equals that of a square with the radius of that sphere as the side.

The *candela* is defined as the luminous intensity of a light source in a certain direction with a monochromatic radiation frequency of 540×10^{12} Hz and a radiant flux of $1/683$ W sr^{-1}.

Colour is defined as the visible colour described by impression, luminous intensity and saturation. Each colour can be composed of red, green and blue.

6.7.3 The interaction of electromagnetic radiation with semiconductors

When a photon with energy E_{ph} coincides with a semiconductor and the photon energy E_{ph} is larger than the bandgap (distance) energy, absorption can occur and an electron can be moved from the valence band into the conduction band. In other words when $E_{ph} > h\nu = E_g$, then an electron–hole pair is generated. This absorption is a function of temperature, the fraction, η, of photons with which they coincide at the surface and the absorption coefficient, α. This absorption can be described with

$$\Phi(x) = \eta\Phi_0 \exp(-\alpha x). \tag{6.18}$$

With increasing temperature the bandgap width decreases and larger-wavelength (smaller-energy photons) radiation can be used to generate electron–hole pairs. For

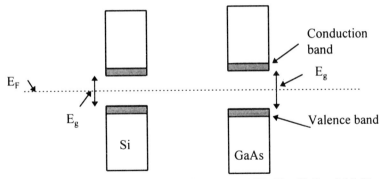

Figure 6.16 Illustration of the bandgap distance E_g (eV). For Si $E_g = 1.15$ eV and for GaAs $E_g = 1.43$ eV. E_F represents the Fermi level.

silicon the bandgap distance is about 1.15 eV and for GaAs it equals 1.43 eV. This is illustrated in figure 6.16.

For intrinsic material (non-doped) there exists an equilibrium in concentration of densities of charge carriers in the valence band and in the conduction band. The Fermi level, E_F, is then found in the middle of the bandgap. For doped material this equilibrium (E_F) shifts towards either the conduction band for n-type doping or the valence band for p-type doping.

At this point we make a very important distinction between *direct-bandgap* materials and *indirect-bandgap* materials. When an electron in a semiconductor is moved from the valence band to the conduction band, or falls back again from the conduction band towards the valence band, then, in general, two things will happen. A match must always be present between energy and wavenumber, which corresponds to the specific transition when an electron moves between valence band and conduction band. For silicon such a transition requires a change in both energy and wavenumber and for GaAs there exists only a change in energy. This is the reason why GaAs is called a direct-bandgap material. For instance, when in GaAs an electron is falling back into the valence band, then the *change* in energy can be observed as generated radiation with no change in wavenumber involved. This is illustrated in a simple diagram in figure 6.17. This characteristic explains the difference in efficiency between Si and GaAs for radiated energy.

6.7.4 The p–n junction in the optical domain

We will not discuss in detail the p–n junction here, but will give only some results relevant for our discussion. For further details reference is made to the literature. When no supply voltage is connected across a p–n junction, there exists a so-called built-in, inner or barrier potential, V_B. This voltage is in equilibrium with the diffusion caused by differences in majority charge carrier concentration densities across the p–n junction and it can be described by

$$V_B = \int_p^n E(x)\, dx \quad \text{or} \quad V_B = \frac{D_p}{\mu_p} \ln \frac{N_A N_D}{n_i^2} \quad \text{or} \quad V_B = \frac{D_n}{\mu_n} \ln \frac{N_A N_D}{n_i^2} \qquad (6.19)$$

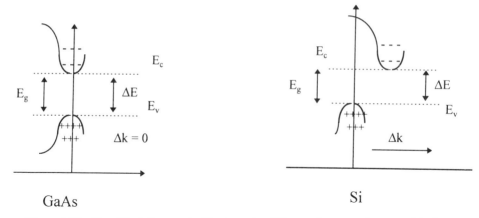

Figure 6.17 Simplified diagram to illustrate the difference between direct- and indirect-bandgap materials.

where $E(x)$ (V m^{-1}) is the field strength in the x direction across the p–n junction, D_n (m^2 s^{-1}) the diffusion coefficient for electrons, D_p (m^2 s^{-1}) the diffusion coefficient for holes, μ_n (m^2 V^{-1} s^{-1}) the electron mobility, μ_p (m^2 V^{-1} s^{-1}) the hole mobility, N_A (m^{-3}) the acceptor density per unit volume and N_D (m^{-3}) the donor density per unit volume.

Further, we have seen previously that this equilibrium is governed by the two transport equations for the two types of charge carrier mentioned in section 6.3.2. For convenience they are repeated here:

$$J_n = q\mu_n n \frac{dV}{dx} + qD_n \frac{dn}{dx} \qquad J_p = q\mu_p p \frac{dV}{dx} - qD_p \frac{dp}{dx} \qquad (6.20)$$

where J_n (A m^{-2}) is the current density of electrons, J_p (A m^{-2}) the current density of holes, q (C) the specific charge for electrons, $dV/dx = E$ (V m^{-1}) the field strength in the x direction, dn/dx (m^{-3}) the density gradient of electrons and dp/dx (m^{-3}) the density gradient of holes. When no external voltage source is connected across the p–n junction, there is no transport of charge carriers and both expressions are null. Then for V_B equation (6.19) is found. In the next section we will see how the temperature and radiation can influence the conductivity of a semiconductor material.

6.7.5 The influence of temperature and radiation on the conductivity

Temperature and radiation influence the generation of electron–hole pairs in a semiconductor and in a p–n junction. The number of electron–hole pairs per unit volume determines the conductivity, σ. So we have to take into account two effects: radiation and temperature. Electron–hole pairs are also generated as a function of temperature when the thermal energy exceeds the bandgap energy. Thermal energy can be described by

$$E_T = kT > E_g \qquad (6.21)$$

where k is Boltzmann's constant $(1.38 \times 10^{-23}$ J K$^{-1})$ and T (K) is the absolute temperature. This influence of temperature can be expressed in general in the conductivity of a semiconductor and is given by

$$\sigma_T = q[n(T)\mu_n(T) + p(T)\mu_p(T)] \tag{6.22}$$

where σ_T (Ω^{-1} m^{-1}) is the conductivity. From this expression it can be found that both the mobility, μ, and the density of charge carriers, and hence the electron–hole pair density, are a function of temperature. Now for *intrinsic* semiconductor material, which means non-doped material, both terms contribute equally to the conductivity. For doped material one of the terms dominates over the other, so for p-type material the contribution of electrons may be neglected and vice versa for n-type material. The doped region often exceeds the number of charge carriers in the undoped region with a factor of at least 1000.

* * *

Example 6.3
An impressive demonstration of this temperature effect can be found in the manufacturing of glass tubes for TV monitors. At room temperature the bandgap distance of glass is larger than 3.5 eV and it is hence a perfect insulator. A TV (CRT) tube is made of two sections: a front plate and a tube. The front and the tube-end are moulded together by first heating both sections up to about 800 °C. Then the two pieces are assembled and two electrodes are connected to the weld area to force a large current through the heated area. The temperature will rise further and a perfect weld is obtained.

* * *

We have also seen that incident radiation on a semiconductor generates electron–hole pairs when the energy exceeds the bandgap distance. This also gives rise to an increase in conductivity and the same expression as (6.22) can be used to describe the increase in conductivity, or

$$\sigma(I) = q[n(I)\mu_n + p(I)\mu_p] \tag{6.23}$$

where $\sigma(I)$ is now the specific conductivity as a function of incident radiation, I. It is not quite clear whether the mobility is a function of incident radiation. However, when a p–n junction is subjected to incident radiation electron–hole pairs are generated as long as $\lambda = hc/E_g$. This gives rise to a photocurrent. When no correlation is assumed we may sum both contributions and find

$$\sigma(t) = \sigma(T) + \sigma(I). \tag{6.24}$$

Four other parameters are sometimes mentioned when these mechanisms are considered:

1 the photocurrent
2 the reflectance
3 the external quantum efficiency
4 the internal quantum efficiency.

We will give the applicable definitions.

The *photocurrent* is the current flowing in an external circuit connected to the photodiode as a direct result of photons incident on the active area of the photodiode. Any current flowing in the absence of incident photons is called the *dark current* and must be subtracted from the total current to determine the photocurrent.

The *reflectance* is the ratio of the rate at which the photons are reflected away from the active area to the rate at which they are incident on the same area.

The *external quantum efficiency* is the ratio of the number of electrons per second (photocurrent) to the rate at which photons are incident on the active area of the photodiode.

The *internal quantum efficiency* is the ratio of the photocurrent to the rate at which photons *enter* the active area of the photodiode.

Finally, we will mention two basic photoelectric effects. In particular the transverse photoelectric effect and a lateral photoeffect have to be considered, which are met in different applications.

The *transverse photoelectric effect* is the phenomenon that electron–hole pairs generated by incident radiation are separated by an internal electric field in the depletion region of a p–n junction. A current can flow from the front side towards the back side of the p–n junction when the right connections are made. A voltage difference is generated, ranging from 0.2 to 0.7 V per p–n junction.

The *lateral photoeffect* is characterized by a generated current across the surface of the p–n junction. With this mechanism position sensitive devices (PSDs) can be fabricated.

6.7.6 Radiant energy in silicon microtransducers

In this section we will give a comprehensive overview of optical semiconductor devices. Some basic optical components are summarized in table 6.16. We follow the classification as outlined in section 6.2 and table 6.4.

Table 6.16 Review of some basic optical components.

Description	Name	Application
Metal–semiconductor rectifying junction	Schottky barrier diode	Detector for UV and visible radiation
Photoconductor	Photoresistor	Detection
p–n junction	Photodiode or bipolar phototransistor	Detection
n-material–intrinsic–p-material	P–I–N diode	Optical communication
Avalanche effect	Avalanche photodiode	Optical communication
Organic (PTCDA) on inorganic semiconductor material	Diode structures	Photodetection
Metal–insulator–semiconductor junction	MIS structure, MOS structure, PMOS, NMOS or CMOS, charge transfer devices (CTDs)	MOS structures are basic for charge transfer components

MI [el, el, ra]

Photoconductor

A photoconductor in Si consists of a n-type diffused strip in an p-type substrate with ohmic contacts at both ends of the strip. An auxiliary electrical energy source is required to detect a change in conductivity due to radiation.

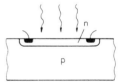

Photodiode

This device is a reverse-biased p–n junction in which electron–hole pairs are generated by incident radiation in or close to the depletion layer. The reverse current is a measure of the incident light intensity. A large variety of photodiodes is manufactured.

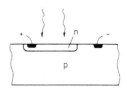

*　　*　　*

Example 6.4

To obtain an impression of used specifications, in table 6.17 an example is given of the technical specification of a photodiode.

Table 6.17   Specification of the BPW 21 photodiode.

Peak spectral response	560 nm
Wavelength range	460–750 nm
Power dissipation	250 mW
Sensitivity (short circuit)	$7\,\text{nA lx}^{-1}$
Open-circuit potential at $E = 1$ klx	280 mV
Dark current (at $V_r = 5$ V)	2 nA
Rise time ($I = 100\,\mu$A, $R_1 = 1$ kΩ)	$3.5\,\mu$s
Junction capacitance ($V_r = 5$ V)	170 pF

*　　*　　*

Phototransistor

This device can be considered as a photodiode connected in parallel with the base collector junction of a bipolar transistor. Electron–hole pairs are generated by incident light in or near the depletion layer. The reverse current in the base collector current is amplified by the transistor resulting in a large change of the collector current due to incident light variations.

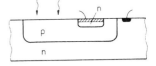

Photo-integrated circuit

Complete light-sensitive integrated circuits with electronics are manufactured in great variety.

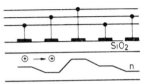

Photodiode arrays

These are arrangements of photodiode arrays in two dimensions. With an x and y line geometry in combination with MOSFET switches, such arrays are manufactured for optical character recognition.

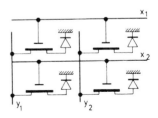

Charge-coupled devices (CCDs)

The devices consist of arrays of closely spaced metal electrodes on top of an oxidized silicon substrate. Minority carriers can be stored in potential wells beneath the electrodes or can be freely moved from one electrode to another electrode by application of suitable voltages to the electrodes.

MO [ra, ra, el]

Digital micromirror device (DMD)

This component consists of arrays of micromirrors 16 μm square. The mirrors are made of a reflective aluminium alloy suspended on two small torsion hinges above a silicon substrate. A voltage of 5 V underneath can tilt the small mirrors over 10° separately with the help of address lines (Texas Instruments). An alternative device is made by Berkeley Sensors & Actuators Center.

Optical switch

The modulator is made by suspending a silicon nitride thin-film membrane a controlled distance over the substrate. Applied electrostatic forces deflect the membrane toward the substrate and by bringing them in contact a reflection mirror is transformed into an antireflection mirror.

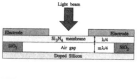

SI [ra, el, 00]

Photovoltaic effect

Radiant energy is converted into electrical energy without the need of an auxiliary energy source. Incident light falls on the depletion layer of a p–n junction and generates electron–hole pairs creating a potential difference across the p–n junction. Every type of radiant energy with a wavelength shorter than 1.1 μm provides enough energy to bridge the bandgap energy of 1.15 eV. For instance photons, γ radiation, x-rays, neutron radiation and so on produce the required energy. The theoretical maximum efficiency is about 18%.

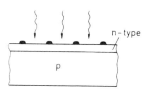

SO [el, ra, 00]

As discussed previously, in direct-bandgap materials, with conservation of momentum, photons can be emitted with high efficiency when recombination of electrically generated electron–hole pairs occurs. The electrical generation of electron–hole pairs can be realized by forward biasing of a p–n junction. Silicon is not a direct-bandgap material and therefore the emission of radiant recombination energy is most inefficient. Furthermore, the emission occurs in the infrared spectrum. Only at very high reverse-biased voltages across a silicon p–n junction is avalanche breakdown realized, producing white light with low efficiency.

6.7.7   Radiant energy in compatible technology

MI [el, el, ra]

Photoconductivity
For certain applications the spectral response of silicon is not the desired one. A customized spectral response can be realized by evaporating suitable materials on top of an oxide layer of SiO_2. For instance CdS with a wavelength of 0.7 μm or InSb with a wavelength of 7 μm are suitable materials. The change in conductivity can be detected with the underlying electronic circuitry.

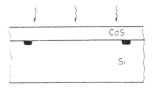

MO [ra, ra, el]

Passive liquid crystal displays (LCDs)
LCDs consist of two flat electrodes of indium oxide between two glass plates between which slightly ordered long molecules of for instance methoxy-benzol-butylaniline (MBBA) are placed. One electrode is made transparent and by means of rather small voltages the optical properties of the MBBA material can be changed. By applying suitable materials and doping, LCDs are made in several colours. A matrix of pixels is turned on and off by the voltage difference between a row line pulse and a column line pulse for that pixel.

Active matrix liquid crystal displays (AMLCDs)
These are based on the same principle as passive LCDs; however, the active colour display pixels are activated by a matrix of transistors manufactured in thin-film transistor technology (TFT).

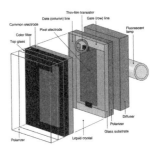

Plasma-addressed LCD (PALCD)

The transistors are replaced here by a matrix of gas plasma contained in channels in the display's glass backplate.

Colour plasma display panels (PDPs)

This is a full-colour plasma display panel. Sizes of 60 in have been built already. Monochrome displays use a matrix addressing scheme to ionize a gas mixture containing neon, and the neon glows with its characteristic red–orange colour. In colour PDPs, the gas mixture is replaced by one containing xenon, which emits in the ultraviolet region. The inner walls of the pixels are coated with red, green and blue phosphors that are sensitive to UV.

Electroluminescence technology

This technology offers colour displays based on the principle that when some phosphors are subjected to a large AC or DC voltage light is generated. For this technology the commercial prospects are modest.

Cathode ray tube (CRT)

This is the most classical one. A Japanese firm (Matsushita Electrical Industrial Co.), has made a flat panel (98 mm thickness) matrix-addressed CRT with almost 10 000 mini-CRTs in one panel. The required control electronics are very complex. This panel is commercially available, although rather expensive.

SI [ra, el, 00]

So far this effect has not been reported in compatible technology.

SO [el, ra, 00]

Light-emitting diodes (LEDs)

Light-emitting diodes are made of direct-bandgap materials such as GaAs with suitable dopants, for instance phosphorus, indium, caesium etc. The complete visible

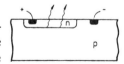

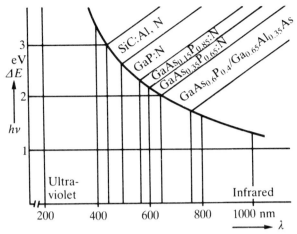

Figure 6.18 Recombination or direct-bandgap energy versus wavelength. The required composite material and epitaxial layer are also depicted.

spectrum can be covered including blue. This is illustrated in figure 6.18 in which the required recombination energy versus wavelength is depicted. In the same figure the composites are shown. The efficiency is high; up to 50% is reported. The above-mentioned materials can be combined with silicon technology; they are applied directly on top of the silicon wafer.

Injection laser
In a p–n junction of GaAs with polished surfaces perpendicular with respect to the junction an optical resonance cavity can be created. Because of the bandgap of GaAs of 1.35 eV the generated wavelength is about 9000 Å. Other suitable materials are GaP, GaSb, ZnS, PbS etc. The efficiency is about 75%.

A whole range of solid-state and gas lasers exists, covering a large bandwidth in the electromagnetic spectrum from UV to the far-infrared. For further reading reference is made to the literature.

6.7.8 Fibre technology
We will present in brief an overview of fibre technology. Basically, optical technology was invented for telecommunications purposes by Alexander Graham Bell as early as 1880. Optical fibre technology has made this viable on a commercial basis. Fibre optics can operate in extremely severe and explosion-prone circumstances.

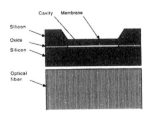

Table 6.18   Classification of optical fibre sensors.

Intensity sensors heterogeneous (AM)		Phase sensors, interferometers (FM)	
Extrinsic and hybrid	Intrinsic	Mach–Zehnder	Sagnac
Different type of field depending on design		Hydroacoustic Magnetic field Electric field Temperature Acceleration	Rotation Position
Industrial and medical applications		Military and industrial applications	

Table 6.19   Review of known physical effects in the radiant domain.

Name of effect	Notation	Macroscopic description
Photovoltaic	[ra, el, 00]	A voltage is generated by incident radiation at the junction of two dissimilar materials
Photomagnetoelectric	[ra, el, ma]	An electrical field is generated by both a magnetic field and incident radiation
Photoconductivity	[el, el, ra]	The increase of electrical conductivity of a material due to incident radiation
Photoelectric	[el, el, ra]	Electrons and holes are generated and separated in a junction area by incident radiation
Photodielectric	[el, el, ra]	The change of a dielectric constant and loss due to incident radiation
Laser	[el, el, ra]	An energy level population inversion is generated by incident radiation
Photoluminescence	[ra, ra, 00]	Radiant energy is emitted by incident radiation with shorter wavelength
Radioluminescence	[ra, ra, 00]	Visible radiant energy is emitted by incident x-rays or γ rays
Radiation heating	[ra, th, 00]	The increase of temperature of a material by incident radiation
Photomagnetic	[ra, ma, 00]	The change of magnetization by incident radiation
Photochemical	[ra, ch, 00]	The change of structure due to incident radiation
Electroluminescence (Destriau)	[el, ra, 00]	The illuminating excitation of a material due to an alternating electrical field
p–n luminescence (Lossev)	[el, ra, 00]	The radiation of recombination energy in a forward-biased p–n junction
Incandescence	[el, ra, 00]	The emission of radiation by thermal movement of atoms activated by an electric current
Kerr electro-optic	[ra, ra, el]	The generation of double refraction of radiation due to an electrical field
Kerr magneto-optic	[ra, ra, ma]	The change of a polarization plane of polarized radiation due to a magnetic field
Faraday	[ra, ra, ma]	As for the Kerr magneto-optic effect
Pockel's effect	[ra, ra, el]	The rotation of polarization of polarized radiation by an electrical field
Cotton–Mouton	[ra, ra, ma]	The generation of double refraction of radiation in a liquid due to a magnetic field

The cables are not sensitive to electromagnetic interference, and there is much less attenuation per unit of distance than in copper cables. Extremely large bandwidth is available in the range of terahertz. Therefore it is easy to multiplex a large number of sensors by simple amplitude modulation on one transmitting fibre. Phase-sensitive detection techniques allow measurements of one millionth of a fringe period.

Fibres for sensors are used in *single mode* and *multimode*. Applications are found in interferometry, fibre optical gyros, fibre optic thermometry and mechanical micro-bend applications. Basically, it can be said that all sensor effects can be produced with fibre optic sensors. Two interesting diagrams can illustrate the fibre optical sensor classification in some more detail, where a distinction is made between intensity sensors and phase sensors (table 6.18), which can be compared with AM and FM modulation. The diagram in figure 6.19 indicates the subdivision of the optical fibre intensity sensors group. A complete discussion of fibre optics would require an entire chapter. The interested reader is referred to the literature.

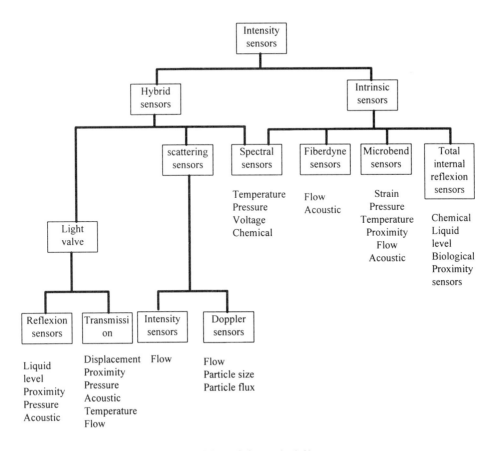

Figure 6.19 Subdivision of the optical fibre sensor group.

6.7.9 Review of known optical effects

In table 6.19 a review is presented of known physical effects on which all discussed photoelectric transducers are based. This concludes our discussion of the optical domain.

This concludes our discussion concerning the optical domain.

6.8 THE MECHANICAL ENERGY DOMAIN

Mechanical transducers are used for a large variety of applications and can be found in the determination of quantities such as

- position and (angle) displacement
- speed, acceleration
- weight and pressure
- surface flatness
- torsion, pull strength
- vibrations.

6.8.1 Physics

Numerous applications exist for measuring mechanical quantities, manufactured in a great variety of technologies. We will focus mainly on semiconductor and compatible technologies. A number of basic physical phenomena are used for mechanical transducers in semiconductor technology; in particular we will mention

- piezoresistivity effect
- piezojunction effect
- piezoelectric effect
- capacitive effect
- surface acoustic waves (SAWs)
- microresonators.

Focusing on silicon semiconductors, table 6.20 presents the main mechanical characteristics of silicon and some other materials used in semiconductor technology. Note the extremely high specific thermal conductivity of diamond, which is sometimes used for heat sinks for electronic devices in specific environments.

Table 6.20 Some mechanical properties of silicon compared with other materials used in the semiconductor industry.

Material	Strain force (10^9 N m^{-2})	Knoop hardness (kgf mm^{-2})	Young's modulus (10^{11} N m^{-2})	Specific density (kg m^{-3})	Specific thermal conductivity (W m^{-1} K^{-1})	Specific expansion (10^{-6} °C^{-1})	Melting point (°C)
Diamond	53	7000	10.35	3500	20 000	1	3900
Silicon carbide (SiC)	21	2480	7	3200	350	3.3	2760
Silicon nitride (Si$_3$N$_4$)	14	3486	3.85	3100	19	0.8	1870
Silicon	7	850	1.9	2300	157	2.33	1420
Aluminium	0.17	130	0.7	2700	236	25	659

Piezoresistivity

In 1954 it was discovered that Si and Ge show a piezoresistivity which is 100 times larger than in metals. We have noted already that silicon does not show piezoelectricity. For a change in resistivity of a conductor we can write

$$R'_\sigma = \frac{l'\rho}{S'} = \frac{l+\Delta l}{S+\Delta S}(\rho+\Delta\rho) \tag{6.25}$$

where R' (Ω) is the resistance under pressure, l (m) the length of the wire, ρ ($\Omega\,$m) the specific resistivity and S (m^2) the cross-section of the wire. This can be written in the form of a normalized differential for R. We then find for dR/R

$$\frac{dR}{R} = \frac{d\rho}{\rho} + \frac{dl}{l} - \frac{dS}{S}. \tag{6.26}$$

From this expression it is seen that the change in resistivity is a function of three different parameters. The temperature is not taken into account, but in practice a kind of temperature compensation will be applied.

In this respect another important factor in elasticity is the *Poisson ratio*, which is defined as the ratio of the relative change in diameter and the relative change in length, or

$$v = \frac{dD/D}{dl/l} = \frac{\varepsilon_D}{\varepsilon_L} \tag{6.27}$$

where v is the Poisson factor, ε_D the relative change in diameter and ε_L the relative change in length (specific strain). For strain gauges a type of *gauge factor K* is defined as

$$K = (dR/R)/\varepsilon_L \tag{6.28}$$

or

$$K = \frac{dR/R}{\varepsilon_L} = 1 + 2v + \frac{d\rho/\rho}{\varepsilon_L}. \tag{6.29}$$

Hence there is change in resistance due to a length change, a change in the cross-section and a change in piezoresistance. The gauge factor can also be understood as the ratio of the relative change in resistance and the relative change in length. Often the resistivity, ρ, is considered a constant and the gauge factor then can be written

$$K = 1 + 2v. \tag{6.30}$$

A review of K factors for silicon is given in table 6.21. For metals the K factor is about 2. Note that n-type silicon shows a negative K factor. Silicon is a very brittle material which is the reason that silicon is not used in strain gauges. Usually, with the help of diffusion or ion implantation, resistors are made in silicon and then fabricated as membranes for pressure transducers. Other disadvantages for silicon

Table 6.21 *K* factors for silicon in different crystal orientations.

Crystal orientation	n-type Si *K* factor	p-type Si *K* factor
[111]	−13	173
[110]	−89	121
[100]	−153	5

are the high temperature coefficient, especially when low-doped silicon is used, and the limited temperature range which can be applied.

Piezojunction effect

Another possibility for transducers is the use of the piezojunction effect. When a p–n junction is subjected to a mechanical strain, then a large change in the *I–V* characteristic will occur. When the current is kept constant then with increasing strain the voltage is reduced (figure 6.20). A mechanical stress reduces the bandgap distance and increases the diffusion length. Real applications are not found owing to the sensitivity to mechanical vibrations.

Piezoelectricity

Piezoelectricity is a reversible effect and is not found in silicon or germanium. Reversible means that when a mechanical force is applied an electrical tension is generated and vice versa: when a voltage is connected to that material a mechanical change can be observed. Piezoelectricity can only be observed under two conditions: that in the material considered no crystal symmetry is present, and ionic bonding dominates over covalent bonding. In other words, when piezoelectricity is present in materials there is no centre of symmetry (table 6.22).

Layers of CdS or ZnO are deposited by low-pressure chemical vapour deposition (LPCVD) on silicon substrates, which offers the possibility for making pressure transducers which show piezoelectricity. A structure designated as PI-DMOS is shown in figure 6.45 and is based on a MOSFET structure.

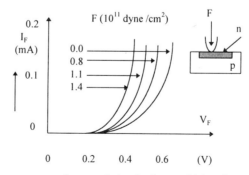

Figure 6.20 Voltage–current characteristic of a forward-biased p–n junction as a function of mechanical force.

Table 6.22 Review of some materials which show piezoelectricity (acentric) and some which show only piezoresistivity.

Crystal structure	I	II	III	IV	V	VI	VII
Centrosymmetric				Si^{4+}			
Centrosymmetric				Ge^{4+}			
Acentric			Ga^{3+}		As^{3-}		
Acentric		Cd^{2+}				S^{2-}	
Acentric		Zn^{2+}				O^{2-}	
Centrosymmetric	Na^+						Cl^-

Capacitive effect

Excellent capacitive transducers can be made because micromachining gives the possibility to manufacture 3D structures. A change in capacitance is measured as a function of stress. This is described by

$$C = \varepsilon_r \, \varepsilon_0 \, \frac{A}{d} \qquad (6.31)$$

where ε_r is the relative permittivity, ε_0 ($8.854\,188 \times 10^{-12}$ F m^{-1}) the permittivity of vacuum, d (m) the distance between the plates and A (m^2) the area of the plates. Materials used are poly-Si, SiO_2 or Si_3N_4.

Surface acoustic waves

Surface acoustic wave (SAW) transducers are able to convert an electrical signal at the input into an acoustic signal and at the output to convert it again into an electrical signal. The generated waves can propagate at a more or less elastic surface. The amplitude is attenuated exponentially with the penetration depth. For this type of transducer the only relevant type of wave is a wave in a Rayleigh mode. The waves are characterized by their specific mode, where their mechanically and electrically equivalent components lie in one plane perpendicular with respect to the surface. Characteristics of Rayleigh waves are as follows:

1. waves can be modulated at the surface
2. the technology for SAWs is compatible with planar technology
3. bandwidth ranges from 20 MHz to 2 GHz
4. a piezoelectric substrate is required to generate Rayleigh waves.

Commonly used materials are STX and YZ lithium niobate. ST and T represent specific crystal orientations and X and Z indicate the direction of propagation. Here, S stands for surface crystal orientation (this is expressed in the so-called Euler angle), and T stands for the transversal crystal orientation (this is also expressed in the Euler angle). ST combines these factors.

In a SAW transducer, phase velocity and amplitude are determined by the following substrate specifications: the elasticity; the piezoelectricity; the dielectric constant; the conductivity; mass of the substrate. When one of these characteristics can be modulated a specific transducer is created. A very common application is the SAW used as an oscillator. The SAW can then be used as a delay line or as a resonance

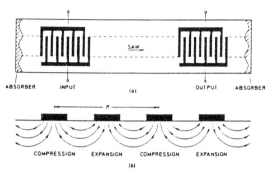

Figure 6.21 Diagram showing the principle with direction of propagation and nodes of compression and expansion.

circuit. Both phase and amplitude can be measured accurately, and recently SAWs have been used as chemical transducers. Note here again that SAWs cannot be manufactured in silicon technology. See figure 6.21.

Silicon microresonators

A resonant microsensor, or microresonator, is a device with a mechanical element at resonating frequency of which the resonance frequency changes its output as a function of a physical or chemical parameter. The conversion from measured quantity to the change in frequency can be accomplished by a change in mass, stress or form of the resonator. Micromachining has made these resonating structures possible. What are used are mechanical cantilevers, micro tuning forks and diaphragms.

A *cantilever* is an elastic tongue or elastic beam suspended at one side (figure 6.22). In practice these beams can be fabricated in silicon and in GaAs with the help of micromachining. On these beams resistors can be integrated by diffusion or ion implantation. A mechanical force applied to this beam will change the resistivity in a reproducible fashion as long as the deformation is acting according to Hooke's law, or

$$\sigma = \varepsilon_L E \qquad (6.32)$$

where σ (N m^{-2}) is the stress per unit area, ε_L the specific expansion coefficient and E (Pa or N m^{-2}) the modulus of elasticity.

Sometimes an array of such cantilevers is used to make an array of resonating tongues. For each cantilever the resonating frequency can be described by

$$f = 0.16 \frac{t}{l^2} \left(\frac{E}{\rho} \right)^{1/2} \qquad (6.33)$$

where f (Hz) is the resonance frequency, t (m) the thickness of the tongue, E (Pa) the modulus of elasticity, ρ (kg m^{-3}) the specific mass and l (m) the length of the tongue. Other possibilities are single or double suspended microbridges and diaphragms. The principle of a microresonator is a type of feedback loop and is

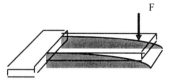

Figure 6.22 Loaded cantilever subjected to a mechanical force.

depicted in figure 6.23. In this diagram we distinguish a vibration excitation unit, a vibrating structure and a vibration detector.

Excitation and detection can be performed in six different ways.

1. *Electrostatic excitation* is based on two electrodes in close proximity to each other. The capacitive changes of this structure are used for detection.
2. *Dielectric excitation* uses a sandwich structure; the detection is also capacitive.
3. *Piezoelectric excitation* uses the built-in dipole charges of a piezoelectric material. The material is deformed when a voltage is applied. This principle can also be used for detection.
4. *Resistive heating excitation* is accomplished by heat dissipation in an integrated diffused resistor; the detection can be performed by the change in piezoresistivity.
5. *Optical heating excitation* is used with a periodically activated laser focused on the resonator. The piezoresistivity is used for detection.
6. *Magnetic excitation* uses the force resulting from the interaction between the electrical current through the structure and a magnetic field. The reverse effect is used for detection.

Temperature stability is a most important parameter and must be as low as possible. It is interesting to check the dimensions for the frequency in (6.33). For diaphragms only bending is allowed within Hooke's law, and hence permanent distortion is not allowed. These membranes can be fabricated with a thickness of a few micrometres.

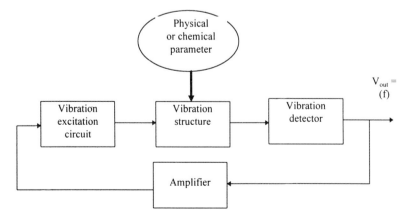

Figure 6.23 Circuit for excitation and detection of a microbridge resonator.

Note that quartz-based resonators have been well known for a long time and can be used for thickness monitoring. However, many other commercial applications are realized already.

In the following sections we will give an overview of known transducers in silicon and in compatible technology.

6.8.2 Mechanical energy in silicon microtransducers

MI [el, el, me]

Piezoresistor

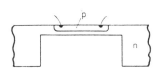

In response to strain a diffused resistor in silicon changes its value. Silicon is therefore used for a large variety of strain gauges. The temperature dependence of silicon strain gauges is rather high. For the measurement of pressures a very thin n-type silicon layer in which four p-type doped resistors are diffused in a Wheatstone bridge configuration is applied as diaphragm.

Piezojunction

When a silicon p–n junction is stressed the current-to-voltage characteristics change. This is caused by the piezoresistive effect and bandgap changes if a mechanical load is applied.

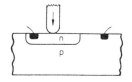

Piezotransistor

A transistor shows the same strain sensitivity as a piezojunction, but as well as this effect it shows a current gain factor which is also strain sensitive. For MOSFETs the same piezoresistivity is reported.

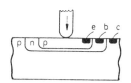

Lateral photoeffect

When a light spot illuminates a silicon solar cell a voltage is generated across the ohmic contacts as a function of the position of the light spot. With this effect large two-dimensional light spot sensitive photodiodes can be made, so-called position sensitive devices (PSDs).

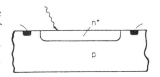

Integrated resistive pressure transducer

A complete Wheatstone bridge combined with temperature corrective circuitry and signal conditioning circuitry is integrated on a silicon chip. The piezoresistive effect is then applied. Very sophisticated etching techniques making strain bars in silicon are applied.

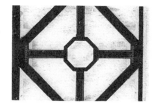

Accelerometer
The sensor consists of a seismic mass with a cantilever beam supported at one side. The piezoresistive device handles linear acceleration with primarily AC response. These accelerometers are being made with three-axis detection characteristics.

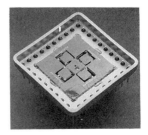

Integrated transistor pressure transducer
A couple of transistors can be used in which one is subjected to stress by a diamond stylus on the emitter surface. If the required electronic circuitry is placed on the same chip, for instance current sources, operational amplifiers and so on, fully integrated pressure transducers are manufactured.

Electret microphone
In silicon an electret microphone is developed, in which a permanent polarized electret is built in, acting as the auxiliary energy source.

Microresonators
This is the collective name for the use of silicon macromachined mechanical resonating microstructures to detect a change of frequency as a function of a physical or chemical parameter. Good examples are accelerometers and pressure sensors.

MO [el, me, el]
So far the electrical modulation of mechanical energy in silicon has not been observed.

SI [me, el, 00]
Silicon does not show the *piezoelectric* effect because of the symmetry in the lattice of silicon. So in silicon no self-generating transducer is available. The piezoelectric effect requires an asymmetric centre of inversion symmetry.

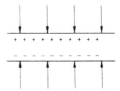

SO [el, me, 00]
So far this effect has not been reported in silicon technology.

6.8.3 Mechanical energy in compatible technology

MI [el, el, me]

Piezoresistor
A great variety of materials are used to manufacture pressure sensitive diaphragms or wires. Composites of platinum–tungsten are reported in a great number of forms. Thin films, wires and thin membranes are used.

Piezo field effect transistor (PI-FET)
The PI-FET is made by depositing a piezoelectric material, for instance ZnO or CdS, on top of a silicon wafer. Very sensitive pressure transducers can be made when the piezoelectric material is deposited between the channel of a MOSFET and the gate electrode. The piezoelectric layer is polarized by applying a mechanical strain and the resulting electric field modulates the channel current. By integration techniques more electronic circuitry can be added. The frequency behaviour is excellent and ranges from 0 to 20 MHz. Basically here a tandem transducer is involved, because the first effect is piezoelectric, and this effect is applied to modulate a MOSFET.

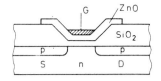

Surface acoustic wave (SAW) filters
When a piezoelectric material such as CdS or ZnO is deposited on a silicon wafer or chip a surface acoustic wave filter can be manufactured. These filters can be combined with other electronic circuitry, resulting in very versatile filters for microwave frequencies up to the gigahertz region. See also figure 6.21.

A capacitive tactile sensor
An array of small-capacitance pressure sensors is integrated on top of a silicon substrate in compatible technology. Using a capacitance-to-phase converter a resolution of 1 fF is obtained. Each taxel is addressed by activating a row, a column and an enable line.

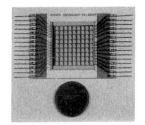

MO [el, me, el]
So far this effect has not been reported in a compatible technology.

SI [me, el, 00]

Piezoelectric field effect transducer (PI-FET)
This effect is combined in a compatible technology with the piezo field effect transistor.

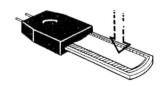

Piezoelectric polymer film

These piezoelectric polymer films are made from polyvinylidene fluoride (PVDF). The material shows a strong piezoelectric and a strong pyroelectric effect. Pyroelectric materials show a strong temperature sensitivity. A large variety of input transducers can be made. The material is also known as *Kynar piezo* film. The piezoelectric coefficient is 15 pC N^{-1} or 14.4 V N^{-1}. The pyroelectric coefficient is 8 V K^{-1}.

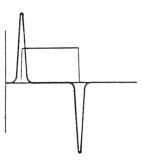

SO *[el, me, 00]*

Because the above-described material shows a piezoelectric effect the material shows also a reversible effect. Hence, an electric voltage change connected to this material gives a mechanical change in dimensions. For instance, loudspeakers can be manufactured from this material.

6.8.4 Review of effects in the mechanical energy domain

A summary is provided in table 6.23. This concludes our discussion about mechanical transducers.

6.9 THE THERMAL ENERGY DOMAIN

We start again with some physics for thermal transducers. We will mainly focus on semiconductors and compatible technologies; however, temperature is one of the oldest and most frequently measured parameters. In power stations thermal energy is converted into electrical energy. Here we mean that thermal information is converted into electrical information.

6.9.1 Physics

Again, firstly we have to look for the physical mechanisms available to make transducers in semiconductor technology. The following basic mechanisms are relevant for our discussion: the Seebeck effect; Peltier effect; Thomson effect; thermoresistance. Then we will discuss in brief the same temperature effects in semiconductors.

The Seebeck effect

The Seebeck effect describes the generation of an electrical voltage due to a temperature difference between a weld of two different materials and the other side of these two wires. This is illustrated in figure 6.24. A voltage difference is only measured when $T_1 \neq T_2$. The Seebeck effect can be described by

$$\Delta V = \alpha_S \Delta T \qquad (6.34)$$

Table 6.23 Review of known physical effects in the mechanical domain.

Name of effect	Notation	Macroscopic description
Piezoelectric	[me, el, 00]	The generation of a surface charge due to a mechanical force
Acoustoelectric	[me, el, 00]	The generation of an electric current by a travelling longitudinal acoustic wave
Triboelectric	[me, el, 00]	The generation of positive or negative surface charges due to rubbing of materials
Piezoresistance	[el, el, me]	The change in conductivity in semiconductors due to a mechanical force
Dember effect	[ra, el, me]	The generation of a voltage across two regions in a semiconductor due to radiation
Lateral photovoltaic	[ra, el, me]	The generation of a lateral voltage across a junction by radiation (PSD)
Lateral photoelectric	[ra, el, me]	The generation of a geometric current division across two contacts at one side of a reverse-biased p–n junction by radiation (PSD)
Thermoelastic	[th, el, me]	The generation of a voltage in two regions of a metal due to a mechanical strain and a temperature difference in the regions
Magnetostriction	[me, ma, 00]	The change in magnetization by a mechanical force
Piezo-optic	[ra, ra, me]	The change in refractive index due to a mechanical force
Photoelastic	[ra, ra, me]	The generation of double refraction by a mechanical force
Magnetostriction	[ma, me, 00]	The mechanical deformation of a material by a magnetic field

where ΔV (V) is the generated voltage difference, α_S (μV K^{-1}) the Seebeck coefficient and ΔT (K) the absolute temperature difference. Basically, the Seebeck effect is a macroscopic description of a number of effects generated when a temperature gradient is present; in particular, the Seebeck coefficient describes a bulk effect, determined by the following effects.

- The temperature difference generates a difference in Fermi level with respect to both sides.
- The bandgap distance changes as a function of temperature.
- The gradient of charge carriers changes for n and p type as a function of temperature.
- The diffusion coefficient is a function of temperature.

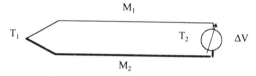

Figure 6.24 Illustration of the Seebeck effect for two different metals M$_1$ and M$_2$.

- Charge carriers move from the heated side to the cold side; this is called thermodiffusion.
- A temperature gradient gives rise to charge carrier transport and thermal transport of phonons. This is called phonon drag.
- An electric field will be generated owing to this transport of charge carriers.

The Seebeck effect is used mainly for the manufacture of thermocouples with which temperature measurements can be performed ranging from -200 to $+1600$ °C. For thermocouples calibration procedures are always required because strict linearity is not present.

Peltier effect

The Peltier effect can be considered as the reverse effect of the Seebeck effect. In 1834 Peltier discovered that when a current flows through a junction of two different metals heat is dissipated or absorbed towards or from the environment. The reason is that a current flow is connected to a heat flow which differs for different materials. Right on the spot of the weld this difference of heat flow can only be balanced when an exchange of heat with the environment occurs. It can be expressed in the following relationship

$$Q = -\Pi_{ab} J_{ab} \tag{6.35}$$

where Q (J m^{-2}) is the heat dissipated or absorbed, Π_{ab} (J C^{-1}) the Peltier coefficient for a junction with materials a and b and J_{ab} (C m^{-2}) the charge carrier density flowing from a to b. Under certain conditions the following relationship for the Peltier coefficient holds:

$$\Pi_{ab} = \alpha_S T. \tag{6.36}$$

The dimensions are consistent. This expression is known as the first Kelvin relation.

The Thomson effect

The Thomson effect describes the mechanism that a current flowing in a wire in which a temperature gradient is present shows a heat exchange with its environment. The effect can be described by

$$Q_{th} = \gamma \boldsymbol{J} \cdot \nabla T \tag{6.37}$$

where Q_{th} (W m^{-2}) is the heat flow, γ (V K^{-1}) is the Thomson coefficient, $\boldsymbol{J}$ (A m^{-2}) is the current density and T (K) is the temperature. Kelvin proved the following relationship between the Seebeck coefficient and the Thomson coefficient:

$$\gamma = T \frac{\partial \alpha_S}{\partial T}. \tag{6.38}$$

Thermoresistance

A macroscopic description for the temperature dependence of resistors can be given as

$$R(T) = R(0)(1 + AT + BT^2) \tag{6.39}$$

where $R(T)$ is the resistance at temperature T, $R(0)$ the resistance at $T = -273$ K, A (K^{-1}) a temperature coefficient and B (K^{-2}) a temperature coefficient. In most metals higher-order terms, such as the B term, may be neglected. The resistivity is expressed by the relation discussed already in section 6.7.5, (6.22). For convenience (6.22) is repeated here in short form, with the condition that the contribution of electrons dominates:

$$1/\rho = \sigma = nq\mu \tag{6.40}$$

where ρ (Ω m) is the specific resistivity, σ ($\Omega^{-1}\,m^{-1}$) is the specific conductivity, n (m^{-3}) is the number of charge carriers per unit volume, q (1.6×10^{-19} C) is the specific charge and μ ($m^2\,V^{-1}\,s^{-1}$) is the electron mobility. In general for metals the resistivity is proportional to the absolute temperature.

6.9.2 Thermal effects in semiconductors

The Seebeck effect in semiconductors
In semiconductor material a number of effects determine the Seebeck effect. In addition, a distinction must be made between p-type and n-type material. This gives the following relations:

$$\alpha_S|_n = -\frac{k}{q}\left(\ln \frac{N_c}{n} + 2.5 + s_n + \phi_n \right) \tag{6.41}$$

$$\alpha_S|_p = -\frac{k}{q}\left(\ln \frac{N_v}{p} + 2.5 + s_p + \phi_p \right) \tag{6.42}$$

where k is Boltzmann's constant (1.38×10^{-23} J K^{-1}), q the elementary charge, N_c (m^{-3}) the state density of the conduction band, N_v (m^{-3}) the state density of the valence band, n (m^{-3}) the doping concentration of electrons, p (m^{-3}) the doping concentration of holes, $s_{n,p}$ an exponent describing the relationship between relaxation time τ and charge carrier energy and $\phi_{n,p}$ the contribution of phonon drag. In practice the Seebeck coefficient is often represented as a function of the specific resistivity:

$$i\alpha_S = \frac{mk}{q}\ln\frac{\rho}{\rho_0} \tag{6.43}$$

where $\rho_0 \approx 5 \times 10^{-6}$ Ω m and $m \approx 2$.

The Peltier effect

For the Peltier effect in semiconductors it suffices to refer to the previous section in which it was discussed.

The resistivity of semiconductors

Here also reference can be made to the previous discussion. A distinction must be made between *intrinsic* or non-doped material and *extrinsic* or doped material. In doped semiconductors one of the energy carriers will always dominate by a factor of at least 1000. The expression then simplifies to one term only. However, the description is rather complicated and will not be discussed here in more detail. For intrinsic semiconductors we find again the relation discussed in section 6.7.5, (6.22):

$$1/\rho = \sigma = n_i q(\mu_n + \mu_p) \tag{6.44}$$

where we have seen that the mobility is also a function of temperature. The resistivity shows a strong temperature dependence.

Thermistors

The noun thermistor is an abbreviation of thermally sensitive resistor. All thermistors are composed of a sintered ceramic semiconductor and metal oxides, such as of manganese, cobalt, copper or iron. The resistance of a thermistor can be described by

$$\rho(T) = \rho(T_0) \exp[-B(1/T - 1/T_0)] \tag{6.45}$$

where $\rho(T_0)$ is the resistivity at $T = T_0$ and B a constant, with order of magnitude 4000 K. The behaviour is exponential and shows a negative temperature coefficient (NTC). The temperature coefficient is defined as:

$$\alpha = d\rho/\rho \, dT = -B/T^2 \tag{6.46}$$

An exact description is complicated and usually two graphs are presented (figure 6.25), one for thermistors with a positive temperature coefficient (PTC) and one for thermistors with an NTC.

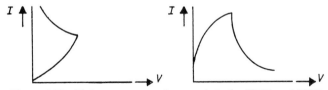

Figure 6.25 Voltage–current characteristic for NTC and PTC.

In the following sections we will present a rather comprehensive overview of transducers in semiconductor and compatible technologies. The development of thermal sensors is growing rapidly and conference proceedings should always be consulted for best results in the state of the art.

6.9.3 Thermal energy in silicon microtransducers

MI [el, el, th]

Thermoresistor

In intrinsic (not doped) semiconductor material the temperature coefficient for conductivity is very large and shows an exponential behaviour owing to an exponential increase in charge carrier density of the electron–hole pairs. This material can be used in temperature measurement applications. Extrinsic (doped) semiconductor

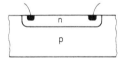

material shows a much less pronounced temperature coefficient for conductivity.

Thermodiode
Reverse- as well as forward-biased diodes show a temperature-dependent behaviour. The change is rather small and appears to be about $-2\,\mathrm{mV\,°C^{-1}}$ for a forward-biased diode.

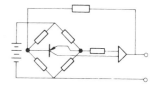

Thermotransistor
The base–emitter voltage of a transistor appears to be a linear function of temperature if the emitter current is maintained constant. Compared with the diode a considerable increase in sensitivity is obtained ranging from $10\,\mathrm{mV\,°C^{-1}}$ to $200\,\mathrm{mV\,°C^{-1}}$.

Integrated temperature sensor
Very complete sophisticated integrated temperature sensors are manufactured by applying for instance a type of current mirror configuration of two matched transistors as depicted. The difference between the base–emitter voltages at different collector currents is proportional to the absolute temperature. An integrated Zener voltage reference and built-in operational amplifiers provide a transducer applicable for the temperature range between $-25\,°C$ and $+85\,°C$, offering a linear output of $10\,\mathrm{mV\,°C^{-1}}$.

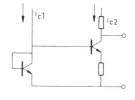

Integrated silicon flow meter
A flow meter can be manufactured by applying a Wheatstone bridge configuration composed of four p-type ion-implanted resistors in an n-type silicon substrate. The four resistors have a large length-to-width ratio of about 85 and are integrated in a square configuration. The bridge current heats the chip well above ambient temperature. When the chip is placed plane-parallel in the airflow the cooling effect of the four resistors is orientation dependent with respect to the direction of the airflow. The signal due to any thermal imbalance that develops is a measure of the flow velocity. The same principle can be applied for a transistor configuration showing more sensitivity but less linearity and stability. An improved version of the resistor configuration applying a double-bridge configuration with thermal feedback offers excellent stability. This principle has been discussed in more detail in section 3.6.

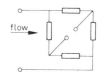

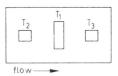

MO [el, th, el]

Heating transistor array
The conversion of electrical energy into thermal energy is very simple indeed. The effect is caused by the interactions of passing current carriers with the lattice. The same effects can be found in diodes and transistors.

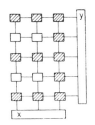

SI [th, el, 00]

Thermopiles
A voltage difference can be generated across differently doped silicon structures joined together when a temperature difference between the two regions is present. This is called the Seebeck effect. When a closed loop is made current will flow. When a series connection of these differently doped regions is made very sensitive thermopiles can be made suitable for long-range thermal measurements. The photograph shows a $Si_{70\%}Ge_{30\%}$ thermopile with a Seebeck coefficient of $75 \, \mu V \, K^{-1}$.

SO [el, th, 00]

Thermal matrix
This effect is well known and has been described already under 'Modulating output' transducers, where a simple resistor can also act as a self-generating output transducer and a transistor in principle can be used as a modulating output transducer. A matrix of such components in combination with heat-sensitive paper can be used to make a thermal matrix dot printer. The addressing electronics can be integrated on the same chip.

6.9.4 Thermal energy in compatible technology

A considerable variety of thermal energy transducers can be made in compatible technology, because it is very easy to deposit a metal layer onto a silicon wafer. With suitable etching techniques, very fine, accurate and stable resistance networks can be fabricated. For this purpose an Ni–Fe alloy is often used. The limiting factor is of course the maximum allowed temperature for silicon before degradation occurs, which will happen above 175 °C.

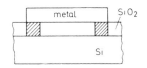

For families of thermocouples reference is made to table 6.24. We mention a few other temperature transducers which are compatible with semiconductor technology.

MI |el, el, th|

Resistive temperature-dependent (RTD) devices
A layer of an alloy of Ni–Fe is deposited onto a silicon substrate. With the help of macromachining, tiny, stable and accurately defined resistor structures can be etched. A maximum operating temperature of 200 °C is allowed.

Thermistors
Thermistors are sintered composites and show a strong temperature-dependent non-linear behaviour. They can be obtained with a positive (PTC) and negative (NTC) temperature coefficient. Applied housings differ considerably.

MO

Heating transistor arrays
These devices have already been discussed and can also be made in compatible technology, for instance for thermal printers.

SI [th, el, 00]

Thermocouples
Thermocouples can be made in thin- and thick-film technology. An overview of frequently used thermocouple materials is given in table 6.24.

Polymers
The pyro effect of some polymers has already been discussed in the section on radiant energy.

Table 6.24 Review of some important thermocouple materials.

Type of material	Sensitivity (μV K^{-1})	Range (°C)	Inaccuracy (% fractional error)
Platinum rhodium–platinum (R&S)	6	0 to +1500	0.25
Chromel–alumel	40	−200 to +1300	0.75
Copper–constantan	42.8	−200 to +350	0.5–0.75
Fe–constantan	54	−150 to +1000	1.0
Chromel–constantan	60	−125 to +800	1.0

SO

Peltier elements

These elements exist and are commercially available for cooling and heating purposes based on the Peltier principle. Specifications are, for instance, a heat pump capacity of 70 W and a maximum temperature difference of 65 °C.

6.9.5 Review of effects in the thermal energy domain

These are summarized in table 6.25. This concludes our discussion about thermal transducers.

Table 6.25 Review of thermal physical mechanisms used for transducers.

Name of effect	Notation	Macroscopic description
Thermoelectric, Seebeck	[th, el, 00]	The generation of an electric current in a closed loop of two dissimilar conductors by different junction temperatures
Pyroelectric	[th, el, 00]	The change in polarization due to a temperature change
Nernst	[th, el, ma]	The generation of an electric field due to a temperature gradient and a magnetic field which are mutually perpendicular
Thermodielectric	[el, el, th]	The change of permittivity of a ferroelectric due to temperature
Thermoconductivity	[el, el, th]	The change in conductivity due to temperature
Superconductivity	[el, el, th]	The conductivity is infinite below a critical specific temperature, T_c
Thermoelastic	[me, el, th]	The generation of a voltage between two regions in a metal due to mechanical strain and their temperature difference
Incandescence	[th, ra, 00]	The emission of radiant energy when a material is heated
Thermoluminescence	[th, ra, 00]	The emission of radiant energy of certain crystals due to temperature
Curie temperature	[th, ma, 00]	The change to paramagnetism of ferromagnetic material at specified temperature
Néel temperature	[th, ma, 00]	As for the Curie temperature but for ferrimagnetic material
Thermochemical	[th, el, 00]	The change of structure due to temperature
Electrothermal	[el, th, 00]	The generation of heat in a conductor by electric current
Peltier	[el, th, 00]	The generation of a temperature difference between two junctions when a current passes through them

6.10 THE MAGNETIC ENERGY DOMAIN

In this section we will discuss magnetic transducers in silicon and compatible technologies. We start again with some basic physics to understand the working principles of magnetic transducers. In particular we will discuss one of the most advanced smart magnetic transducers, the Hall element available as a complete integrated transducer.

6.10.1 Physics

In this section we will discuss the following topics:

- the cause of magnetism
- superconductivity
- the Hall effect in conductors
- the Hall effect in semiconductors
- magnetoresistivity.

Magnetism is one of the fundamental forces we find in nature and in many technical applications. Here we will focus on transduction of information into and from the magnetic energy domain. For instance, in electrical power stations we find applications for energy conversion. In general we can distinguish four types of magnetism.

Table 6.26 Survey of types of magnetism and related forces.

Type of magnetic material	Examples	Resulting force as a function of current in a magnetic field with gradient (N)	Resulting magnetic dipole moment when $H_{uitw} = 0$ A m^{-1}
Diamagnetic	Cu, salt, water	F quadratic (I^2); repulsive	0
Paramagnetic	Al, Na, liquid oxygen	F quadratic (I^2)	$\neq 0$
Ferromagnetic	Ni, Fe	F linear (I)	$\gg 0$
Ferrimagnetic	Fe$_3$O$_4$	F linear (I)	$\gg 0$

This is summarized in table 6.26. In general forces are exerted on moving charge carriers in a magnetic field which are described by the Lorentz force law

$$F = -q(v \times B) \qquad (6.47)$$

where q is the elementary charge of an electron (1.6×10^{-19} C), v is the electron velocity and B is magnetic induction. Note that a vector product is implied here. From table 6.26 it can be seen that above a certain strength of the magnetic field the force is quadratic with the current. For ferri- and ferromagnetic materials the force is linear with the current. On a microscopic (atomic) scale we distinguish three different possible contributions from the moving charges to the magnetic field:

1 the *magnetic orbital moment* due to the movement of the electrons in their orbitals
2 the *magnetic spin moment* due to the fact that electrons are spinning around their axes, the so-called Bohr magnetons
3 the *magnetic core moment*, because the core, with its positive charge, also spins around an axis.

In diamagnetic materials the net magnetic moment is zero and the force acting in a strong magnetic field is repulsive. For paramagnetic materials a net magnetic moment can result and for these materials the so-called *Langevin* function is valid. The

Langevin function is

$$M = Np \left[\coth(x) - \frac{1}{x} \right] \tag{6.48}$$

with

$$x = \frac{\mu_0 p H}{kT} \tag{6.49}$$

where M (A m^{-1} or A m^2 m^{-3}) is the total magnetic moment per unit volume, N (m^{-3}) the total number of magnetic dipoles per unit volume, p (A m^2) the magnetic dipole moment of the atom, μ_0 (1.256 637 × 10^{-6} H m^{-1} or V s A^{-1} m^{-1}) the permeability of vacuum, k the Boltzmann constant (1.38 × 10^{-23} J K^{-1}) and T (K) the absolute temperature. This behaviour is illustrated in figure 6.26. Above a certain critical temperature, the so-called Curie temperature, ferromagnetism disappears and the material becomes paramagnetic.

Remnant magnetism occurs in ferromagnetic materials and it is the phenomenon that there is still magnetism at zero external magnetic field. In ferromagnetic materials a kind of Weiss field H_{weiss} is present, which is similar to a kind of molecular magnetic field located in so-called magnetic domains. These domains can be oriented in parallel permanently. The internal Weiss field is described by $H_{\text{weiss}} = \lambda M$ and is added to the magnetic field. Then (6.49) can be written as

$$x = \frac{\mu_0 p H_a + \lambda M}{kT} \tag{6.50}$$

where H_a is the external applied field. The ratio M/H is called the magnetic susceptibility and represents the resulting magnetization caused by the external applied field strength, H_a. In a ferromagnetic material at or above the *Curie temperature*, T_C, the magnetic polarization is completely randomized, and magnetism will vanish. This temperature is specific for a given material. For instance, this mechanism is applied in soldering irons to maintain a constant temperature. Ferrimagnetic materials are of less interest for transducers and therefore will not be discussed here.

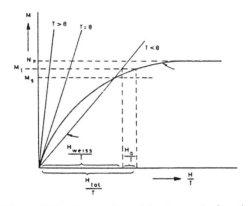

Figure 6.26 Illustration of the Langevin function.

Superconductivity

In this respect we also want to discuss the phenomenon of superconductivity. This is characterized by an infinite conductivity below a certain temperature. However, a critical field strength also exists at which the magnetic polarization vanishes. Certain transducers are also based on this principle. This can be described by

$$H_c(T) = H_0\left(1 - \frac{T^2}{T_c^2}\right)$$ (6.51)

where $H_c(T)$ (A m^{-1}) is the critical field strength at which magnetic polarization disappears, H_0 the critical field strength at $T = 0$ K, T (K) the temperature of the material and T_c (K) the critical temperature at which superconductivity disappears.

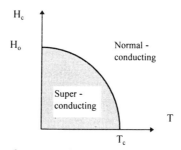

Figure 6.27 Illustration of superconductivity as a function of temperature and field strength.

This function is illustrated in figure 6.27. Many materials show superconductivity, in particular certain alloys. It is noticed that in another type of superconducting material the graph is smoothed to a large extent with regard to temperature. These are called type II superconductors and they can be used for making superconducting coils. Recently, in 1986, superconductivity was discovered in certain sintered alloys at the temperature of liquid nitrogen. So far, commercial applications for these new high-temperature superconductors have not been reported.

6.10.2 Hall effect

In many applications the Hall effect is used and therefore it will be discussed in some detail. We make a distinction between the Hall effect in conductors and in semiconductors.

The Hall effect in conductors

The Hall effect was discovered by Edwin Hall in 1879. It describes the effect that, when a current is flowing in a (semi)conductor and this conductor is placed in a magnetic field, not in parallel with the current direction, an electric field will be generated perpendicular with respect to the current and the magnetic field direction.

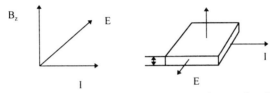

Figure 6.28 Illustration of the Hall effect. In a semiconductor, for electrons and holes the generated voltage difference, E, opposes each other.

This is illustrated in figure 6.28. In semiconductors the generated Hall voltages, V_H, for electrons and holes oppose each other, because of the different transport directions for electrons and holes. Note that the force acts in the same direction for both types of charge carrier. The Hall voltage can be described by

$$V_H = \frac{R_H I_x B_z}{t} \tag{6.52}$$

where V_H (V) is the Hall voltage, R_H (m^3 C^{-1}) is the Hall constant, I_x (A) is the current in the x direction, B_z (V s m^{-2}) the magnetic induction in the z direction and t (m) thickness of the considered plate. The force executed on the charge carriers is described by (6.47). In practice the movement can be periodic or circular. The Hall voltage is inversely proportional to the thickness, t, of the applied material. It can also be shown that the Hall constant $R_H = 1/nq$, where the electron density is n and q the elementary charge. A large density gives a small Hall coefficient. Some Hall coefficients are given in table 6.27.

Table 6.27 Some numeric values for Hall coefficients.

Material	Hall coefficient (m^3 C^{-1})
InSb	-116×10^{-6}
InAs	-112×10^{-6}
InP	-1.4×10^{-3}

The Hall effect in semiconductors

As said previously, in semiconductors we have to deal with two types of charge carrier. To describe the Hall effect, we have to take into account both charge carriers. It can be shown that in this case we find for the Hall coefficient that

$$R_H = \frac{p\mu_p^2 - n\mu_n^2}{q(p\mu_p + n\mu_n)^2} \tag{6.53}$$

where p (m^{-3}) is the hole density per unit volume, n (m^{-3}) is the electron density per unit volume, $\mu_{n,p}$ (m^2 V^{-1} s^{-1}) is the mobility for electrons and holes respectively and q (C) the specific charge for electrons. For n-type material where $n \gg p$ the Hall coefficient becomes

$$R_H = -1/nq \tag{6.54}$$

and for p-type material where $p \gg n$ the Hall coefficient becomes

$$R_H = 1/pq \tag{6.55}$$

For most semiconductors the mobilities of holes and electrons differ considerably. For instance, for silicon we find that $\mu_p = 600 \times 10^4 \, \text{m}^2 \, \text{V}^{-1} \, \text{s}^{-1}$ and $\mu_n = 1600 \times 10^4 \, \text{m}^2 \, \text{V}^{-1} \, \text{s}^{-1}$. Note that different authors give different values for the mobility. However, this is the reason that intrinsic material also shows a Hall effect. The Hall coefficient then is

$$R_H = \frac{\mu_p - \mu_n}{n_i q (\mu_p + \mu_n)}. \tag{6.56}$$

In practical circumstances the Hall coefficient differs from the theory by a factor of 2. For this reason a factor r is introduced. The Hall effect is often used to determine the mobility of semiconductors. The product of the Hall coefficient and the specific density equals the mobility, or

$$R_H \sigma = \mu_H(n) \qquad R_H \sigma = \mu_H(p) \tag{6.57}$$

This is sometimes called the *Hall mobility*, because in these expressions the factor r is included.

Magnetoresistivity

Magnetoresistivity is the increase in resistance due to a magnetic field which exerts a force on the charge carriers. The cause can be (1) physical or (2) geometrical. The effect of this second possibility is large and can be used for making magnetoresistors. For instance, the influence on the mobility is small for silicon but large for InSb, because for the mobility of InSb $\mu = 80\,000 \times 10^4 \, \text{m}^2 \, \text{V}^{-1} \, \text{s}^{-1}$ is reported.

Magnetoresistivity in magnetic layers is much larger than in semiconductors. This gives the possibility to apply a magnetic field on magnetically sensitive layers which can cause a change in the direction of the current in that layer.

The resistivity of a magnetic resistor, when B is perpendicular with respect to the current, can be written

$$\rho = \rho_0(1 + k_m B^2) \tag{6.58}$$

where ρ (Ω m) is the specific resistivity, ρ_0 is the resistivity when $B=0$ and k_m is a magnetic sensitivity factor. For InSb $k_m = 38 \, \text{T}^2 \, \Omega^{-1} \, \text{m}^{-1}$ at $B=1$ T. For the resistance of a magnetic film as a function of the angle between magnetic field and current we find

$$R(\Theta) = R(0) - [R(0) - R(90)] \sin^2(\Theta) \tag{6.59}$$

where $R(0)$ is the resistance of the magnetic layer when the directions of the magnetic field and current are parallel, $R(90)$ the resistance of the magnetic layer when the directions of the magnetic field and current are perpendicular and Θ the angle between magnetic field direction and current. The highest efficiency is obtained when current and magnetic field are perpendicular, which is realized in a so-called *barber-pole* structure (figure 6.29). Layers of Ni–Fe and Ni–Co are used.

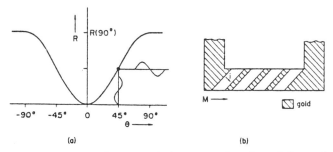

Figure 6.29 The resistance of a magnetic layer as a function of the angle between *B* and *I*; a barber-pole structure.

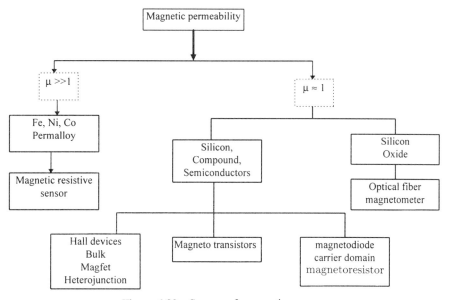

Figure 6.30 Survey of magnetic sensors.

A survey of magnetic sensors is given in figure 6.30. In the following section a survey of magnetic transducers in silicon and in compatible technology is given.

6.10.3 Magnetic energy and silicon microtransducers

MI [el, el, ma]

Magnetoresistor
The resistance of a carrying conductor will change in a magnetic field. The effect is known as the Gauss effect and is rather small, depending strongly on the applied geometry of the diffused resistors. The largest effect can be obtained with a so-called Corbino disk.

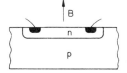

Hall effect

When a current-carrying semiconductor is subjected to 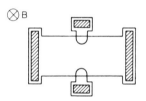 a magnetic field a so-called Hall voltage will be generated across the semiconductor at right angles to both current and magnetic field direction. The polarity of the Hall voltage is dependent on type of semiconductor material (p type or n type), the direction of the current and the direction of the magnetic field.

Magnetodiode

In semiconductors the charge carriers can be deflected from their straight path applying a magnetic field, caused by the Lorentz force. Magnetodiodes show a high sensitivity to magnetic fields when they are manufactured with large differences in the recombination rates between the two regions. The diode characteristics change when a magnetic field is applied. Diodes made in SOS (silicon on sapphire) technology have a high sensitivity but are less linear than a Hall plate.

Magnetotransistor

The magnetotransistor consists of a bipolar transistor structure and is composed of one emitter and two collectors. Without a magnetic field the collector current divides equally between both collectors. When a magnetic field is applied in the plane of the silicon chip perpendicular to the line connecting both collectors an unequal current division will occur. If two load resistors in the collectors are connected the difference in current is converted into a voltage difference and appears to be a linear function of the magnetic field.

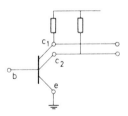

MAGFET

The MAGFET consists of a MOSFET with two drain contacts analogous to those of the magnetotransistor and proves to have a comparable sensitivity.

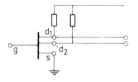

Hall MOSFET

A Hall MOSFET consists of a MOSFET with a rather 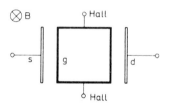 wide channel area and additional Hall contacts. The Hall voltage is inversely proportional to the charge carrier density and the thickness of the plate, but proportional to the current through the plate and the magnetic field. Sensitivities of $100 \, \text{mV} \, \text{mT}^{-1}$ are obtained.

Integrated magnetic transducers
Hall plates can be integrated together with signal-processing circuitry on the same chip. Applications are magnetic recording heads, brushless motors, displacement transducers and so on.

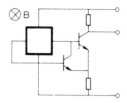

Magnetobridge
Four sensitive magnetoresistors are deposited on a single silicon chip in a Wheatstone bridge configuration and this appears to have a sensitivity of 1 mV at a magnetic field variation of 100 A m^{-1} at a supply voltage of 5 V. They can be used for speed and direction control of electric motors.

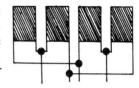

MO [el, ma, el]
So far applications of this effect are not known.

SI [ma, el, 00]
This effect can be realized when a silicon coil made by diffusing a p-type winding in an n-type substrate is subjected to an alternating magnetic field. Because of Faraday's induction law a relatively small voltage will be generated. Because of the technological limits only a small number of windings can be integrated.

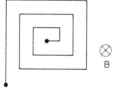

SO [el, ma, 00]
Of course this effect can be realized but no application has so far been reported.

6.10.4 Magnetic energy in compatible technology
In some cases the temperature sensitivity of magnetic transducers is unacceptably large. In that case the magnetic effect of other magnetically sensitive materials can be used, for instance layers of Ni–Co or Ni–Fe being deposited on top of a silicon dioxide layer with underlying silicon substrate containing the signal processing circuitry.

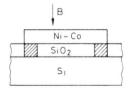

Magnetic resistors
A layer of thin magnetic material is deposited on a substrate. A current perpendicular to the magnetic field direction influences the resistivity. A nanotesla resolution can be obtained.

Optoelectronic magnetic sensor
Here light is used as the information carrier. The sensor is based on the Faraday
rotation of the polarization plane of linearly polarized light due to the Lorentz force
unbound electrons in insulators. Fibre optic sensors can be used for this purpose.

SQUID
SQUID stands for superconducting quantum interference device and is based on the
principle of current tunnelling through a sandwich of two extremely thin superconducting and one insulating layer. For instance, an $Nb/Al_2O_3/Nb$ sandwich conducts
below 20 K, It can also be used as a kind of switch known as the Josephson junction.
Based on this principle with these junctions, a supercomputer was once built. With
SQUID magnetic fields of picoteslas can be measured. The device is used in medical
application measuring brain activities.

6.10.5 Physical effects in the magnetic energy domain
These are summarized in table 6.28. This concludes our discussion about magnetic
transducers.

Table 6.28 Survey of known magnetic transducer effects.

Name of effect	Notation	Macroscopic description
1 Hall	[el, el, ma]	The generation of an electric field in a conductor due to a current and a magnetic field which all are mutually perpendicular
2 Magnetoresistance	[el, el, ma]	The change in resistivity of a material caused by a magnetic field
3 Suhl	[el, el, ma]	the change in conductivity of a semiconductor surface caused by a magnetic field
4 Superconductivity	[el, el, ma]	The change to normal state of a superconductor at a specific magnetic field and/or temperature
5 Photomagneto-electric	[ra, el, ma]	The generation of an electric field due to a magnetic field and incident radiant energy
6 Nernst	[th, el, ma]	See T3
7 Magnetostriction	[ma, me, 00]	A mechanical deformation is generated in a ferromagnetic material by a magnetic field
8 Faraday	[ra, ra, ma]	The rotation of a polarization plane of polarized radiation by a magnetic field
9 Kerr magneto-optic	[ra, ra, ma]	See R16
10 Cotton–Mouton	[ra, ra, ma]	See R19
11 Ettinghausen	[el, th, ma]	The generation of a temperature gradient due to an electrical current and a magnetic field
12 Maggi–Righi–Leduc	[th, th, ma]	The change in thermal conductivity caused by a magnetic field
13 Righi–Leduc	[th, th, ma]	The generation of a temperature gradient due to a heat flow and a magnetic field
14 Electromagnetic	[el, ma, 00]	The change in magnetization due to an electric current
15 Curie and Néel	[th, ma, 00]	See T10 and T11

6.11 THE CHEMICAL ENERGY DOMAIN

The discussion of this type of energy conversion in semiconductors is the most difficult to classify, because it also includes effects which are less easy to classify in one of the above classes and which are not purely chemical in their effects. A very serious problem concerning packaging of this class of sensors is also a major complicating factor in the manufacture and lifetime of chemical sensors. Some types have an MTTF (non-repairable) of a couple of hours, for instance the oxygen sensor. In general it is possible to apply the following classification:

- chemical sensors
- sensor for biological systems
- miscellaneous sensors.

Table 6.29 illustrates the different possibilities under which a potential difference is generated. With the help of this table all well known generated potentials can be characterized.

Table 6.29 Survey of electrochemical potentials.

Medium/separation	Name of potential
Semipermeable membrane	Cell membrane potential
Electrolyte–electrolyte	Liquid junction potential
Metal–metal	Contact potential
Semiconductor p–n junction	Contact potential
Metal–oxide semiconductor	
Polymer–metal	
Metal–electrolyte	Half-cell potential

6.11.1 'Physics'
In the following sections we will present some of the chemical phenomena used in chemical transducers by which a potential difference is generated in different circumstances. We also have called it 'physics' to make the presentation consistent. We start with the well known basic Nernst equation for concentration differences.

The Nernst equation
In general, it can be said that when two compartments, a and b, which differ in chemical composition (ion concentration) are brought together, an exchange of charge carriers can occur. We have seen already that in this transport mechanism of charge carriers two basic forces will balance each other. In the equations for the p–n junction we met the forces of diffusion and the force of the electrostatic field. Here,

the type of charge carrier must be extended towards positively and negatively charged ion groups. The generated voltage difference between the two compartments a and b can be described with help of the *Nernst* equation, or

$$E = V_1 - V_2 = -\frac{RT}{z_i F} \ln \frac{[C]_{i,a}}{[C]_{i,b}} \qquad (6.60)$$

where E (V) is the voltage difference between a and b, R (8.31 J mol^{-1} K^{-1}) the universal gas constant, T (K) the absolute temperature, z_i the valence of the ion involved, F Faraday's constant (9.6487×10^4 C mol^{-1}), $[C]_{i,a}$ (m^{-3}) ion density i, in compartment a and $[C]_{i,b}$ (m^{-3}) ion density i, in compartment b. The Faraday constant represents the charge on a mole of electrons. This equation holds for one type of ion. A potential difference is always present when two different conducting media are in contact with each other. This equilibrium is the so-called Donnan equilibrium. In practice the situation can be much more complicated when more than one type of ion is involved. In biocell membranes we may have to deal with ions such as K$^+$, Na$^+$, Cl$^-$ and many other types of ion group.

Some cells have the property that, when they are stimulated appropriately, they depolarize and can even generate an action potential difference. At depolarization the cell membrane potential changes sign. For instance, tangible or palpable sensorial organs respond in skin by pressure deformation with changing sign, after which this signal is transported with constant speed across the cell membrane. This change of polarization can occur within a very short time (several milliseconds). The distance of change is of the order of nanometres.

The metal–electrolyte junction

Metal–electrolyte junctions represent a type of interface which also gives rise to the generation of a so-called half-cell potential. When we assume that the interface consists of a metal electrode with atoms type C, and an electrolyte with anions type A$^-$ and metal cations type C$^+$, then the following reaction will occur:

$$C \rightleftharpoons C^{n+} + ne^- \qquad A^{m-} \rightleftharpoons A + me^- \qquad (6.61)$$

where n represents the valence of the cations and m represents the valence of the anions. When the current equals zero then at the interface oxidation and reduction balance each other and the net electron current is zero. When a current passes the interface, which reaction dominates depends on the current direction. The magnitude and the stability of the reaction are determined by the type of material, the type and concentration of the electrolyte and whether an electrical current passes the interface. Standard half-cell voltages are measured with reference to a hydrogen electrode which by definition is 0 V. In table 6.30 some of these half-cell potentials and their reactions are given. The half-cell potential can be described by the expression

$$V = V^\circ + \frac{RT}{F} \ln(a_{C^{n+}}) \qquad (6.62)$$

Table 6.30 Survey of some standard half-cell potentials.

Metal	Reaction	Potential (V)
Al	$Al^{3+} + 3e^-$	−1.705
Zn	$Zn^{2+} + 2e^-$	−0.763
H_2	$2H^+ + 2e^-$	0.000
$Ag + Cl^-$	$AgCl + e^-$	+0.223
Ag	$Ag^{1+} + e^-$	+0.799
Cu	$Cu^{2+} + 2e^-$	+0.340
Cu	$Cu^{1+} + e^-$	+0.522

where V (V) is the half-cell potential, V° (V) is the standard half-cell potential referred to H_2, R the gas constant ($8.31 \, J \, mol^{-1} \, K^{-1}$), T (K) the absolute temperature, F the Faraday constant ($9.6487 \times 10^4 \, C \, mol^{-1}$), $a_{C^{n+}}$ (mol l^{-1}) the ion activity of cation C^{n+} and n the valence of the electrode material. The ion activity can be interpreted as the availability of the ions in the solution to participate in the reaction. Up to a certain limit the activity equals the concentration C_i. In larger concentrations $a_i = \gamma_i C_i$, where γ_i is the coefficient of activity where $0 < \gamma_i < 1$. These voltages are also called *potentiometric* voltages.

Polarization

The half-cell potential across a metal–electrolyte junction is not a constant but a function of temperature, the activity, the type of metal and the current flowing through it. The change in voltage of a half-cell potential as a function of current is called the *overpotential* and is due to *polarization of the electrode*.

Three different mechanisms contribute to this polarization: (1) Ohm's effects; (2) change of concentration; (3) change of activity. The main cause of this overpotential is the energy required to convert an electron current into an ion current at the interface of a metal–electrolyte junction. (1) Ohm's overpotential is due to the dissipation in Ohm's resistance at the interface. (2) The concentration overpotential is caused by the change in ions at the metal–electrolyte interface when a current is flowing through the interface. (3) The activation overpotential represents the difference in energy between oxidation and reduction. These three mechanisms may be added to the polarization which then becomes

$$V_p = V_r + V_C + V_a \tag{6.63}$$

where V_p is the polarization potential, V_r is Ohm's overpotential, V_C is the concentration overpotential and V_a is the activation overpotential. It is strongly preferred that the polarization potential, V_p, is kept as low as possible. A non-polarizable electrode is called a *perfect non-polarizable electrode*. The other way round is the existence of *perfect polarizable electrodes*, where an infinite energy is required to perform a transfer of electrons across the interface. This can be interpreted as an ideal capacitor. Electrodes made from gold or platinum are almost ideal polarizable electrodes. Also, extremely high-input-impedance amplifiers can provide this characteristic, minimizing currents through the interface as much as possible.

Metal–oxide junctions

These types of junction are the most complex to describe. Often a noble metal such as Pt, Pd or Ag is used as a catalyst to reduce the activation energy needed. In the automotive industry an oxygen sensor is used to monitor the air–fuel mixture for optimum combustion. It also provides information about the exhaust gas mixture for CO, H_2S, methane, NO_x and other noxious gases. The conductivity of a metal oxide such as SnO_2 changes with the gas concentration. Because of regulations NO_x is an important area in monitoring emission control.

For the automotive industry the oxygen sensor is made from a hollow-shaped zirconia ceramic, ZrO_2, covered with a porous layer of platinum on both sides of the tube. This creates an electrochemical cell that develops a potential difference between the two electrodes. A high operating temperature is required, $>350\,°C$, in order to keep this sensor functioning properly. Metal-oxide semiconducting thin films on aluminium oxide substrates are used to make small sensors for each cylinder separately.

Polymers

Polymer materials can be used for making sensors. They can be fabricated highly selectively for a specific type of gas. It is noted that polymer transistors can also be manufactured. We will not discuss this in further detail, but reference is made to the literature.

Miscellaneous sensors

Miscellaneous sensors are for instance humidity sensors. They are difficult to classify in one of the energy domains discussed. Usually, a kind of capacitive structure is used to measure the change in conductivity due to a change of relative humidity between the two electrodes. These structures can be made in silicon technology, and can also be combined with small Peltier elements underneath the silicon substrate.

In the following sections examples are given of transducers in the chemical domain.

6.11.2   Chemical energy in silicon microtransducers

MI [el, el, ch]

Ion electrolytic conduction
A simple monolithic structure with two electrodes can be made such that the conductivity of an ionic solution on top of the silicon substrate can be determined. Here a major problem concerning packaging and corrosion of such a type of transducer arises.

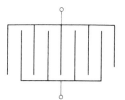

Humidity sensors
A humidity sensor can be made by applying a silicon chip on top of a Peltier cooling element. On top of the chip a polycrystalline cam-structured geometry is implemented whose capacitance will change abruptly when water begins to condense on top of the silicon chip. A thermal energy classification might be justified here.

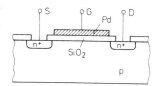

Hydrogen-sensitive MOSFET

A MOSFET with a palladium gate appears to be sensitive for hydrogen (H_2), showing a threshold voltage as a function of the hydrogen concentration. The sensitivity proves to be 10 ppm H_2. It is also shown that this device is sensitive for H_2S and NH_3.

Ion-sensitive field effect transistor (ISFET)

When the gate electrode of a MOSFET is not metallized but covered with an ion-selective layer of Si_3N_4, Al_2O_3 polymer the device appears to be sensitive for ion concentrations in an electrolyte, providing the possibility to manufacture pH sensor elements. Here again the packaging problem strongly limits the MTTFF. A reference electrode is needed.

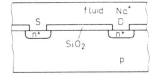

Oxygen sensor

Here the silicon substrate is used as a carrier only in which a spiral channel is etched providing a chamber for an electrolytic solution of AgCl. The spiral chamber is covered with a membrane suitable for oxygen diffusion. Two silver electrodes are deposited on the substrate for use as anode and cathode and when a voltage is applied the charge carrier flow is a measure of the oxygen pressure in the chamber. The MTTFF is a serious problem and is a mere couple of hours owing to packaging problems. For the automotive industry special emission control sensors made from ZrO_2 have been developed.

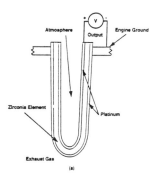

Smart nose transducer

An array of selective membranes is deposited on an array of ISFETs. This eight-by-eight array sensor has been developed to be able to discriminate between 64 different concentrations of gases.

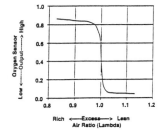

MO [el, ch, el]

So far no chemical output transducers in silicon are known.

SI [ch, el, 00]

Galvanic cells can be realized by bringing suitable electrodes and solutions of electrolytes together, generating a voltage. The same effect can be obtained when a drop of an electrolyte is put on top of a silicon chip with appropriate electrodes. The generated voltage is a measure of the ionic concentration. A limited lifetime is involved because of corrosion problems. Integrated chemical sensors are available today. A good example is the electronic nose where selective membranes can discriminate between different species.

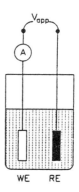

A very special family of chemical transducers is being developed now with the help of fibre technology; see the survey in section 6.7.8. The SAW sensors are another category which are also appropriate for measuring chemical parameters. These have been discussed already in the section on radiant energy

6.11.3 Review of chemical effects

In table 6.31 we present an overview of known physical effects for making chemical transducers. Some transducers show a tandem effect which means that more than one energy conversion is involved in one transducer element. For instance in a flow transducer the gas flow generates a temperature difference and the temperature difference generates a resistive difference which is converted into a voltage difference, so at least two energy conversions are involved: me–th and th–el.

This concludes our discussion about chemical transducers.

Table 6.31 Survey of known chemical effects for making transducers.

Name of effect	Notation	Macroscopic description
Volta	[ch, el, 00]	The generation of a voltage between two dissimilar metals
Galvano-electric	[ch, el, 00]	The generation of a voltage between two dissimilar metals in an electrolyte
Chemical–dielectric	[el, el, ch]	The change in permittivity caused by a gas or liquid
Chemical–electric	[el, el, ch]	The change in conductivity of a semiconductor surface caused by an electrolyte
Chemical–magnetic	[ch, ma, 00]	The change in magnetization of a magnetic material caused by absorption of a gas
Electrochemical	[el, ch, 00]	The change of structure caused by an electrical current
Chemical–radiant	[ch, ra, 00]	The change in reflectivity or refraction caused by a chemical reaction
Thermochemical	[th, ch, 00]	The change of structure caused by temperature

6.12 FUTURE TRENDS IN TRANSDUCERS

In discussing future trends for transducers and especially for silicon microtransducers two different directions can be distinguished. The first trend is a purely technological one whereas the second appears to have a more fundamental character.

6.12.1 Technological trends

In the development of transducers an increasing trend is present to apply silicon technology because of the advantages already discussed in section 6.6. Of course, some problems and disadvantages exist and should be mentioned to put silicon technology in the right context. From this consideration some trends can be recognized of what may be expected in future.

1   Starting the development of an integrated sensor requires considerable effort and time before a marketable sensor is available. For numerous sensors the time of development has been ten years or more. At a first glance this seems an unbelievable time to invest, but the development of a sensor is really a complicated affair. Many problems have to be solved in a multidisciplinary environment. A sensor is not pure electronics and, although proving a physical effect is not too difficult, manufacturing a reliable sensor with linear characteristics and good response time is a formidable task. Furthermore, often for every application an appropriate packaging must be developed. All these factors must be translated in a price–performance ratio and it is not too difficult to understand that this will start to be worthwhile when large series can be fabricated. It is a lucky circumstance that the need for sophisticated transducers is still increasing, because we are becoming more and more dependent on electronic systems and environmental pollution legislation demand for improved control and hence more sensors. Transducers are the sense organs for every system communicating with the outside world. It will be clear that when large series are manufactured the result is a considerable cost reduction.

2   The application of silicon is limited to the allowed applicable temperature range for silicon. The applicable temperature range lies between $-100\,^{\circ}\mathrm{C}$ and $200\,^{\circ}\mathrm{C}$. Below $-100\,^{\circ}\mathrm{C}$ the applied dopant materials in silicon are hardly activated and above $200\,^{\circ}\mathrm{C}$ degradation of silicon will occur. Here GaAs can provide new possibilities because the bandgap of GaAs is $1.43\,\mathrm{eV}$ which is $0.31\,\mathrm{eV}$ larger than that of silicon. The consequence is a higher allowed operational temperature. It is likely that GaAs may also be applied for a great variety of transducers, as will occur for other III–V semiconductor components as long as they are compatible with planar silicon technology concerning cost aspects.

3   The bonding technique used for connecting the chips with their outer environment is very sophisticated and is performed fully automatically and can be executed, for instance, with a speed of eight bonds per second. However the gold or aluminium wire bonds have a thickness of 1 mil or $25\,\mu\mathrm{m}$ which is thinner than a hair, and so are highly vulnerable when they are exposed to open air which is required for some transducers. Smart packaging can solve this problem.

4   Other challenging trends can be found in further integration of transducers with their processing electronics towards completely integrated systems or smart sensors. It may be expected that in the future more and more fully integrated sophisticated transducers for several physical parameters combined with computing power on one single chip will be fabricated. However, this development may be considered as a purely technological development.

5   Micromachining makes it possible to make three-dimensional transducers performing new and improved and unexpected tasks.

6   Fibre technology offers also completely new sensor technology in all types of energy domain.

7   Finally, polymer technology and the development of *biosensors* give new insights and possibilities of which the consequences are difficult to foresee. There is one simple statement applicable here: it is a never-ending story for everybody who is interested in this field.

PROBLEMS

6.1   Describe a general layout of a data-collecting system.

6.2   Describe the main reasons why integrated silicon transducers can be important.

6.3   What is the definition of an intensive and an extensive quantity?

6.4   What is the difference between a cross-effect and a direct effect?

6.5   What is the difference between the internal and the external photoelectric effect?

6.6   What is understood by a direct and indirect bandgap material?

6.7   What is the characteristic difference between a GaAs and a GaP diode?

6.8   Two physical effects can be applied for pressure transducers. What are they?

6.9   Why is the temperature coefficient of the piezoresistance effect in   rinsic silicon larger than in extrinsic silicon?

6.10 What are thermoelectric effects?

6.11 Which effect(s) can be used for manufacturing an output transducer denoted as [el, ma, 00]?

6.12 An error signal of 5 mV is caused by an interference signal of 25 V in a transducer. What is the CMRR?

6.13 (a) Over a period of 16 h the measured zero drift at the output of a transducer system is 500 mV. There is a gain factor of 500. What is the drift of the transducer per hour? (b) The signal output drift over the same period of time is 100 mV. What is the drift of the transducer per hour now? (c) How can this difference be explained when no temperature influences are taken into account?

6.14 Check for both terms at the right-hand side of equations (6.12) and (6.13) that a current density $A\ m^{-2}$ is involved.

6.15 In the range of 10^5 Pa a pressure sensor has a sensitivity of $1\ V\ Pa^{-1}$ and a temperature coefficient of $0.05\ mV\ K^{-1}$. What is the cross-sensitivity for this transducer as a function of temperature?

6.16 In a Wheatstone bridge consisting of four equal resistors of 50 Ω one of them has a differential sensitivity of 2 mV Ω^{-1}. Calculate the relative sensitivity when the supply is 10 V.

6.17 (a) Give a general expression for an arbitrary transducer's output signal. (b) Explain why it is not possible to eliminate transducer drift by electronic means. (c) Why is drift more severe than offset?

6.18 A capacitive difference transducer has membrane dimensions of $1500 \times 10^{-6} \times 1000 \times 10^{-6}$ m^2. Because of a pressure change the plate distance varies by 3 μm. What is the change in capacitance? (Answer: 4.4 pF.) The measured change in charge across the transducer is 0.22 pC; what is the change in voltage?

6.19 A strain gauge is fed by a current source of 1 mA. The length l of the strain gauge is 2 m, and the diameter is 0.5 mm. The specific resistivity is 45×10^{-8} Ω m. Calculate the change in voltage when a pulling force causes a change in length of 4 mm and in diameter of 0.4 mm.

6.20 The output of a four-sensor bridge strain gauge configuration is connected to an amplifier. The strain gauge is subjected to torsion. The bridge is fed by a voltage source of 10 V. The required inaccuracy corresponds to a relative change in resistance of 10^{-6}. Calculate the minimum CMRR for the amplifier.

6.21 Calculate the Hall constant R_H in a semiconductor material with a doping profile of $n = 2 \times 10^{17}$ cm^{-3} and $p = 5 \times 10^{14}$ cm^{-3}.

6.22 A semiconductor sensor has a thickness of 0.25 mm and a width of 1 mm. A current of 0.1 A is flowing through it. In a magnetic field of 1 T a Hall voltage of 0.8 V is measured. Calculate the Hall constant. (Answer: 2×10^{-3} m^3 C^{-1}.)

6.23 A transducer shows a transfer function with air velocity as input variable with a maximum deviation from a straight line of 3×10 cm s^{-1} air flow velocity. The maximum allowed air speed is 15 cm s^{-1}. Calculate the non-linearity error as a percentage.

6.24 When the principle of operation in 6.23 is based on thermal anemometry discuss the form of the transfer characteristic.

6.25 Calculate the absolute temperature of a silicon sensor with a surface area of 5×5 mm^2 when the radiated power is 50.5 mW. The chip may be considered as a grey body. The emittance coefficient is 0.92.

6.26 Calculate the mobility of electrons for a copper conductor when the number of electrons in a copper conductor is given as $n = 10^{22}$ cm^{-3}.

6.27 For which wavelength λ_g does complete absorption exist for pure silicon at 300 K?

6.28 A photodiode is connected to a circuit as is illustrated in figure 6.31. The luminous intensity is 2000 lx and it generates 0.002 mA. $R_s = 100$ kΩ. Determine the resistor values R_1 and R_2 for an output voltage of 5 V.

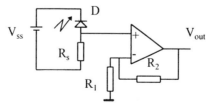

Figure 6.31 Photodiode circuit.

6.29 Discuss why fibres can be used for temperature measurement.

6.30 A linear resistive position sensitive device is connected to a load resistance R_l as is shown in figure 6.32. Find an expression for the output voltage and the non-linearity in terms of R and R_l.

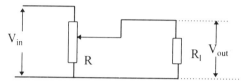

Figure 6.32   Circuit for problem 6.30.

6.31 (a) Derive an expression for the temperature behaviour of a single diode with a bias current I_D. (b) Calculate the ambient temperature in degrees Celsius when the bias current is 1 mA and the saturation current is 1 nA.

6.32 An integrated square Wheatstone bridge configuration consists of four equal temperature-dependent resistors with a temperature coefficient $\alpha = 0.022 \, \mathrm{K}^{-1}$. The bridge supply voltage is $V_{SS} = 5$ V. The bridge is used as a thermal anemometer. Find the geometric layout for maximum performance.

6.33 Find for the configuration discussed in 6.32 an expression for the output voltage when the supply voltage is V_{SS}. Show that when the flow direction changes, the output signal changes polarity.

6.34 Calculate the on-chip temperature difference when the bridge output voltage is 1 mV.

BIBLIOGRAPHY

Here only a selection of references is made; the literature about transducers is overwhelming and it is always efficient to consult the biannual proceedings of Transducer's Solid-State Circuit Conference. Also *Sensors and Actuators* should be consulted.

[1] Thomson R B 1994 Fiber optic sensors, advance chemical analysis from afar *IEEE Circuits Devices* **10** 14–21

[2] Forest S R 1989 Organic-on-inorganic semiconductor heterojunctions: building block for the next generation of optoelectronic devices? *IEEE Circuits Devices* **5** 33–7

[3] Leheny R F 1989 Optoelectronic integration, a technology for future telecommunications systems *IEEE Circuits Devices* **5** 38–41

[4] Kapon E 1992 Quantum wire lasers *IEEE Proc.* **80** 398–410

[5] Suematsu Y, Iga K and Arai S 1992 Advanced semiconductor lasers *IEEE Proc.* **80** 383–96

[6] Yariv A 1989 Quantum well semiconductor lasers are taking over *IEEE Circuits Devices* **5** 25–8

[7] IMEC Leuven 1994 *Scientific Report*

[8] Lambeck P V (ed) 1994 *Sensor Technology 1994, Proc. Dutch Conf. on Sensor Technology* (Twente)

[9] van Putten A F P and Middelhoek S 1974 An integrated silicon anemometer *Electron. Lett.* **10** 425–6

[10] Sze S M 1994 *Semiconductor Sensors* (New York: Wiley)

[11] van Putten A F P 1990 Thermal feedback drives sensor bridge simultaneously with constant supply voltage and current *Trans. IEEE (IMTC/89)* 48–51

[12] van Putten M H P M, van Putten M J A M, van Putten A F P and van Putten P F A M 1993 Drift Eliminatie bij Sensoren *Dutch Patent Application* 9301399

[13] van Putten M J A M, van Putten M H P M and van Putten A F P 1994 Full additive drift elimination in vector sensors using the alternating direction method (ADM) *Sensors Actuators* A **44** 13–7

[14] van Putten M H P M, van Putten M J A M, van Putten A F P and van Putten P F A M 1995 Method for drift elimination in sensors *US Patent* 5426969

[15] van Putten M H P M, van Putten M J A M, van Putten A F P and van Putten P F A M 1994 Method for drift elimination in sensors *European Patent Application*, 94202293.0

[16] *van Dijk R, van Putten A F P and van Bokhoven W M G 1995 Silicon semiconductor sensor for seasurement of respiratory airflow* Proc. *ProRisc/IEEE Benelux Workshop on Circuits, Systems and Signal Processing (Mierlo, 1995)*

[17] Mullins A, Medler A, Court D, Bayford R, van Putten A F P, Patel D and Sinnadurai N 1995 Research on novel sensors on silicon substrates *Microcircuits Electron. Packaging* March 1995

[18] Mullins A, van Putten A F P, Bayford R and Butcher J B 1995 Potential for a smart sensor based on an integrated silicon anemometer 1995 *Sensors Actuators* A **46–47** 342–348

[19] Mullins M A, van Putten A F P, Bayford R and Butcher J B 1994 Potential for a smart sensor based on an integrated silicon anemometer *Eurosensors VIII (1994)*

[20] Pascoe K J 1972 *Properties of Materials for Engineers* (New York: Wiley)

[21] van Putten A F P 1988 Integrated silicon anemometers *PhD Thesis* Leuven

[22] Middelhoek S and Audet S A 1989 *Silicon Sensors* (New York: Academic)

[23] Bouwstra S 1990 Resonating microbridge mass flow sensors *PhD Thesis* (Twente University of Technology)

[24] Middelhoek S and Van der Speiegel J (eds) 1987 *Sensors and Actuators, State of the Art of Sensor Research and Development* (Lausanne: Elsevier Sequoia)

[25] Hedberg A (ed) 1990 *Trends in Sensor Technology* STU-info 788 (Swedish National Board for Technical Development)

[26] *IEEE Solid-State Sensor and Actuator Workshop SC, (1992)*

[27] Dagenais M, Leheny R F and Craw J (eds) 1995 *Integrated Optoelectronics* (New York: Academic)

[28] Brignell J and White N 1994 *Intelligent Sensor Systems* (Bristol: Institute of Physics)

[29] Wilson J and Hawkes J F B 1989 *Optoelectronics. An Introduction* 2nd edn (New York: Wiley)

[30] Dakin J P (ed) 1990 *The Distributed Fibre Optic Sensing Handbook* (IFS)

[31] Pal B P 1992 *Fundamentals of Fibre Optics in Telecommunications and Smart Systems* (New York: Wiley)

[32] Culshaw B and Dakin J 1989 *Optical Fibre Sensors, Systems and Applications* vol 2 (Artech)

[33] Ristic L (ed) 1994 *Sensor Technology and Devices* (Artech)

[34] Déjous C 1994 Capteurs acoustiques á modes de plaque transverses horizontaux. Application á la détection en milieux liquides et gazeux. *These de Docteur* L'Université Bordeaux I

Seven

OFFSET AND DRIFT

7.1 INTRODUCTION

For two reasons it can be said that linear amplification is one of the most important operations in the field of electronics. Firstly, a correct matching to the signal source can be achieved and secondly for reasons of sensitivity linear amplification will always be necessary. The importance of this sensitivity aspect can easily be illustrated by the very low signal source levels which are always encountered in the medical and biological field. Signal sources with millivolt and microvolt voltage level are usual. Other examples are thermocouples which deliver usually a voltage change of $40\,\mu\text{V K}^{-1}$.

In the problem of linear amplification four different aspects can be distinguished which mainly determine the final accuracy:

1 offset and drift
2 guarding and shielding
3 noise
4 the application of universal active elements.

In the following chapters the first three items will be emphasized and discussed in detail, and in less detail the application of active elements, because they can be found in every textbook about electronics. Generally, it can be said that ultimately the overall accuracy of a measurement set-up is determined by the offset, drift, guarding and noise characteristics of all components involved. Guarding and noise will be discussed in chapters 8 and 9 respectively; this chapter considers offset and drift. Firstly, we will discuss the principal concepts of offset voltage and offset current in active components. Secondly, the drift behaviour of active components is treated.

A special section will be devoted to drift in transducers; in particular the input transducer is sensitive for drift. Basically, we have to deal with *multiplicative* drift and *additive* drift in sensors. These phenomena complicate the description and the context of drift, because here drift can originate from other energy domains than the electronics domain exclusively, when the required physical parameter shows an anisotropic character and is superimposed on an isotropic signal originating from the same energy domain. In other words sensor asymmetries can cause a common-mode signal from another energy domain to be converted into a differential-mode

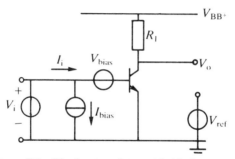

Figure 7.1 Bipolar transistor with biasing sources.

signal. When the required signal shows an asymmetric behaviour and the sensor shows directional sensitivity for the desired parameter then it can be eliminated.

This type of drift cannot be eliminated by electronic means. Then other precautions must be taken. This will be discussed in more detail in section 7.10. For instance, it can be realized with the ADM principle or with another sensitive element whose output is also a function of the same physical parameter.

7.2 PRINCIPLES OF OFFSET VOLTAGE AND CURRENT

It is well known that every active element requires a certain biasing current and voltage in order to make this element operative in the desired region. With the input signal source put to zero, the effect of this biasing will result in a certain output voltage level. This is illustrated in figure 7.1 for a single bipolar transistor and in figure 7.2 for an amplifier with a differential input stage. In figure 7.1 a bias current I_{bias} and a bias voltage source V_{bias} are depicted. In the case of the differential amplifier it makes sense to talk about differences in biasing voltages and *differences* in biasing currents at both input terminals, because these are the resulting parameters. We will return to that later. By definition, the required input voltage to be applied to make the output voltage exactly zero is called the input offset voltage of the active element. In the same way, by definition the required input current to be applied to make the output voltage exactly zero is called the input offset current of that active element.

In both definitions it is assumed that the signal source is put to zero. Both conditions must be fulfilled to make the output zero. The required voltage and current to make the output voltage zero can be considered to be delivered by two independent

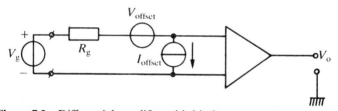

Figure 7.2 Differential amplifier with biasing sources I_{bias} and V_{offset}.

Table 7.1 Example of input offset voltage and input bias current of a high-current amplifier (BB 3573).

	Conditions	Minimum	Typical	Maximum	Units
Input offset voltage	$-25\,^\circ$C	n.s.	± 5	± 10	mV
versus temperature	$\leq T_{\text{case}} \leq$		± 10	± 65	μV K^{-1}
versus supply voltage	$85\,^\circ$C		± 35		μV V^{-1}
Input bias current	$-25\,^\circ$C	n.s.	15	40	nA
versus temperature	$\leq T_{\text{case}} \leq$		± 0.05		nA K^{-1}
versus supply voltage	$85\,^\circ$C		± 0.02		nA V^{-1}

n.s. = not specified.

sources V_{bias} and I_{bias} in the single-stage case and V_{offset} and I_{offset} for the differential stage. These sources are often interpreted as interfering sources, which can be drawn outside the amplifier at the input's side. This concept allows one to talk about a bias-free amplifier, which is, of course, an imaginary, idealized amplifier. Usually, for operational amplifiers the (input) offset voltage and the (input) bias current are specified by the manufacturer. In table 7.1 the specifications for a high-current amplifier are given.

Figure 7.3 illustrates the differential amplifier in more detail. In this figure for symmetry reasons, two bias voltage sources and two bias current sources can be recognized, from which figure 7.2 is derived. In figure 7.3(b) the bias- and offset-free amplifier is drawn with the corresponding offset and bias sources. This concept will be met again in chapter 9. Methods of shifting (noise) sources in circuits, and some rules will be presented in chapter 9.

7.3 CALCULATION OF THE OFFSET VOLTAGE AND CURRENT

In this section the expressions for the offset voltage and current of the bipolar transistor and the field effect transistor will be derived.

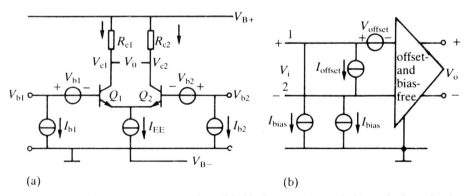

Figure 7.3 (a) Differential amplifier with biasing sources and (b) equivalent circuit with offset sources.

7.3.1 The bipolar transistor

To calculate the offset voltage and current for the bipolar transistor we use the following expression:

$$I_C = K \frac{A_E}{W_B} \exp(-qV_{GAP}/kT) \left[\exp(qV_{BE}/kT) - 1\right] \tag{7.1}$$

where I_C (A) is the DC collector current, K a temperature-dependent quantity proportional to T^3, A_E (m^2) the emitter surface, q the specific electronic charge (1.6×10^{-19} C), V_{GAP} (V) the bandgap between valence band and conduction band, T (K) the absolute temperature, V_{BE} (V) the base–emitter voltage, W_B (m) the base thickness and k the Boltzmann constant (1.38×10^{-23} J K^{-1}).

Neglecting the -1 term, taking the natural logarithm of (7.1) and assuming both $V_i = 0$ and $V_o = 0$, an expression for the base–emitter voltage is found:

$$V_{BE} = \frac{kT}{q} \ln\left(\frac{I_C W_B}{KA_E}\right) + V_{GAP}. \tag{7.2}$$

Because of the large value of K the first term will be negative and to find the magnitude of V_{BE} with $V_{output} = 0$ the actual value for I_C must be substituted. Then the bias current with $I_i = 0$ and $V_o = 0$ equals

$$I_{bias} = I_C/\beta \tag{7.3}$$

where β is the current gain factor. The order of magnitude for V_{BE} is 0.6–0.7 V and that for the current appears to be 1–5 μA. Note that for a single stage V_{BE} equals V_{offset} and I_B equals I_{bias}. Of course in a single stage these values are always present and no precautions are available to compensate them effectively other than by external sources.

7.3.2 The field effect transistor

The offset behaviour of a field effect transistor can be subdivided into four operational regions and is determined by the region in which the transistor is operative. Apart from this we have to distinguish between two types: the junction and the isolated gate field effect transistor. This last type is usually called a MOS-type transistor. In figure 7.4 the characteristic of a FET is drawn with the four regions to be distinguished. The first operative region is called the *linear region* and is characterized by a strong linear behaviour between I_D and V_{DS}. The second region is called the *weak inversion region* and shows a strong exponential behaviour. The third region is called the *boundary region* between weak and strong inversion and is characterized by a quadratic behaviour. The fourth region is known as the region for *strong inversion* and shows a quadratic behaviour. This makes the discussion of the offset behaviour a little more complicated. We shall confine the discussion to the usually most applied region for strong inversion.

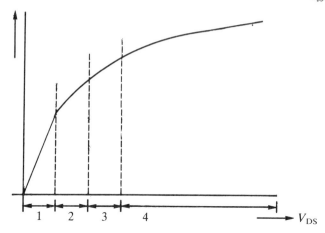

Figure 7.4 Typical characteristic for a field effect transistor.

1 The junction field effect transistor

For the junction field effect transistor the offset behaviour can be derived from the equation

$$I_D = \Gamma \frac{W}{L} V_p \left[1 - 3 \frac{V_{GS}}{V_p} + 2 \left(\frac{V_{GS}}{V_p} \right)^{3/2} \right] \qquad (7.4)$$

where I_D is the drain current, Γ a geometrical and physical constant of the junction FET, V_p the pinch-off voltage, V_{GS} the gate source voltage, W the channel width and L the channel length. This expression can be approached by

$$I_D = \Gamma \frac{W}{L} V_p \left(1 - 3 \frac{V_{GS}}{V_p} \right). \qquad (7.5)$$

If V_{GS} is made explicit we find the expression for the offset voltage for a junction FET:

$$V_{offset} = V_{GS} = \frac{1}{3} V_p \left[1 - \left(\frac{I_D}{\Gamma V_p W/L} \right)^{1/2} \right]. \qquad (7.6)$$

2 The MOS field effect transistor

For a MOS field effect transistor in the same region, the expression for strong inversion is

$$I_D = \frac{K_p W}{2nL} (V_{GS} - V_p)^2 \qquad (7.7)$$

where I_D is the drain current, K_p a geometrical and physical constant of the MOSFET, W the channel width, n a slope factor related to the actual inversion region, L the channel length, V_{GS} the gate–source voltage and V_p the pinch-off voltage. The pinch-off voltage, V_p, is defined as the gate–source voltage V_{GS} required to cause pinch-off of the channel to zero width. Note the difference with the threshold voltage V_t, which is defined as the gate-to-source voltage required to cause an inversion layer. Solving equation (7.7) for V_{GS} delivers the desired expression for the offset voltage of the MOS field effect transistor:

$$V_{\text{offset}} = V_{GS} = V_p + \left(\frac{I_D \, 2nL}{K_p \, W}\right)^{1/2}. \tag{7.8}$$

Characteristic values for the offset voltages for both the junction and the MOSFET are several volts. For the (input) bias current these values are of the order of 10–100 pA.

7.4 OFFSET BEHAVIOUR OF DIFFERENTIAL AMPLIFIER STAGES

Here again we have to make a distinction between the bipolar and the field effect transistor configuration. We refer to the layout of figure 7.3, where by definition the offset voltage of the differential stage can be written

$$V_{\text{offset}} = |V_{\text{off.1}} - V_{\text{off.2}}| \tag{7.9}$$

or

$$V_{\text{offset}} = |V_{\text{bias1}} - V_{\text{bias2}}|_{V_{\text{out}} = 0} \tag{7.10}$$

and for the offset current by definition

$$I_{\text{offset}} = |I_{b1} - I_{b2}|_{V_{\text{out}} = 0}. \tag{7.11}$$

The referred configurations are so-called coupled pairs.

7.4.1. The bipolar differential input stage
Applying equation (7.2) for the bipolar differential input stage we find

$$V_{\text{offset}} = \frac{kT}{q}\left(\ln\frac{I_{C1}\,W_{B1}}{KA_{E1}} - \ln\frac{I_{C2}\,W_{B2}}{KA_{E2}}\right) \tag{7.12}$$

in which it is assumed that both transistors have the same temperature and equal K values. The condition $V_{\text{out}} = 0$ gives the expression

$$I_{C1}\,R_{C1} = I_{C2}\,R_{C2}. \tag{7.13}$$

Combining equations (7.12) and (7.13) yields

$$V_{\text{offset}} = \frac{kT}{q}\ln\frac{R_{C2}\,W_{B1}\,A_{E2}}{R_{C1}\,W_{B2}\,A_{E1}}. \tag{7.14}$$

Typical values for the offset voltage of the bipolar differential stage are 0.5–5 mV. This offset is easily nulled via R_{C1} or R_{C2}, but as can be seen from the expression for the offset *symmetry* for all components involved plays a key role in the offset behaviour of the differential stage. This is true for the geometrical dimensions as well as for temperature coefficients and doping concentration levels for the transistors and resistors involved. The offset current can be derived from the expressions (7.10) and

$$V_{out} = R_{C1} I_{C1} - R_{C2} I_{C2} = 0. \tag{7.15}$$

Equation (7.15) with (7.3) yields

$$\beta_1 R_{C1} I_{b1} - \beta_2 R_{C2} I_{b2} = 0 \tag{7.16}$$

or

$$I_{b1}/I_{b2} = \beta_2 R_{C2}/\beta_1 R_{C1}. \tag{7.17}$$

Rearranging (7.11) yields for the offset current

$$I_{offset} = I_{b2}(I_{b1}/I_{b2} - 1). \tag{7.18}$$

Substituting (7.17) in (7.18) gives

$$I_{offset} = I_{b2}\left(\frac{\beta_2 R_{C2}}{\beta_1 R_{C1}} - 1\right). \tag{7.19}$$

Here again, it is obvious that the symmetry of β and R_C stipulates the offset current level.

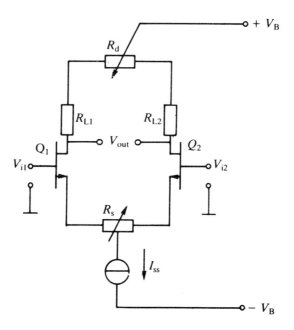

Figure 7.5 Junction FET differential input stage. R_s provides offset compensation for V_d; R_d provides offset current compensation for I_d.

7.4.2 The junction FET differential input stage

This configuration, often called a source-coupled pair, is depicted in figure 7.5. If the same definitions (7.10) and (7.11) are applied for the offset voltage of a junction FET differential input stage we find

$$V_{\text{offset}} = V_{\text{GS1}} - V_{\text{GS2}}$$

or

$$V_{\text{offset}} = V_{p1} - V_{p2} + \left(\frac{I_{D2}}{\Gamma V_{p2}(W/L)_2}\right)^{1/2} - \left(\frac{I_{D1}}{\Gamma V_{p1}(W/L)_1}\right)^{1/2}. \tag{7.20}$$

The offset current of a FET is generally neglected because of the very low value of the gate current. The gate currents for junction FETs are of the order of nanoamperes to picoamperes, and for the offset current some tens of picoamperes are reported. This will be discussed again in a later section. If desired, the expression for the offset voltage can be rearranged in relative ratios in W/L and the load resistances R_L.

7.4.3 The MOSFET differential input stage

Applying again the definitions for the offset voltage, the following expressions can be derived:

$$V_{\text{offset}} = V_{\text{GS1}} - V_{\text{GS2}} = V_{p1} - V_{p2} + \left(\frac{I_{D1} 2nL_1}{K_p W_1}\right)^{1/2} - \left(\frac{I_2 2nL_2}{K_p W_2}\right)^{1/2}. \tag{7.21}$$

This is equivalent to

$$V_{\text{offset}} = \Delta V_p + \frac{(V_{\text{GS}} - V_p)}{2}\left(-\frac{\Delta R_L}{R_L} - \frac{\Delta(W/L)}{W/L}\right) \tag{7.22}$$

where

$$\Delta V_p = V_{p1} - V_{p2} \tag{7.23}$$

$$V_{\text{GS}} = \tfrac{1}{2}(V_{\text{GS1}} + V_{\text{GS2}}) \tag{7.24}$$

$$\Delta R_L = R_{L1} - R_{L2} \tag{7.25}$$

$$R_L = \tfrac{1}{2}(R_{L1} + R_{L2}) \tag{7.26}$$

$$W/L = \tfrac{1}{2}[(W/L)_1 + (W/L)_2] \tag{7.27}$$

$$\Delta W/L = (W/L)_1 - (W/L)_2. \tag{7.28}$$

The same remark is applicable to the offset current as to the case of the junction FET. In the next section we will come to that again. In most cases it is easy to compensate for any offset in order to obtain a nulled output when no signal is present. An example of an offset nulling procedure for a differential input stage for current and voltage is depicted in figure 7.5. This nulling can be performed also by trimming techniques (by laser) on the integrated circuit itself. However, much more important are drift phenomena which will be discussed now in the next section.

7.5 OFFSET IN TRANSDUCERS

When a physical non-electrical parameter must be measured sensors are used. In various aspects this is often the most crucial part in electronic systems. In chapter 6 we have presented an comprehensive discussion about transducers. Using modulating sensors, often a Wheatstone bridge configuration is used. Integrated circuit technology makes it possible to integrate all four resistors on to one chip. Partly integrated bridge configurations where the configuration is completed with external resistors are also used. However, here again, in general an output signal can be measured, when no input signal is present. For the Wheatstone bridge circuit depicted in figure 7.6 the output voltage yields

$$V_{\text{offset}} = V_{\text{AB}} = V_{\text{CC}} \left(\frac{R_3}{R_1 + R_3} - \frac{R_4}{R_2 + R_4} \right). \tag{7.29}$$

Each asymmetry in this circuit results in an output offset voltage whatever the cause may be. To illustrate this, causes may be all resistors have a positive, but not precisely equal temperature coefficient, or the resistor values in magnitude are not equal, etc. It will be obvious that these asymmetries easily can result in a change in the output voltage. We will come to that again when we discuss the drift of sensors.

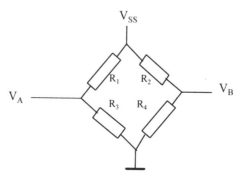

Figure 7.6   Resistive Wheatstone bridge configuration.

7.6 DRIFT BEHAVIOUR OF OFFSET CURRENT AND VOLTAGE

Drift behaviour of the offset current and offset voltage is much more crucial and complex than offset itself. In most cases, good measurements are available to prevent or compensate for any offset of either current or voltage. Drift phenomena are characterized by slow variations of the offset sources due to several causes and finally, in connection with noise, they determine the lowest limit which can be measured with specified accuracy. Drift is more complex to solve and is still one of the most discussed subjects in analogue electronics. As discussed previously we will present our discussion of drift phenomena in electronic components and drift phenomena in transducers separately, of which the sensor is the most crucial one.

In general four main factors generating drift can be mentioned:

1 stochastic variations in electrical conductivity
2 aging, humidity, mechanical vibrations, radiation, etc
3 power supply variations
4 temperature changes affecting parameters.

For transducers we find all possible combinations of these, especially for the input sensor.

7.6.1 Stochastic variations in electrical conductivity

Especially at low frequencies stochastic variations may influence the drift behaviour. This is especially the case where $1/f$ noise comes into scope, which may influence the drift behaviour considerably. Examples of components which have to be considered as large $1/f$ drift sources are carbon composite resistors, electrolytic capacitors and Zener diodes. Possible precautions are frequency shifting to a higher frequency band, filtering and the choice of components with little $1/f$ noise. More details can be found in section 10.4.

7.6.2 Aging, humidity and vibrations

The electrical conductivity can be influenced considerably by aging and humidity.

1 Aging can be considered as the changing of electrical properties of components with passing time.
2 Chemical and physical changes in the material characteristics from which the components are made are likely to occur as time passes. This can be considered as a wear-out process. Possible precautions are, for instance, derating in load conditions. More details about aging are found in chapter 5.
3 Humidity will also have an influence on conductivity. To prevent this, effective measures are available, such as hermetic sealing and slightly raising the temperature of humidity sensitive components.
4 Vibrations are mainly caused by hazardous environmental conditions as in missiles, aircraft, ships and mobile electronic equipment. These conditions can easily influence the electronic parameters and so the working performance of the electronic components. Suitable stable and rigid constructions, especially of connectors and plugs, can eliminate these problems. With regard to reliability, connectors, soldered connections and plugs show highest failure rates.

7.6.3 Power supply variations

In general, power supply variations will affect either the output voltage or the output current for which two factors can be mentioned related to voltage and current. By definition the *voltage power supply rejection ratio* (PSRR) is the equivalent input voltage which must be connected to the input to keep the output voltage constant when the supply voltage changes an amount V_B, because the relationship between input and output voltage is the gain factor. For a single bipolar amplifier stage as shown in figure 7.1, taking the derivative of

$$V_B - V_{out} = g_m V_i R_c \qquad (7.30)$$

this yields

$$\frac{1}{H} = \left. \frac{V_i}{V_B} \right|_{V_{out}=const} = \frac{\partial V_i}{\partial V_B} = \frac{1}{g_m R_c} = \frac{r_e}{R_c} \qquad (7.31)$$

where $r_e = kT/qI_C$, k is Boltzmann's constant (1.38×10^{-23} J K^{-1}), T (K) the absolute temperature, q (1.6×10^{-19} C) specific charge, I_C (mA) collector current, R_c collector resistance and V_B power supply voltage (V). In the case of a single stage, the supply voltage will directly influence the output voltage in a one-to-one relationship. Usually, this figure of merit is expressed in decibels.

*   *   *

Example 7.1
For instance, assume $I_C = 1$ mA, and $R_c = 2.5$ kΩ. From $I_C = 1$ mA we find $r_e = 25 \, \Omega$; then $1/H = 25/2500 = 0.01$, and the PSRR $= 10$ mV V^{-1} or 40 dB, which is a poor figure of merit.

*   *   *

For the power supply rejection ratio concerning the current we find

$$V_{out} = V_B - I_C R_c \qquad (7.32)$$

or

$$V_{out} = V_B - I_C R_c = 0 \qquad (7.33)$$

and with $I_C = \beta$, I_i we find

$$\frac{1}{H} = \left. \frac{I_i}{V_B} \right|_{V_{out}=const} = \frac{1}{\beta R_c} \qquad (7.34)$$

in which β is the current gain factor. A large value of β will deliver a large value for the rejection ratio in decibels. This value is often not specified by the manufacturer but may be of importance when current sources are implemented.

It will be clear that when the same definitions are applied for balanced amplifier configurations as in a differential stage, the power supply rejection ratio for the voltage and the current will increase considerably and a value of 120 dB can easily be obtained. Here too, symmetry aspects determine the final rejection ratio and for the current the tail current source involved will be dominant. Suppose for a differential amplifier a PSRR of 10 μV V^{-1} is specified; this equals 100 dB.

7.7   THE INFLUENCE OF TEMPERATURE

The influence of temperature producing drift is caused by the fact that most electronic parameters show a temperature-dependent behaviour. This temperature drift is one of the main causes of offset drift and, as said already, is much more cumbersome to compensate for. It should be emphasized here that, normally, offset nulling does not mean drift nulling. In general this offset nulling is only valid for one temperature. Firstly, we will demonstrate the causes of temperature drift with several examples.

*   *   *

Example 7.2
Thermocouple effects caused by temperature gradients across dissimilar metals can generate voltages proportional to the difference between the junction and the measurement end of the metal. This voltage can range from zero to hundreds of microvolts per kelvin depending on the materials involved. Small temperature differences can easily exist across small distances of several millimetres. It will be clear that this implies considerable problems when low signal levels are present. Temperature gradients can be found across resistors and package-to-printed-circuit-board interfaces. (a) The thermoelectric output of a wirewound resistor is zero if both ends have the same temperature. If not then the thermal EMF may be $\pm 40 \, \mu V \, K^{-1}$. (b) Also, natural and forced cooling effects may generate temperature differences across resistors. This is illustrated in figure 7.7.

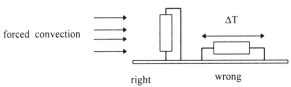

Figure 7.7   A correctly and a wrongly positioned resistor in the case of forced convection. In the case of natural convection only, the right and wrong positions must be interchanged (chimney effect).

Whole families of resistors exist with their specified temperature coefficients provided by the manufacturer. Carbon composite, oxide and some metal-film resistors can cause large thermocouple errors. Wirewound resistors manufactured from manganese are best, since they only generate about $2.0 \, \mu V \, K^{-1}$ with copper as reference. Finally, also radiation can generate thermal differences which induce natural convection. It has been shown that temperature differences can exist easily on silicon integrated circuits as well, despite their relative good specific thermal conductivity of 80–$150 \, W \, m^{-1} \, K^{-1}$. The thermal conductivity of silicon is also a doping-dependent parameter. This effect is used as an advantage in thermal anemometers.

Special software programs are available for thermal simulations of circuits at printed circuit boards (PCBs) and at IC level. For instance, with HSPICE three-dimensional (3D) thermal circuit model simulations can be performed at IC and PCB level.

*   *   *

Example 7.3
Suppose a low-drift amplifier is monitored over a 10 min period and during this period it appeared to have an input-referred offset variation of $\pm 5.0 \, \mu V$. When the

circuit is shielded from draught the offset variation is reduced to $\pm 0.5\,\mu$V, which results in a factor of improvement of 10.

$$* \qquad * \qquad *$$

Example 7.4
Suppose the gain setting resistors of an amplifier cannot track with temperature; then a gain error will result. For instance, if a gain of 1000 amplifier with a constant input of 10 mV appears to have a temperature mistrack of 0.5% over the desired operating temperature range, the error at the output will be 50 mV, which referred to the input is 50 μV. It is important that the gain setting resistors have the same temperature coefficient and have a close temperature tracking.

$$* \qquad * \qquad *$$

7.8 TEMPERATURE DRIFT IN BIPOLAR TRANSISTORS

Single-stage amplifier
The temperature drift behaviour of a bipolar transistor can be calculated by differentiating equation (7.2) with respect to T. This results in

$$\left.\frac{\partial V_{BE}}{\partial T}\right|_{I_C} = \frac{k}{q}\ln\frac{I_C}{A}. \tag{7.35}$$

Multiplying the right-hand term with T/T and substituting again equation (7.2) yields

$$\frac{\partial V_{BE}}{\partial T} = \frac{1}{T}(V_{BE} - V_{GAP}). \tag{7.36}$$

For a single bipolar transistor $V_{BE} = 0.6$–0.7 V, the bandgap voltage, $V_{GAP} \approx 1.2$ V, and, substituting for $T = 300$ K, the temperature coefficient of the offset voltage of the bipolar transistor is approximately -2 mV K^{-1}. For the temperature dependence of the offset current at the input we find

$$\left.\frac{\partial I_B}{\partial T}\right|_{I_C} = \left.\frac{\partial(I_C/\beta)}{\partial T}\right|_{I_C} = \frac{1}{\partial T}\left(\frac{I_C\,\partial\beta}{\beta^2}\right) \tag{7.37}$$

where I_C is considered to be a constant. Substituting again $I_C = \beta I_B$, the temperature coefficient for the offset current becomes

$$\left.\frac{\partial I_B}{\partial T}\right|_{I_C} = -\frac{I_B}{\beta}\frac{\partial\beta}{\partial T}. \tag{7.38}$$

This value appears to be smaller than 1% of I_B per kelvin.

The balanced amplifier

It is very easy to find an expression for the temperature dependence of the offset voltage of the *balanced* bipolar amplifier stage. Differentiating expression (7.14) delivers the desired result, which is

$$\frac{\partial V_{\text{offset}}}{\partial T} = \frac{k}{q} \ln \frac{R_{\text{C2}} W_{\text{B1}} A_{\text{E2}}}{R_{\text{C1}} R_{\text{B2}} A_{\text{E1}}} \tag{7.39}$$

and multiplying the right-hand term with T/T and substituting the original expression for the offset we find

$$\frac{\partial V_{\text{offset}}}{\partial T} = \frac{V_{\text{offset}}}{T}. \tag{7.40}$$

* * *

Example 7.5
Suppose the offset voltage for a balanced amplifier is 2 mV and $T = 300$ K; then the temperature coefficient of the offset voltage is

$$\frac{\partial V_{\text{offset}}}{\partial T} = \frac{2 \times 10^{-3}}{300} = 7.7 \ \mu\text{V K}^{-1}.$$

* * *

Theoretically it is possible to null this coefficient by controlling the ratio $R_{\text{c2}}/R_{\text{c1}}$ and making the logarithmic term 1, but in practical circumstances other temperature-dependent parameters remain, reducing the effect to several microvolts per kelvin. One of the most important reasons for this is the small temperature differences which still can be generated on a single piece of silicon, despite the relative good thermal conductivity for silicon as mentioned previously. This can be caused by dissipating differences on the chip itself, so it might be important to design the input stage and the output stage by calculating the thermal symmetry axis and placing the transistors around it. A type of temperature cross-matching of transistors is performed.

The temperature coefficient of the offset current is found by differentiating expression (7.11) and regarding (7.38); we find

$$\frac{\partial I_{\text{off}}}{\partial T} = \frac{\partial}{\partial T} (I_{\text{B1}} - I_{\text{B2}}) = I_{\text{B2}} \frac{\partial \beta_2}{\beta_2 \, \partial T} - I_{\text{B1}} \frac{\partial \beta_1}{\beta_1 \, \partial T} \tag{7.41}$$

and if we assume that

$$\frac{\partial \beta_2}{\beta_2 \, \partial T} = \frac{\partial \beta_1}{\beta_1 \, \partial T} = \frac{\partial \beta}{\beta \, \partial T} \tag{7.42}$$

then

$$\frac{\partial I_{\text{off}}}{\partial T} = I_{\text{off}} \frac{\partial \beta}{\beta \, \partial T} \tag{7.43}$$

where again $I_{\text{off}} = I_{\text{B2}} - I_{\text{B1}}$. This expression is a factor $I_{\text{off}}/I_{\text{bias}}$ smaller than in the case of the single-stage amplifier.

We still want to consider the case when the two transistors composing the input stage do not have the same temperature. Then the offset becomes, with all other parameters equal,

$$V_{\text{off}} = V_{\text{BE1}} - V_{\text{BE2}} = (T_1 - T_2)\frac{k}{q}\ln\frac{I_C W_B}{KA_E}. \tag{7.44}$$

Rearranging equation (7.2) gives

$$\frac{kT}{q}\ln\frac{I_C W_B}{KA_E} = V_{\text{BE}} - V_{\text{GAP}}. \tag{7.45}$$

Multiplying equation (7.44) with T/T and substituting equation (7.45) in (7.44) the result for V_{off} as a function of temperature T is

$$V_{\text{off}} = \frac{\Delta T}{T}(V_{\text{BE}} - V_{\text{GAP}}) \tag{7.46}$$

where

$$\Delta T = T_1 - T_2.$$

*   *   *

Example 7.6
Suppose a chip temperature is 353 K and the resident temperature difference ΔT on the chip is 1 °C, then the temperature coefficient of the offset voltage becomes

$$V_{\text{off}}|_T = 1 \times (0.6 - 1.15)/353 = -1.56 \text{ mV K}^{-1}$$

which is a considerable and rather poor value. Reducing the temperature differences between components and transistors will improve this value to a large extent and can be realized by applying integrated circuit technology only, which technology guarantees a much better temperature tracking of all components involved. Nowadays, drift specifications of less than 1 μV K^{-1} can easily be obtained.

*   *   *

7.9 TEMPERATURE DRIFT IN FETs

Here again we have to deal with two types of transistor, which require different descriptions.

7.9.1 The junction field effect transistor
Two main causes are responsible for the temperature drift in junction FETs.

1 With increasing temperature the channel resistance will increase.
2 The thickness of the depletion layer will decrease when temperature increases. This results in a lower channel resistance.

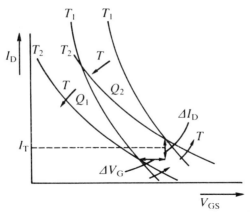

Figure 7.8 Behaviour of the transconductance of two junction FETs at two different temperatures.

As can be seen, both causes oppose each other, but a complete cancellation is obtained only in the region well above the pinch-off voltage. Figure 7.8 shows the transconductance of two junction FETs at two different temperatures T_1 and T_2. To calculate the temperature coefficient (TC) the expression for a *reverse-biased* diode is applicable; thus from (7.1)

$$I_D = I_{ss} = -K \frac{A_E}{W_B} \exp\left(-\frac{qV_{GAP}}{kT}\right)$$
(7.47)

where I_{ss} is the saturation current, and all other parameters are as mentioned before. Differentiating (7.47) with respect to T gives

$$I_D = I_{ss} = -K \frac{A_E}{W_B} \left(-\frac{qV_{GAP}}{k}\right) \exp\left(-\frac{qV_{GAP}}{kT}\right) d(1/T).$$
(7.48)

Substituting again equation (7.47) in (7.48) gives

$$\frac{dI_D}{dT} = +I_D \frac{q}{kT^2} V_{GAP}.$$
(7.49)

* * *

Example 7.7
If we suppose $T = 300$ K and $V_{GAP} = 1.15$ V, then substituting these values in (7.49) the temperature coefficient equals

$$\frac{dI_D}{dT} = 0.147 I_D \qquad (\text{A K}^{-1}).$$
(7.50)

Note again that here I_D is the reverse-biased diode current and not the drain current. This is a rather high value ($\pm 15\%$) but, owing to the very low absolute value of the gate current of a few nanoamperes, the influence remains rather low.

*   *   *

Offset voltage drift in junction FETs is not easy to calculate because several terms in the expression for the offset have different temperature coefficients and partly cancel each other. This is especially the case at one value for the drain current, I_D. The value of the temperature-dependent offset voltage for a source-coupled junction FET pair appears to be some microvolts per kelvin.

7.9.2 The MOS field effect transistor

Determining the TC of a MOSFET is much more complicated than in the case with the junction FET, but here a very low value of the offset current is involved, picoamperes, and so it makes little sense to calculate this value which has a negligible influence. As will be seen the $1/f$ noise is more important and therefore, in general, applications of MOSFETs in DC input stages are not found. However, because of recently obtained large improvements in CMOS technology more and more linear DC CMOS amplifiers can be met in a great variety of integrated (sensor) applications.

It is interesting to compare several types of amplifier with their state-of-the-art specifications. This is depicted in table 7.2 for three different types of amplifier in

Table 7.2 Comparison of three types of amplifier.

Description	Bipolar (741)	FET amplifier (OPA100 (BB))	Chopped amplifier (7650 (Intersil))	Units
Input offset voltage (25 °C)	1000–5000	250	1	μV
V_{off} temperature coefficient	15	± 10	± 0.01	$\mu V\,K^{-1}$
Input bias current (25 °C)	80 000	± 1	35	pA
Input offset current (25 °C)	20 000	± 0.5	0.5	pA
I_{off} temperature coefficient	$0.5\,nA\,K^{-1}$	NS	doubles every 10 K	
CMRR	90	82	130	dB
PSRR	96	NS	130	dB
Offset voltage versus time	NS	NS	0.100	$\mu V\,month^{-1}$

NS, not specified.

which the well known 741 is mentioned for reference reasons. In the fourth column, a so-called chopped amplifier is mentioned, in which the offset is automatically nulled at regular time intervals. This is realized with an internally generated frequency and two added capacitors by which the fault voltage is nulled. A possible principle of operation is illustrated in figure 7.9, in which an internal oscillator, a main amplifier and a so-called null amplifier are shown. The oscillator drives two electronic switches A and B. Both amplifiers can be nulled, whereas the main amplifier directly connects the input to the output. The null amplifier measures its own offset when the A switches are closed and B switches are open. When the B switches are closed and the A switches are open the nulling amplifier measures the output of the main amplifier. The required A and B switches are analogue MOSFET switches and are controlled by the local integrated oscillator for which only two capacitors should be added externally. The two capacitors store the required compensating voltages. The

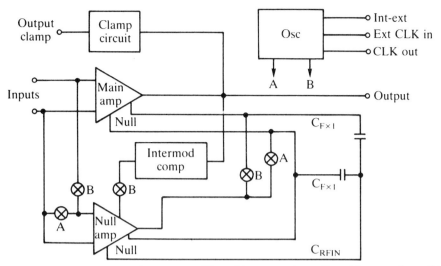

Figure 7.9 Principle of a chopped amplifier with extreme low offset. (Courtesy Intersil.)

extreme low offset voltage of less than $1 \mu V$ is realized by comparing the output voltages of both amplifiers by an algebraic subtraction via the two capacitors.

Some other measures are available to eliminate offset voltage, such as isolating the interfering source. This can imply power supply stabilization, maintaining a constant temperature for all temperature-sensitive parts and the choice of very low sensitive materials and components. Of course you will have to pay for it, but it can be worthwhile.

7.10 DRIFT IN SENSORS

Drift in sensors has become a very real problem because of an increase of potential for more accuracy and sensitivity. The sensitivity of sensors has become extremely high and it is easily possible to detect signals at the microvolt level. However, at the same time this low signal level requires questions to be answered: what is the real signal contribution and what is the noise contribution? At the input of a system noise and temperature effects determine final accuracy. This is part of a conversion process in the transducer itself, because the measurand is converted by the transducer into an electrical quantity. We will state that the main errors and/or noise contributions originate from the transducer's site and not from the electronic part. In this respect a very important factor is also that IC technology has made it possible to integrate together sensor part and processing part on to one single chip. We have called this type a smart sensor; see also chapter 6. In general, for transducers we can say that we have to deal with

- an offset-to-signal ratio, $V_{\mathrm{offs}}/V_{\mathrm{s}}$
- a drift-to-signal ratio, $V_{\mathrm{drift}}/V_{\mathrm{s}}$
- a noise-to-signal ratio, $V_{\mathrm{noise}}/V_{\mathrm{s}}$.

Table 7.3   Review of errors in transducers.

Type of error	Characteristic cause or effect	Dimension
Offset	Non-zero output signal due to asymmetries in circuitry or layout at zero signal input	V or I
Multiplicative drift	Scaling error change, or sensitivity change	e.g. $\partial V/\partial T$
Additive drift	Slowly varying output as a function of time, aging, pollution, temperature or other external influences	e.g. $V\,h^{-1}$ or $V\,K^{-1}$
Interference	External influences, e.g. EMC (up to megahertz)	V
Noise	Physical effects of materials of components; frequency and/or temperature dependent	e.g. $nV\,Hz^{-1/2}$

In all three cases we want a minimum contribution of these effects. However, in many applications offset-to-signal ratio and drift-to-signal ratio are a more serious problem than noise-to-signal ratio.

By definition an offset error is just a fixed quantity and just is the level in which the output signal deviates from zero when the input signal is zero. Causes are usually asymmetries in the design or applied circuitry. When the output changes slowly and no input signal is present we are talking about *drift*. For instance, when the offset is influenced by an undesirable parameter, such as temperature, pressure, radiation, a magnetic field, aging, pollution or another slowly varying external physical influence, it is drift. When the output is influenced by internal physical effects in the transducer elements itself, we talk about noise. This will be discussed in chapter 9. Noise limits the final sensitivity of any transducer. Finally, in this respect we have to mention *interference* which is characterized as any other external electromagnetic cause of disturbance of the output signal. These causes of error can cover a large range in frequency up to megahertz. This is summarized in table 7.3.

We have introduced the concept of drift already in chapter 4, figures 4.1 and 4.11, where we have discussed the difference between multiplicative or scaling errors and additive or offset error changes. For convenience we have depicted this again in figure 7.10 in a slightly different form. Scaling or multiplicative errors result in a sensitivity change, whereas offset errors only give a level shift of the signal.

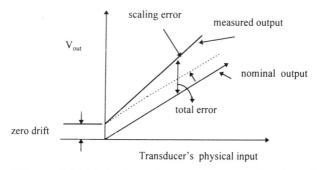

Figure 7.10   Offset and drift behaviour of a transducer as a function of an arbitrary physical input.

For instance, assume that the drift behaviour is a function of time and temperature. Then we can specify the indicated value as the measured value of the offset at a certain time, or at a certain temperature; hence

$$V(t)_{\text{zerodrift}}|_{V_s=0} = \frac{V_{\text{out}}}{t} \qquad (7.51)$$

or

$$V(T)_{\text{zerodrift}}|_{V_s=0} = \frac{V_{\text{out}}}{T}. \qquad (7.52)$$

* * *

Example 7.8
A transducer at zero input, $V_s = 0$, shows a temperature drift of 44 mV over a temperature change of 80 °C; then the specific zero drift per degree is $44/80 = 0.55$ mV K^{-1}.

The same transducer shows a drift in time of 15 mV over 10 h; then the zero drift in time is $12/10 = 1.5$ mV h^{-1}.

The transducer shows a positive sensitivity drift of 22 mV over 40 °C; then the scaling error γ is $12/40 = 0.3$ mV K^{-1}. The scaling error $\partial V(s)/\partial s$ is not necessarily a constant and then it must be determined by an extensive calibration procedure over a specified range.

* * *

We have assumed so far that effects are independent of each other; in practice this is often not true and we have seen already that cross-effects complicate the description considerably. If we assume that all parameters are independent, so that no cross-effects are present, then we can describe the drift behaviour with the following expression:

$$V_{\text{out}} = \gamma V(S) + V_{\text{offset}} + E(V_1, V_2, \ldots, V_n) \qquad (7.53)$$

where V_{out} is the output signal, $V(S)$ the desired signal voltage, γ the multiplicative coefficient or scaling factor defined as $\partial V(S)/\partial s$, V_{offset} the magnitude of output signal when no input signal is present, E the total of additive error voltages as a function of undesired physical parameters and V_i the error voltage caused by the physical input i. This basic description makes it possible to discuss some methods of drift elimination. It is easy to see that when $V(S) = 0$ the offset as output signal remains. In general we can eliminate drift by compensation and by applying a type of feedback, which is always better than compensation. Compensation requires an exact knowledge of the drift phenomenon; only under this condition can a reversed signed signal be added to the input to compensate precisely for the error signal. It will be clear that eliminating possible drift causes, before the signal can enter the transducer, is the best way to do so. In the next sections we will discuss several methods to eliminate drift.

7.10.1   Drift elimination by modulation

A well known method to reduce drift influences is the application of *modulating* the measurand. This can be performed in three different stages at the input: (1) before the measurand enters the transducer, applicable for self-generating transducers, (2) in the transducer itself, applicable for modulating transducers, and (3) directly after the transduction of the measurand. It will be clear that modulation of the measurand before transduction is the most effective one.

The first possibility arises when the process has a self-generating physical quantity. Here, we can apply a *mechanical chopper disc* between the physical quantity and the transducer. An example is the measurement of the wall temperature by measuring its IR radiation. This principle is illustrated in figure 7.11(a).

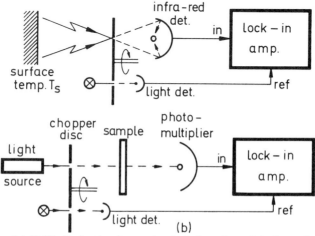

Figure 7.11   (a) Drift elimination by mechanical chopping of the incoming IR radiant energy. (b) Drift elimination applying mechanical chopping for a modulating transducer.

Secondly, when we have to deal with modulating transducers, which need a power supply, an electrical form of modulation can be used, applying a type of AC signal to excite the transducer. Quite a number of variations exist on this concept. One example is given in figure 7.11(b) where the absorption energy of a sample is measured. An auxiliary (light) energy source is needed.

These two modulation principles deal with modulating the incoming energy before entering the transducer. A second possibility is to apply modulation in the transducer itself. A third possibility is to chop the electrical output of the transducer using a so-called chopper amplifier. In this case the modulation is performed after the transduction. An example of this principle of a chopped amplifier has been given already in figure 7.9.

It is obvious that modulation must be performed preferably as early as possible. Once a disturbing signal has become part of the desired signal, no distinction can be made any longer. In all these circumstances a kind of discrimination is generated between the desired signal and the undesired signal. Filtering, or the application of a lock-in amplifier, performs the final required operation to produce the measurand in an electrical form.

Another completely passive measure to reduce errors is the prevention of disturbing signals entering or influencing the transducer. This can be performed by proper guarding, shielding and filtering of the input transducer with the system the transducer is connected to. This will be discussed in more detail in chapter 8.

However, recently a newly patented approach has been developed to eliminate additive drift in sensors. This will be discussed in the next section.

7.10.2 The alternating direction method (ADM)

Referring to (7.53), in this discussion we focus on additive drift only, and assume $\gamma = 1$. In the ADM, the basic idea is to construct *vector* sensors that are *anisotropic* for the measurand S and *isotropic* for the parameters V_i. When the measurand S is a vectorial quantity, for instance flow, an electric field, acceleration or velocity, which we shall henceforth refer to as S, we shall then have the ideal situation that

$$V_{\text{sensor}} = V(S) + \varepsilon E(V_1, V_2, \ldots, V_n) \tag{7.54}$$

with ε arbitrarily small. The ADM implies that two separate partial measurements are performed, in which the anisotropic characteristics of the vector sensor are used. Then the anisotropic characteristics of the sensor make it possible to measure two different sensor signals at the time t with $V^A_{\text{sensor},t}$ and at the time $t + \Delta t$ with $V^B_{\text{sensor},t+\Delta t}$, depending on the two relative orientations A and B of the sensor with respect to the vector S. This is illustrated in figure 7.12. When we further may assume without loss of generality that

$$V^A_t(S) = -V^B_{t+\Delta t}(S) \tag{7.55}$$

and again assuming further a constant input signal S, it will then hold that in measuring A we have

$$V^A_{\text{sensor},t} = V^A_t(S) + E_t(V_1, V_2, \ldots, V_n) \tag{7.56}$$

and in the second measurement

$$V^B_{\text{sensor},t+\Delta t} = V^B_{t+\Delta t}(S) + E_{t+\Delta t}(V_1, V_2, \ldots, V_n). \tag{7.57}$$

Note now that the two partial measurements introduce two different states of the sensor in the input energy domain S, but similar states in the energy domain V_i.

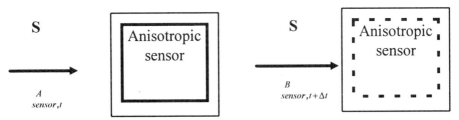

Figure 7.12 Principle of the ADM using an anisotropic sensor and performing two measurements in two sequential short time intervals, t and $t + \Delta t$.

When we subtract (7.56) and (7.57) and we take into account equation (7.55) we obtain the ADM signal:

$$V_{\text{sensor},t}^{A} - V_{\text{sensor},t+\Delta t}^{B} = V_t^A(S) = -V_{t+\Delta t}^B(S) + \varepsilon \qquad (7.58)$$

or

$$V_{\text{ADM}} = 2V_t^A(S) + \varepsilon \qquad (7.59)$$

where the influence of the drift term reduces to $\varepsilon \approx (\mathrm{d}E/\mathrm{d}t)\,\Delta t$. By choosing the time interval sufficiently short relative to the time interval over which the drift significantly affects the output signal, the drift influence can be made arbitrarily small. We also see that any offset value has also become irrelevant because of the applied subtraction. The application of a vector sensor requires that either we reverse, in the two indicated time intervals, the sensor with respect to the measurand S, or we reverse the direction of S.

The use and power of this ADM are best demonstrated with long-term high-accuracy flow and volume measurements.

*   *   *

Example 7.9
The basic vector sensor configuration has been described already in chapter 3, figure 3.41. The sensor consists of a silicon integrated double-Wheatstone-bridge config-uration, heated to a constant mean temperature in a thermal feedback loop. The temperature gradient $T(x, y)$ that is generated by the air flow defines our measurand S and yields a voltage difference as a measure of the air flow velocity. This flow sensor is bidirectional in the sense that the sign of its output signal changes when the relative orientation of the flow with respect to the sensor changes by 180°. Owing to imperfections, measurements made using this sensor represent errors due to asymmetries in a component that are a function of the total power dissipation of

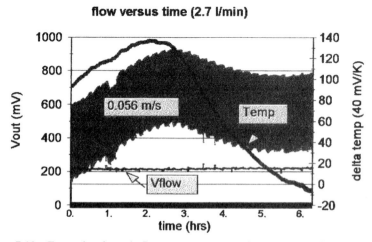

Figure 7.13 Example of an air flow measurement using the ADM for an average flow velocity of 0.056 cm s^{-1} over a 6 h time period.

the sensor itself, but also errors due to drift in electronics. In this example the chip is rotated over 180° in the indicated time intervals.

ADM cannot be realized by electronics alone, or by changing interconnections, or by using two sensors, since from the output signal of the sensor it is not possible to know the contribution of the drift since V_{sensor} as defined in (7.53) does not define the causes of its components. Hence, replacement using two similar sensors also does not guarantee perfect cancellation where in general $E_1 \neq E_2$.

It can be said that all sensors work; however, when accurate long-term measurements must be performed the ADM appears to be very powerful. Note that often long-term measurements are not specified by the manufacturer at all. An example of a flow measurement is given in figure 7.13.

In figure 7.13 we observe the change in the average position of the flow measurement over a 6 h time period from 400 mV to 700 mV that represents a drift of over 300% with respect to the true output signal of about 200 mV. Further, note that the absolute value of the offset voltage has become totally irrelevant, because, by subtraction, the offset is also cancelled. The *'bandwidth'* of the measurand is a constant under constant flow conditions. The average flow velocity of 0.056 cm s^{-1} is an extreme low value and can normally only be measured with a laser Doppler anemometer. Note also the minor influence of the change of the temperature change over 3 K. At larger medium temperature changes, the influence of temperature is revealed from a decrease of the sensitivity of the sensor when the medium temperature increases. This is in accordance with the theory for thermal flow sensors.

$$* \quad * \quad *$$

From this example it can be seen that the ADM using vector sensors is a very powerful method for which it is worthwhile to investigate the viability for other physical parameters.

PROBLEMS

7.1 What is meant by offset and what by drift? Explain why it is not always true that a realized nulling involves no drift.

7.2 The temperature coefficient of a differential amplifier is 3 μV K^{-1}. The gain setting factor is 1000 and the offset voltage is 1.5 mV at 25 °C. By what value will the output voltage change when the ambient temperature changes by 15 K?

7.3 In figure 7.14 a temperature measurement set-up is drawn, in which $R(T)$ is a temperature-dependent platinum resistance. The temperature behaviour of $R(T)$ is described with $R(T) = R_0(1 + \alpha T)$, where $R_0 = 100\ \Omega$, $\alpha = 3.2 \times 10^{-3}$ K^{-1} and T is the resistance temperature in degrees Celsius. The resistance of the supply leads is represented by r, and the reference voltage source $V_{\text{ref}} = 7.25$ V. If the offset voltage is 100 μV and $I_{\text{bias}} = 100$ nA, calculate the output error in degrees Celsius if $T = 0$ °C.

7.4 Calculate the offset voltage change for a single-junction FET when the drain current changes 12%.

7.5 Derive an expression for the offset voltage for figure 7.5 when there is a mismatch of 10% in W/L and all other parameters are supposed to be balanced.

7.6 Perform the same procedure as in problem 7.5 for a source-coupled pair composed of MOSFETs.

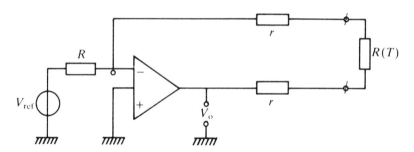

Figure 7.14 Temperature measurement set-up of problem 7.3.

7.7 Derive an expression for the output of the vector sensor in example 7.7 as a function of the flow velocity, v.

BIBLIOGRAPHY

[1] Sudar T 1980 *Digital Instrument Course* vol 3 (Philips Gloeilampen fabrieken)
[2] Solomon 1974 The monolithic op-amp, a tutorial study *IEEE J. Solid State Circuits* **9** 314
[3] Gray and Meyer R G 1980 *Analysis and Design of Analog Integrated Circuits* (New York: Wiley)
[4] Horowitz and Hill W 1986 *The Art of Electronics* 2nd edn (Cambridge: Cambridge University Press)
[5] van Putten M H P M, van Putten M J A M, van Putten A F P and van Putten P F A M 1993 Drift Eliminatie bij Sensoren *Dutch Patent Application* 9301399
[6] van Putten M J A M, van Putten M H P M and van Putten A F P 1994 Full additive drift elimination in vector sensors using the alternating direction method (ADM) *Sensors Actuators* A **44** 13–7
[7] van Putten M H P M, van Putten M J A M, van Putten A F P and van Putten P F A M 1995 Method for drift elimination in sensors *US Patent* 5426969
[8] van Putten M H P M, van Putten M J A M, van Putten A F P and van Putten P F A M 1994 Method for drift elimination in sensors *European Patent Application* 94202293.0

Eight

GUARDING AND SHIELDING

8.1 INTRODUCTION

This chapter includes a comprehensive discussion of guarding and shielding techniques which are available in electronic measuring systems in order to avoid any kind of interference. The causes and the nature of interference will be outlined and it will be seen how important guarding and shielding are with respect to the required common-mode rejection ratio (CMRR). The influence on the claimed accuracy of the applied measuring instrument will be investigated. We shall discuss pitfalls, what to do and what not to do, and will conclude with a collection of problems for practical training. Sometimes noise is also considered as interference, but because the nature of noise is a more fundamental physical process of the components and devices itself, we will treat noise in a separate chapter.

Guarding can be defined as a collection of measures and precautions taken to prevent stray currents entering sensitive electronic parts by connecting the appropriate parts of a circuit to earth or ground.

Shielding is the technique of placing electronic systems in a metal casing to prevent electrostatic and/or magnetic fields entering sensitive components. We will start our discussion with a few examples of how interference can occur and demonstrate some causes of interference.

8.2 THE NATURE AND CAUSES OF INTERFERENCE

Interference can be defined as any kind of physical influence on a given system which reduces the quality and performance of that system. This influence may cause a reduction either in accuracy or in CMRR, as will be seen later. If very low signal levels are present, the influence of noise will play a vital role but, as said already, because of the character of noise, this requires a separate treatment and will be discussed in chapter 9.

Two different types of source of interference can be distinguished. Interference can be caused either by (1) electrostatic fields or by (2) magnetic fields. Both causes will be treated in detail. Since most instruments obtain their operating power from the mains supply, one of the most general sources of interference is the mains (voltage) coming from the power supply with a mains frequency of 50 or 60 Hz depending on the country.

In most cases we will assume a fixed frequency of 50 Hz throughout this chapter. The mains can easily induce a voltage by parasitic capacitive coupling to the measuring system or to the signal source itself. This is shown in the following examples.

$$* \quad * \quad *$$

Example 8.1

Figure 8.1 illustrates the situation where a person touches the input of an oscilloscope which has an input impedance of ≥ 1 MΩ. Assume that the transition impedance of the person is $R_{tr} = 500$ kΩ and the capacitance between the mains (50 Hz) and R_{tr} is $C \approx 5$ pF, which is a typical value for normal rooms; then an input signal V_i will appear on the CRT of which the magnitude can be calculated as follows:

$$V_i = V_m \frac{R_i}{R_{tr} + R_i + 1/j\omega C} \tag{8.1}$$

where V_m is the mains voltage (assumed to be 220 V), R_{tr} is the transition impedance of the person, R_i is the oscilloscope's input impedance and C is the parasitic capacitance.

It can be calculated that $|1/j\omega C| \gg R_{tr} + R_i$; hence with neglect of these terms the denominator becomes

$$|1/j\omega C| = \frac{1}{2\pi 50 \times 5 \times 10^{-12}} \approx 600 \text{ M}\Omega \tag{8.2}$$

and the resulting interfering input voltage will be

$$V_i = \left| \frac{R_i}{1/j\omega C} V_m \right| = |j\omega C R_i V_m| = 0.38 V_{ef}. \tag{8.3}$$

So a considerable input voltage is induced at the input of the oscilloscope as a result of a capacitive coupling of the mains and a high input impedance, R_i. As will be seen, capacitive coupling will be one of the main causes of interference.

$$* \quad * \quad *$$

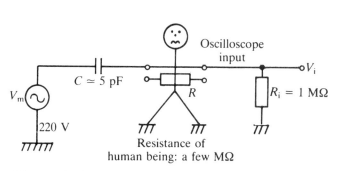

Figure 8.1 Direct mains interference of a human being touching the input terminals of an oscilloscope.

Example 8.2

Another example where interference has been a problem is the determination of an electroencephalogram (EEG) of a human being. This means monitoring the brain activities by the measurement of external voltages on the skin. For this purpose at least two electrodes must be attached to the patient's head skin and, via a small parasitic capacitance of, say, 2 pF, the patient is connected to the mains. This is illustrated in figure 8.2. In the same figure the electrical equivalent of the disturbing source is depicted.

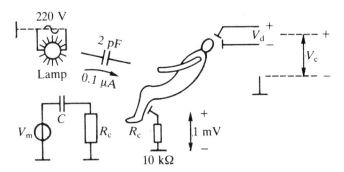

Figure 8.2 Recording problem of an electroencephalogram (EEG) and its parasitic small-signal equivalent circuit.

To reduce the interference, the patient is connected to ground with a separate electrode and, via the parasitic capacitive coupling, a current can flow through the remaining resistance, R_c, of, for instance, 10 kΩ. Despite a carefully made connection to common, a rest impedance will often remain. The magnitude of the interference current can be calculated from

$$I_i = \frac{V_m}{R_c + 1/j\omega C}. \tag{8.4}$$

Substituting the given values and, again, assuming $|1/j\omega C| \gg R_c$, this yields, with $V_m = 220$ V for the current

$$I_i = |j\omega C| V_m = 0.14 \ \mu\text{A} \tag{8.5}$$

and the generated voltage difference over R_c can be calculated from

$$V_c = I_i R_c. \tag{8.6}$$

Substituting the values results in $V_{cm} = 1.4$ mV, which is a factor of 1000 higher than the source voltage level to be measured, which is normally in the range 1–10 μV. Hence the interference completely obscures the required signal. Again, from this example, the conclusion can be drawn that, in circumstances where relatively high frequencies are present, stray capacitances play a vital role.

*   *   *

Apart from this capacitive coupling via the mains, there are still other tedious sources of interference present. The frequency of the mains (50 or 60 Hz) is in the middle of the frequency band of the brain's activities, which is in the range 1–100 Hz and, as a consequence, is sensitive to those frequencies. As well as this, the heart muscle and other muscles generate common-mode voltages of about 1 mV. Both sources can influence the accuracy considerably.

These sensitive measurements require an amplifier with a balanced input configuration. In figure 8.3 two alternative representations of a complete incremental

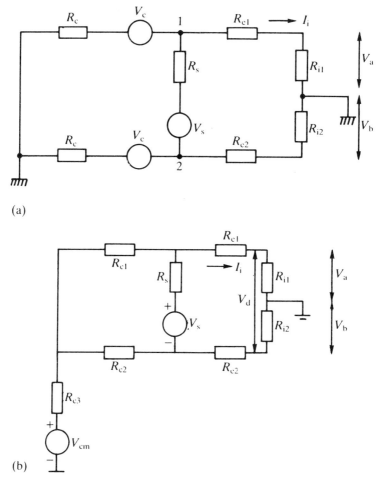

(a)

(b)

Figure 8.3 (a) Interfering signal source at the inputs of the applied balanced amplifier monitoring an EEG; (b) alternative diagram for interference circuit.

model of such a balanced amplifier input configuration are depicted. The signal source V_{cm} with internal resistance R_c represents the interfering source which is present between both electrodes and ground. Because this signal acts simultaneously on both electrodes 1 and 2, with equal sign, it is called a common-mode signal source.

The voltage source V_s, with internal resistance R_s, represents the measurand and is present between both electrodes 1 and 2. V_s is called the desired differential-mode signal V_d. R_{e1} and R_{e2} represent the two electrode impedances which very often fluctuate in magnitude and are of considerable level.

The input impedances of the amplifier's circuitry are denoted by R_{i1} and R_{i2} and should be much larger than R_{e1} and R_{e2} to avoid any attenuation of the small signal. If R_{e1} is not precisely equal to R_{e2} it is easily seen that V_{cm} induces a differential-mode signal, which will be added to the desired signal. We will denote both input signals of the amplifier with respect to ground by V_a and V_b.

Referring to figure 8.3, a definition for the differential-mode signal and the common-mode signal can be found by writing V_a and V_b as a sum and difference of two voltages as follows:

$$V_a = \tfrac{1}{2}(V_a - V_b) + \tfrac{1}{2}(V_a + V_b) \tag{8.7}$$

and

$$V_b = -\tfrac{1}{2}(V_a - V_b) + \tfrac{1}{2}(V_a + V_b). \tag{8.8}$$

In these equations, by definition the component $\tfrac{1}{2}(V_a - V_b)$ is called the differential-mode signal because this signal appears with opposite sign at both inputs of the amplifier. In the same way, $\tfrac{1}{2}(V_a + V_b)$ is called the common-mode signal because this signal appears with the same sign at both inputs of the amplifier.

As we have seen already, it is important to recognize possible interferences due to small parasitic stray capacitances, especially when signal sources can supply low power only. In the past the only good protection against interference was placing the patient in a Faraday cage which then prevents an external electrical field passing through it. However, because of the availability of excellent differential amplifiers, this has become obsolete and is no longer required.

*   *   *

Example 8.3
When a measuring system is used where a digital voltmeter or a recorder is implemented and accuracies are claimed of about 0.1% or better for both instruments, great care must be taken in order to maintain this accuracy. Suppose a recorder has a guaranteed accuracy of 0.25% and is used with a full-scale deflection of 1 mV. This corresponds to 2.5 μV, or 1% corresponds to 10 μV. Suppose further that a parasitic voltage of 20 μV is present. Then an error of 2% arises which completely destroys the claimed accuracy.

*   *   *

8.2.1 Sources of interference
Every part of the system in which small signals are present is basically sensitive to interference. Directly or indirectly the mains is one of the most common sources of interference but other sources may exist as well. The following list gives nine other potential sources of interference. Interference can be caused either by sources which

can produce an electromagnetic field or by sources which have a thermal or mechanical influence on the system.

1   A temperature coefficient can be assigned to every part of a system, so the magnitude of certain parameters of a component is always a function of temperature and temperature changes. Temperature can influence the biasing of active components considerably. In very sensitive parts we have to reckon with temperature gradients caused by dissipation differences in components. This may cause thermopile voltages because of the Seebeck effect. Especially large temperature differences can be present as a result of dissipation differences of components mounted on printed circuit boards. Seebeck voltages are of the order of 40 μV K^{-1}. This can produce a constant error (offset) or a slowly varying error (drift).

2   Mechanical shocks can influence poorly installed and soldered connections.

3   Ignition equipment in cars produces a broad bandwidth of frequencies. The ground frequency lies between 50 and 200 Hz but higher frequencies are also present, ranging from 30 to 300 MHz and, for these frequencies, the spark-plug cables can act as a transmitting antenna.

4   In digital equipment, pulse-shaped signals are applied and the present frequency bandwidth is a function of the slope of the applied pulses and ultimately determines the magnitude of the influence.

5   Switches in power distribution systems can produce large pulses and then act as an interfering source. This is also the case in large production equipment systems in which a large number of switches must be controlled and activated. A very popular type of power controller producing a large amount of interference is the thyristor. The steeper the slope, the greater the interference. Switching at the precise moment the voltage crosses the zero level can reduce this type of interference.

6   The collector of an AC series electromotor can be interpreted as a fast switch, operating with a frequency between 1 and 10 kHz determined by the number of lamellae and the number of revolutions per minute. This mechanical switch produces pulses with steep slopes and thus a broad frequency spectrum will result.

7   Very high voltages can produce corona effects which cause a continuous interfering spectrum with a large bandwidth.

8   In a gas discharge lamp a periodic discharge is generated producing an interference noise signal level above 1 MHz. A filtering element such as an induction coil can reduce this influence.

9   Finally, interfering sources with a very small frequency bandwidth should be mentioned, such as HF generators, magnetrons, welding equipment and all types of transmitter.

Interference of these kinds can be eliminated by two measures:

1   the application of an isolating transformer
2   the use of shielding leads and cabinets.

You will not be surprised that a lot of mistakes can be made when applying these two methods, and that is what we will discuss in the following sections. At first glance it may seem very simple, but large mistakes can be made. Let us look at the mistakes and the solutions, in this sequence.

8.3 ASPECTS OF GUARDING

Applying a type of guard and/or shield is a good method to prevent interference. In this section we will discuss aspects of guarding qualitatively before going into more detail and making calculations. Guarding can be considered as the technique of applying one conductor only as the common, which may be grounded or earthed.

Many instruments have three input terminals, marked with a 'high' or '+'; a 'low' or '−' sign; and an 'earth' or '⊥' mark. The terminal marked with 'earth' is a safety earth to which all metal parts of an instrument are connected and which in its turn may be connected to earth or ground of the mains cable. In a proper installation this safety earth prevents any of the conducting parts of the system which can be touched from rising in potential above 42 V. A voltage of 42 V with respect to earth is the commonly accepted safety voltage for the human body.

Basically, current is the limiting factor here and figures 8.4 and 8.5 show this sensitivity as a function of frequency and current, and the consequences of the current on a human body.

In many instruments, a metal link or lead, the so-called 'low', can be connected to the safety earth terminal, linking the common of the system to ground. If this is the only connection to earth the measurement can be performed straightforwardly, but great care must be taken to ensure that the system under test has no other connections to earth because this can introduce errors in sensitive measurements. In

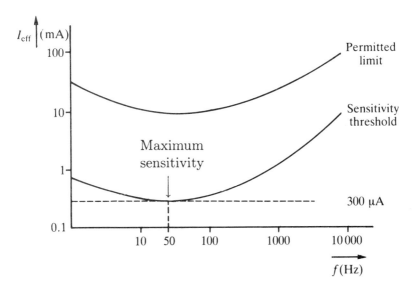

Figure 8.4 Current safety margins for the human body as a function of the frequency. Note the dip between 40 and 60 Hz.

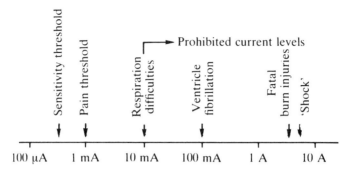

Figure 8.5 Survey of the consequences of a 50 Hz alternating current flowing from hand to hand in a human body.

this case the 'low' and earth have to be disconnected, resulting in a 'floating' common preventing ground loops.

8.3.1 Ground and earth

Another aspect to be considered in earth connections is the very common pollution of earth terminals in the mains plug. In many real circumstances the earth terminal is not clean at all, but polluted with large interference and can even have several volts with respect to ground level.

Ground can be defined as any reference conductor that can be used for a common return. Any ground should be an equipotential for the whole system involved; in other words, it ideally should have infinite conductivity. *Earthing* is just a particular case of grounding. On many instruments, there is a metal link that provides the possibility to make a connection between low and ground to ensure safety earthing.

That a ground is not always an equipotential and that care that must be taken before using a common can be seen from the following examples. In a ship several hundreds of volts can exist between decks, superstructures and rigging. Across an aircraft up to 100 V difference can exist between structural points such as nose, wing and tail parts. In motor vehicles several volts can exist between different points. On PCB level, 5 cm of standard PCB copper track of 0.25 mm width and 0.025 mm thickness can cause 1 LSB drop. Other causes of grounding errors are when no distinction is made between signal ground, digital ground, safety ground and power supply ground. Note that any current flowing out must also return. For instance, when a bias current for a transducer of 50 mA flows back then a common ground can cause a 5 mV level shift in a 0.1 Ω PCB track.

When lightning occurs, earth stray currents can generate several kilovolts between different ground connections. A good earth connection can be made by driving copper- or cadmium-clad iron rods into the ground, sometimes to a depth of up to 20 m, depending on the transition impedance of the rods to earth. In most cases, this transition impedance should not exceed 0.2 Ω. Different values may be required and are application dependent. The transition impedance is to a large extent a function of the specific conductivity and humidity of the soil. It is not unlikely that more than one rod is necessary to obtain the required transition impedance. Used as a prescribed safety earth, a voltage of 42 V between the metallic housing and earth may not be exceeded.

8.3.2 A grounded system

With this in mind, consider the case when two points in a measuring system are connected to ground as depicted in figure 8.6. It will be clear that a type of ground loop is created if the transducer and the measuring instrument are both connected by relatively long cables. You may think of thermocouples or strain gauges for the transducer and a chart recorder for the measuring instrument. In this circuitry, R_1 and R_2 represent the respective cable resistances. In general there are always stray or erratic currents in earth and these currents combined with the soil resistance form an imaginary voltage source V_{cm} causing a current to flow in R_1 and R_2.

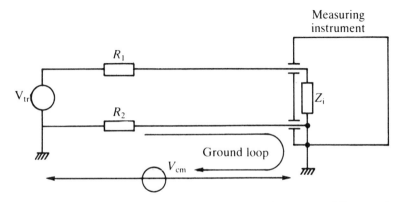

Figure 8.6 The creation of a ground loop by earthing at two different points.

It is obvious that the current in R_2 will be larger than in R_1, because of the relatively high input impedance Z_i of the measuring instrument; hence, because of V_{cm}, unequal series voltages are generated in R_2 and R_1, and these will be added to or subtracted from the differential voltage to be measured. A common-mode voltage, V_{cm}, is converted into a differential-mode voltage as a result of asymmetries in the input stage of the measuring instrument. The introduced error is called a *series-mode voltage error*.

The same situation arises when a collection of instruments is mounted in a metal rack and each instrument is electrically connected to it. A current can flow through the rack via the signal ground terminals, no matter whether the instruments are connected to ground or not. This current can flow because in large racks several hundreds of millivolts may exist between different drawers. Another aspect is the feedback mechanism of interfering signals to the input likely to be coupled back into the circuit by capacitive or electromagnetic induction, and again a series mode error is introduced in the measuring system. The feedback problem will be discussed in more detail later.

Suppose now that the cable resistances R_1 and R_2 are neglected and a thermocouple transducer with given internal resistances R_{tr1}, R_{tr2} is connected to the input of an

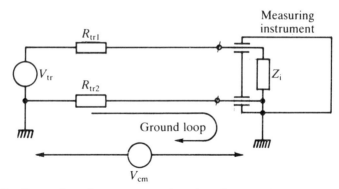

Figure 8.7 Conversion of a common-mode voltage into a series-mode voltage caused by the thermocouple impedance only.

amplifier with an input impedance Z_i, then the configuration in figure 8.7 will result. A thermocouple is a welded connection of two different metals with different specific resistances and, if long thermocouple wires are involved, large differences in resistance can occur. Hence in this configuration the electrical equivalent of the thermocouple is considered as a voltage source with two equally divided, but unequal, resistors.

When this system is earthed at two points as shown, because of V_{cm} and Z_i two different ground loops are created, resulting in two currents different in magnitude and developing two different series-mode voltages across the R_{tr}. This results in a differential-mode voltage across Z_i, demonstrating again the conversion of a common-mode voltage V_{cm} into a series-mode voltage seen by the amplifier's input as a differential-mode signal.

In many measuring applications a type of bridge circuitry is applied which can be fed by a DC or AC voltage source V_{cc}. Very often, one of the bridge nodes is earthed and it will be obvious that the voltage levels of points A and B, with respect to earth, are $\frac{1}{2}V_{cc}$ and both are determined by the accuracies of the bridge branches. Hence, a bridge configuration will always deliver a so-called transducer common-mode voltage to the input of the measuring system. This is illustrated in figure 8.8.

In most cases we are not interested in this common-mode voltage, and the differential-mode voltage generated by the bridge imbalance is the required signal. Hence a

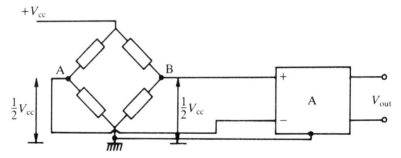

Figure 8.8 Bridge configuration acting as a common-mode voltage source.

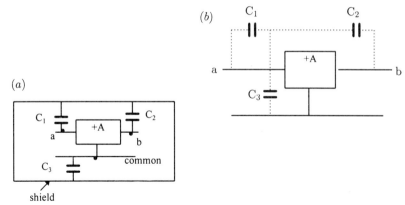

Figure 8.9 (*a*) Parasitic feedback of a shielded amplifier; (*b*) its electrical equivalent.

bridge measurement requires also a differential input stage to eliminate the common-mode voltage.

Great care must be taken because, if a fixed imbalance of the bridge is present, this results in a fixed differential signal, which should be avoided. A bridge-balancing procedure is easy to accomplish and a variety of good compensation techniques are available. However, when the bridge signal varies with temperature or time, when a pressure should be measured for instance, considerable errors can be introduced and other precautions must be taken. (See also chapter 7.)

8.4 ASPECTS OF SHIELDING

As we said in the introduction, shielding can be defined as the placing of sensitive electronic parts and components in a metal casing to prevent electric and magnetic fields entering that casing.

8.4.1 A shielding cabinet
Suppose we have placed a non-inverting amplifier in a shielding cabinet as illustrated in figure 8.9. Inside the cabinet three parasitic capacitors C_1, C_2 and C_3 can be distinguished, of which C_1 and C_2 form a parasitic feedback path from output to input. This can result in undesired oscillations or jamming of the amplifier. The feedback path can be eliminated by connecting the shield to the common and thus short-circuiting C_3 as is shown in figure 8.10.

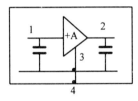

Figure 8.10 Eliminating the parasitic feedback path by short-circuiting C_3.

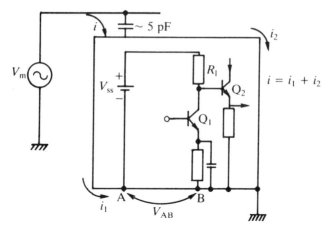

Figure 8.11 A shield as part of an electronic circuit.

From this example it will be clear that a shield is only effective when the shield is connected to the common or zero-signal reference potential of the shielded circuitry, eliminating the undesired feedback path.

8.4.2 A shielding cabinet as part of an electronic circuit

Figure 8.11 illustrates the use of a shielding cabinet as a part of an electronic system. The shielding cabinet is coupled to the mains via a parasitic capacitor which will result in two currents i_1 and i_2 flowing through the cabinet to the earth connection. If within the cabinet different common connections are made, suppose at A and B, a voltage difference V_{AB} between A and B can be developed by V_{cm}. This results in a series-mode voltage with V_{ss} forming an input signal for transistor Q_2 via R_1 and clearly distortion will occur. The solution to this problem is applying one common connection only to eliminate V_{AB}, as is illustrated in figure 8.12. Hence one point only of the circuitry's common should be connected to the shield.

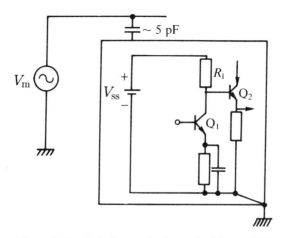

Figure 8.12 Solution to the hazard of figure 8.11.

It is interesting to make the following remark connected to this. Suppose you are walking outside in an open field and there is going to be a thunderstorm and you cannot find a hiding-place: there is one thing you should do and one thing you should never do. When lightning strikes the ground, large stray currents can be induced. If this is likely to happen, make one point of your body common as effectively as possible, hence sit on the ground with your legs drawn up and close together, preventing large voltage differences from being developed between your legs and resulting in stray currents passing your legs. The thing you should never do is to hide under a large tree. When lightning hits a tree, the tree will probably be split into two parts. This will be due to the huge amount of water, which especially in summertime is always present in a tree, and hence a tremendous steam explosion can occur. Note that the developed power involved in a stroke of lightning can exceed 1 000 000 V by 20 000 A, which represents a power generation of 20 000 MW in a split second.

8.4.3 The use of more than one shielding cabinet

Suppose two instruments are required, both of which need shielding; more than one shielding cabinet is involved and it is recommended that only one earth connection is used for the two cabinets. However, sometimes a problem can arise, as illustrated in figure 8.13. If the right-hand cabinet is earthed and both cabinets are connected to the mains via parasitic capacitors, a current i_1 can flow through the parasitic capacitor C_1, the cabinet 1 and the shield of the coaxial cable to the ground connection of cabinet 2. When a rather long coaxial cable connects both cabinets a considerable voltage difference can be developed. This voltage difference V_{AB} can be considered as a series-mode voltage source with the voltage across R_1 and thus will be a part of the input signal for Q_2.

It will be clear that interference and consequently errors will result. The same situation arises when the left-hand cabinet is earthed, when a current is generated from right to left.

Now suppose we ground both cabinets: a voltage difference between the cabinets can still be developed because of current flows i_1, i_2 and i_3, different in magnitude in the loop ABA. However, as well as this mechanism another source of interference

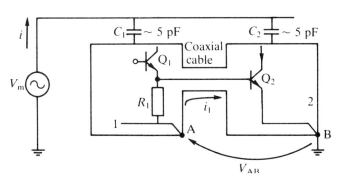

Figure 8.13 The cause of interference when two shielding cabinets are connected in the wrong fashion.

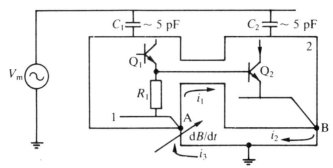

Figure 8.14 The grounding of both cabinets creates a magnetically sensitive loop.

is introduced, as is depicted in figure 8.14. A type of loop is made in which an extra current can be generated if a changing magnetic field dB/dt is present. This problem can be solved by implementing a two-wire shielded coaxial cable and applying a differential input stage at the input of the shielded circuitry in the second cabinet. This solution is depicted in figure 8.15, from which you can see that no ground loop exists and parasitic currents cannot influence any input or create a series-mode voltage.

In the following section another example of creating a ground loop by a wrong connection of the common is discussed.

8.4.4 The connection of a common in a multiple-cabinet configuration

The system to be investigated is shown in figure 8.16 and consists of a three-cabinet configuration interconnected via two-wire shielded cables. At a first glance you may think the common connection seems to be right, but a more careful consideration shows otherwise. This is clarified in figure 8.17 where the equivalent electrical circuit is drawn. It is assumed that rather long cables are installed, of which R_{1c} represents the equivalent resistance. A voltage difference will be generated over R_{1c}, which

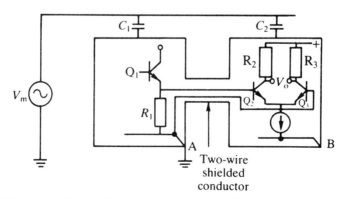

Figure 8.15 A multiple-cabinet configuration avoiding the magnetically sensitive ground loop and the influence of parasitic capacitors C_1 and C_2.

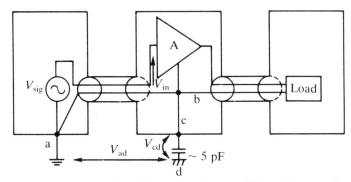

Figure 8.16 The connection of a common in a multiple-cabinet configuration.

equals $V_{ad} + V_{dc}$. Then for the input signal V_{in} connected to the amplifier's input we have

$$V_{in} + V_{cd} + V_{da} - V_{sig} = 0 \tag{8.9}$$

and then for V_{in} we find

$$V_{in} = V_{sig} + V_{ad} - V_{cd}. \tag{8.10}$$

Note the exchange in the indices of V_{ad}.

From this incremental model it can be seen what has to be done to avoid this source of interference. If we disconnect bc the influence of R_{1c} is eliminated and a suitable common connection for the circuit will remain. This is illustrated in figure 8.18. From this example we can make a general statement.

The common should be connected only to the shield at the ground connection of the zero-signal reference potential.

Hence this earth connection is made at the input of the system, but if this is not possible, the common connection may be made at the output as is often performed

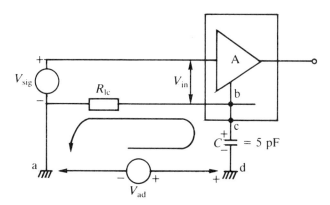

Figure 8.17 Equivalent incremental model for figure 8.16.

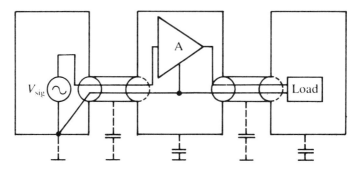

Figure 8.18   The correct common connection for the system.

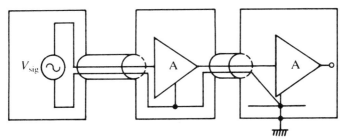

Figure 8.19   An alternative connection for the common when an oscilloscope is used as load.

when the load is an oscilloscope. This is shown in figure 8.19, where the common of the CRT involved is connected to its own cabinet.

Another example is a digital voltmeter (DVM) conceived as a load for the signal source to be measured. This type of instrument can have a large capacitance between its common and the shield, e.g. about 10 nF. The situation is drawn in figure 8.20, where a grounded signal source is connected to the input terminals via a two-wire shielded cable. When the shield of the DVM is also grounded a current I_1 will flow into the common. If the ohmic resistance R_{ac} is of considerable magnitude, a large series-mode voltage V_{ac} can be generated across R_{ac}. This series-mode voltage can be eliminated by bypassing R_{ac}, applying a conductor between points 1 and 2. The shunted current will flow now via the loop a–1–2–c–a with a smaller ohmic resistance, and a reduced interference will result.

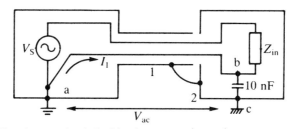

Figure 8.20   Signal source loaded with a large capacitance between common and shield.

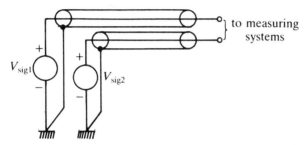

Figure 8.21 A shielding configuration when more than one independent signal source is present.

When in a measuring system more than one independent signal source is present, each signal source should have its own shield, without any interconnection to other shields in the system except for a common reference ground potential. This principle is illustrated in figure 8.21.

Another example is given in figure 8.22, where in a multiple-PCB-card system it can be seen that a complete distinction is made between a *digital ground*, DGND, an analogue ground, AGND, and a power supply ground, GND, except for one common reference point which can act as a ground potential. Note that when large ground currents flow this system must be modified into a more carefully designed ground system. In particular, signal paths must be traced to make them low impedance at the signal frequency used.

It will be clear that the shield and the shielded conductor are always capacitively coupled by a (small) parasitic capacitor. If the shield has a voltage difference with respect to the signal source conductor, a current can flow through the parasitic capacitor between shield and conductor. This introduces an error signal and means that the shield is not allowed to be at a potential with respect to the signal source. To prevent this occurrence, shield currents should be avoided at all times. This problem is illustrated in figure 8.23.

We will return to this point when active guarding is treated. We can summarize our discussion so far with the following basic statements.

1 An electrostatic shield is only effective when it is grounded to the common of the circuitry.
2 The common should be connected to the shield at the signal ground connection only.

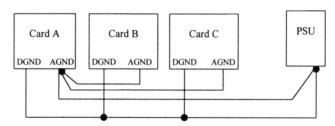

Figure 8.22 Grounding system for a multiple-card system.

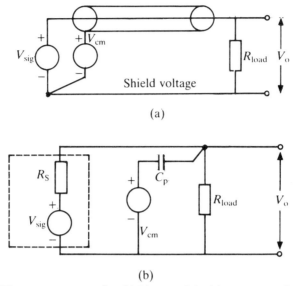

(a)

(b)

Figure 8.23 The consequences of a shield potential with respect to the signal source, as a result of shield currents.

3   When in a system more than one shield is applied, the shields have to be series connected and ultimately to the signal ground and/or common.
4   In general, shield currents must be avoided at all times.
5   A common which is used as a ground must be an 'equipotential'.

8.5   INTERFERENCE CAUSED BY A MAGNETIC FIELD

Interference caused by a magnetic field is much more difficult to eliminate than interferences coming from electrostatic fields. The main reason is that magnetic fields can penetrate conducting materials, and the moment that there are changing magnetic fields with respect to time, induced voltages caused by the $d\Phi/dt$ or dB/dt mechanism are generated in conductors, resulting in currents flowing in the conductors. For this type of interference two causes can be distinguished. Firstly, if wires in the neighbourhood of a system are conducting high alternating currents, the wires will act as a transmitting antenna and are the source of a magnetic field. Secondly, if within a system wires are conducting high currents, the source is already present inside the system. In the first case precautions should be taken to prevent the magnetic field entering the system. In the second case compensation techniques can be applied. However, look first at the penetration depth of a magnetic field in three different kinds of very common material. The penetration depth z of a magnetic field in a conductor can be described by the exponential equation

$$B_z = B_0 \exp(-\alpha z) \tag{8.11}$$

where B_0 is the field strength at the conductor's surface, α is the absorption coefficient which in its turn is a function of the frequency and the applied material and z is the

Table 8.1 Penetration depth as a function of material and frequency.

Frequency (Hz)	Penetration depth (z) (mm)		
	Copper	Aluminium	Steel
50	8.97	11.5	0.91
60	8.5	10.9	0.86
100	6.6	8.5	0.66
1000	2.1	2.7	0.20
10 000	0.66	0.84	0.08
1 000 000	0.08	0.08	0.008

penetration depth in the material. In table 8.1 the influence of material and frequency is demonstrated for three different materials at different frequencies. From this table it can be understood that it is not so easy to shield against magnetic fields. There is one other material to be mentioned: that is mumetal, which is very effective at low frequencies. However, at 100 kHz mumetal is less effective than steel, copper or aluminium. If this material is saturated by very strong fields it provides no shielding at all. Figure 8.24 shows the effectiveness of magnetic shielding of several materials versus frequency.

The best way of minimizing magnetic interference is to minimize loop areas. When the direction of the magnetic field is known, the interception of the magnetic field can be avoided by a correct mounting of the loop area involved.

A good example of a compensation technique is illustrated in figure 8.25, where a correct and an incorrect compensation technique are depicted. If two equal and opposite currents flow through two twisted conductors, two equal and opposite fields are generated, producing a zero net field. This is illustrated in figure 8.25(c). Note

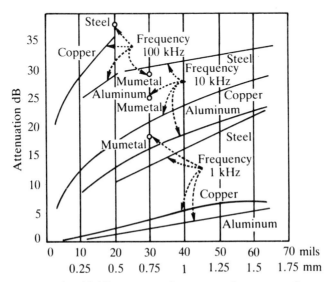

Figure 8.24 Magnetic shielding attenuation versus frequency of several materials (reprinted by permission of John Wiley & Sons Inc.).

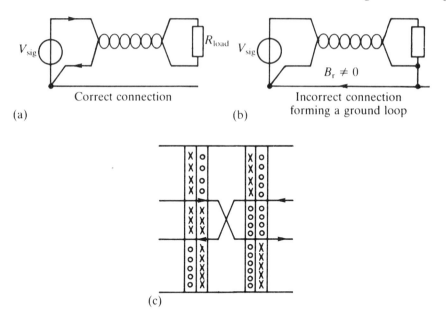

Figure 8.25 Applying a compensation technique with twisted wires.

again that wires can act either as a receiver or as a transmitter of interference. The compensation cancels the magnetic fields present under the condition that the conductors are completely symmetric and no new ground loop is introduced. Another example of a good symmetrical conductor is a coaxial cable, which for reasons of symmetry is completely insensitive to magnetic interferences. Here are some basic rules for minimizing magnetic interference.

1 Use twisted-pair cable for conductors carrying large currents. In this case, the wires are the source of the magnetic field.
2 Shield the circuit with an appropriate material depending on the applied frequency and field strength.
3 Avoid any wire running parallel to the magnetic field and crossing magnetic fields at right angles.
4 Locate magnetically sensitive parts as far as possible from magnetic field sources and at right angles.
5 Avoid ground loops and reduce any area which might be sensitive to magnetic fields.
6 Minimize cable length as far as possible.

In the following section we shall discuss some coupling techniques and related problems you may encounter when different measuring systems are interconnected.

8.6 TYPES OF COUPLING MECHANISM

In many circumstances, systems are interconnected by a kind of coupling technique. We shall discuss several coupling techniques and possible errors which can occur when applying these techniques. In general, coupling between different measuring systems can be performed by different techniques, each of which has its own advantages and disadvantages. Coupling of measuring systems can be accomplished in the following different ways:

1 direct coupling
2 magnetic coupling
3 r.f. coupling
4 microwave coupling
5 opto-coupling.

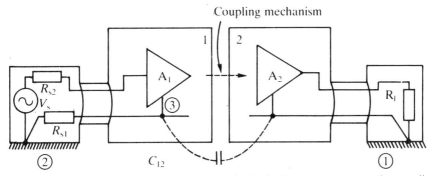

Figure 8.26 Types of coupling mechanism. The dashed arrow represents the coupling mechanism to be discussed.

Suppose we have a system as illustrated in figure 8.26 consisting of two stages A_1 and A_2 and interconnected in one of the five ways mentioned above, denoted by the dashed arrow. Suppose further that the signal source V_s and the load R_1 are at different ground potentials, then the parasitic capacitor C_{12} allows a current to flow through R_{s1}. It will be clear that a common-mode voltage represented by the ground potential difference is converted into a series-mode voltage in the same fashion as we have seen before. This results in a differential-mode voltage for the first stage. We will discuss the consequences of several coupling mechanisms. Of the five methods of coupling mentioned we will only discuss direct coupling and magnetic coupling, because they are the most frequently used coupling methods.

8.6.1 Direct coupling
Direct coupling is a galvanic connection between different parts of the system to which for instance a differential amplifier is applied as illustrated in figure 8.27. In this circuit a parasitic capacitor, C_{24}, between one lead and the shield 4, is present, over which a voltage difference is generated when a ground potential difference V_{56} is present. This causes a current to flow through R_{s2}, when two ground connections are made. The leakage resistor R_{23} allows current to flow from point 5 to 5 via 2, the common 3 and point 6. In this figure R_{13} and R_{23} are the so-called leakage resistances with respect to common of both amplifier inputs. These leakage resistances can be interpreted as the isolation resistances of both inputs with respect to the common, and are present in every input stage.

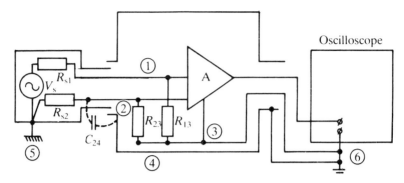

Figure 8.27 Principle of direct coupling by applying a differential amplifier.

As will be shown, in a DVM these resistances should exceed a value of 1000 MΩ. The capacitor C_{24} can be understood as a cable or shield capacitor and, for instance, can have a capacity of 60–100 pF m^{-1}. When high common-mode voltages are involved a balanced attenuator must be installed, because the maximum rating of the input voltage of a standard operational amplifier is usually limited to 30 V. This is illustrated in figure 8.28.

In this way the common-mode voltage can be attenuated by a required factor. The second stage will reject the remaining common-mode voltage and reamplify the desired differential-mode signal. Most amplifiers cannot handle common-mode voltages exceeding 30 V, so when a common-mode voltage of 1000 V must be handled, in the attenuator an attenuation factor of 40 must be implemented. For this purpose, a floating amplifier for the first stage should be installed.

For direct coupling we can mention the following advantages.

1   The coupling mechanism is accurate and stable.
2   There are no limitations in bandwidth.
3   A completely symmetrical input can be realized.

Of course there are some disadvantages.

1   The common-mode voltages have to be processed by the circuitry.
2   The level of the common-mode voltage is confined by the level of the power supply voltage.
3   The common-mode rejection ratio CMRR is strongly frequency dependent.

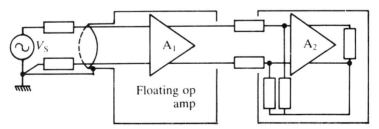

Figure 8.28   Direct coupling of high common-mode voltages via a balanced attenuator.

8.6.2 Magnetic coupling

Magnetic coupling is one of the most widely used although the coupling mechanism has several disadvantages such as costs and bandwidth limitations. Applying a coupling transformer can serve two purposes.

1   In numerous cases the transportation of energy is involved from the mains towards the system to supply.
2   Information can be transported. In this case a type of modulation and demodulation technique is involved by which the information is added to a carrier wave with a relatively high frequency.

The principle of operation of a transformer coupling for transporting information is drawn in figure 8.29 in which a voltage difference V_{14} due to a ground loop and a leakage capacitor C_{23} can be recognized. This can cause a current to flow round the input when an imperfect or wrong shielding technique is applied. As will be shown, this can require an input impedance of more than 1000 MΩ for accuracy.

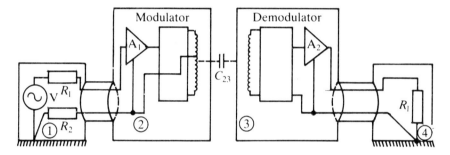

Figure 8.29   Principle of operation of magnetic flux coupling.

For magnetic coupling we can mention the following advantages.

1   The maximum common-mode level is determined by the transformer's insulation quality.
2   No external voltage can influence the shielded input, so the signal source may be floating.
3   Common-mode signals are not amplified and are no longer a function of frequency and/or amplitude. The CMRR will remain independent of gain after the first stage.
4   If at the input a type of modulation is used, within the shielding no electronic component is required. This means that at the input no power is necessary which might cause parasitic currents to flow into the input and shielding.
5   A complete galvanic insulation between two stages in a system can be realized.

The disadvantages are as follows.

1   There is an increase in weight and costs of the system.
2   There is a limitation in bandwidth determined by the quality of the transformer.

In the following section we shall discuss the interference of power supply transformers in more detail.

8.7 INTERFERENCE CAUSED BY POWER SUPPLY TRANSFORMERS

Most instruments use a mains supply transformer to deliver the required electric power. This transformer also provides a good electrical insulation from the mains, because only a magnetic coupling via the core material is involved, by which the electrical energy is transferred from the primary to the secondary coil via a magnetic field. Of course this is an important safety feature but this inherently involves a disadvantage too, as will be seen.

In normal circumstances the instruments are connected to the mains via a three-core cable, of which one lead carries the power voltage and the other is the neutral or null lead. This null lead is connected to earth at the power distribution station where the starpoint is always earthed at the secondary coil of the step-down transformer. However, simultaneously the safety guard wire is also earthed at the same point to which all metal housings are connected via the third lead of the mains cable.

Suppose no isolation transformer is installed at the instrument side; then the situation is as illustrated in figure 8.30. As can be seen from this figure, an extra low-ohmic path to earth is created for all currents forced to flow by an isolation breakdown or an accidental short circuit. Under normal conditions there is also a small current via a capacitive coupling of the conductors. This means that a common-mode voltage will be formed between this power safety ground and the instrument ground and a lot of 'hum' at the input terminals will be the result in all circumstances. So, a most important reason for installing an isolation transformer is for safety.

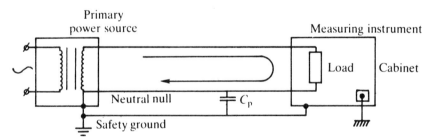

Figure 8.30 Circuit without a power isolation transformer.

When a power isolation transformer is used in a measuring system, another interference problem is introduced. In most countries the primary of this transformer is earthed. This is illustrated in figure 8.31 in which the stray capacitance C between primary and secondary coil is shown.

This capacitance exists between the primary and the secondary windings and is represented by one lumped capacity C ranging from several hundred picofarads up to 5 nF. The common of the circuitry at the secondary is normally grounded by a conductor which represents a certain impedance Z. A current, i, can flow from the

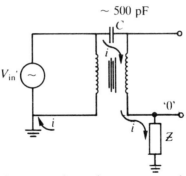

Figure 8.31 Stray capacitance in a power supply transformer.

primary to the secondary conductor whose magnitude is determined by $1/\omega C$ which represents the largest impedance value in the loop to be considered.

The result will be a voltage $V_p = iZ$ above ground level for the common of the secondary side. This need not be too serious, but when two or more instruments are used, each with its own power supply transformer, the commons can have different voltage levels which can cause large currents to flow when they are connected together.

This stray capacitance can be diminished considerably applying two shields for each transformer coil at both sides. The fact that, in a transformer, only one shield cannot eliminate stray capacitances is illustrated in figure 8.32. It can be seen that capacitive currents will flow through the ground connection at the secondary side in both cases.

The solution is shown in figure 8.33, with two shields, one around the primary side and one around the secondary side which are open windings connected to one side only of their windings. If closed windings are used, large eddy currents will flow, totally spoiling the desired effect.

Applying both shields results in a large capacitance between the shield and the cable core, giving a low-impedance path for the common-mode generator V_{cm}. In

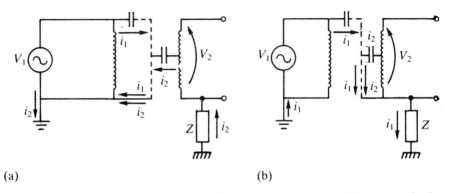

(a)   (b)

Figure 8.32 The application of one shield does not eliminate parasitic currents in the secondary common.

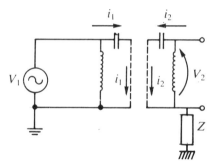

Figure 8.33 Method for avoiding large stray capacitances in a power supply transformer.

this way, the stray capacitance between the primary and the secondary can be reduced to a value of 0.1 pF.

Of course this technique has some disadvantages such as the high costs of manufacturing these types of transformer. The shields are hand made and careful manufacturing is required to prevent short circuiting.

Finally, we will give some other examples in which such a type of transformer is applied.

*   *   *

Example 8.4
Figure 8.34(a) illustrates the application of the shielding principle. The shields around the primary and the secondary windings are connected to one side of the winding,

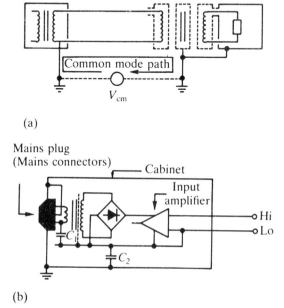

(a)

(b)

Figure 8.34 (a) The shielding principle applying a power isolation transformer; (b) its use in a digital multimeter input configuration (courtesy Philips).

so that stray capacitances between the windings and the shields are eliminated at one end. Moreover, a large capacitance is achieved between the shield and the cable core, resulting in a low-impedance path for the common-mode generator V_{cm}. The secondary is often connected to a centre tap of the secondary winding, reducing the voltage difference. This also reduces the stray capacitance between the primary and secondary windings and a value of 0.1 pF can be obtained, while without shielding the stray capacitance between the two windings may exceed 2000 pF.

This principle is applied in a digital multimeter as is illustrated in figure 8.34(b), for which the specifications are as follows: capacitance from mains to Lo is $C_1 = 50$ pF; capacitance from Lo to chassis is $C_2 = 1.5$ nF. It will be clear that 50 pF is quite a good separation between the mains and the measuring circuitry.

* * *

When more stringent input specifications are required, a solution such as that in figure 8.35 can be applied, which is discussed in the following example.

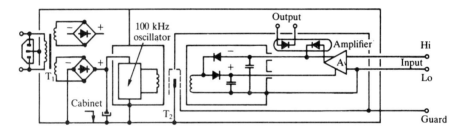

Figure 8.35 Application of the shielding principle applying a second transformer and a type of opto-coupling (courtesy Philips).

* * *

Example 8.5
In order to achieve a further reduction of the influence of the mains, an extra transformer T_2 can be added. One of the secondary windings of T_1 is used to feed a 100 kHz oscillator which forms a part of the measuring system. The output of amplifier A is fed to the remaining part of the instrument's amplifier via a light-coupled diode device. Hence a perfect galvanic separation is obtained. A floating guard which may be connected to an extra terminal on the instrument is placed around the input amplifier, A, and its supply unit to eliminate the influence of external parasitic voltages. With this type of configuration an AC common-mode rejection of 126 dB for the mains frequency can be obtained. The principle of this set-up is depicted in figure 8.35.

* * *

In the next example, a multiple-cabinet configuration is discussed.

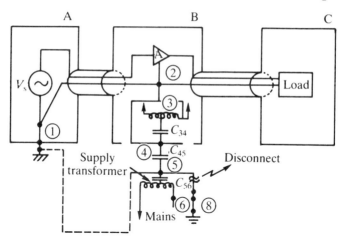

Figure 8.36 Earthing the primary of the shielding transformer in a multiple-instrument configuration.

*   *   *

Example 8.6
Suppose a multiple-instrument configuration is designed as illustrated in figure 8.36. The instruments A and B are connected to the signal source V_s as shown. Suppose further that $C_{45} \approx 1$ nF and a ground potential of 10 V is present; then the current in the loop 1, 2, 4, 5, 8, 1 can be calculated as

$$I = 10/(1/\omega C) = 10/(3 \times 10^6) = 3.2\ \mu\text{A}.$$

If the common conductor has a 1 Ω impedance, a voltage difference between point 1 and 2 is developed of about 3 μV. This parasitic voltage can seriously destroy the performance of a sensitive measurement set-up.

The solution to prevent this parasitic current involves grounding the primary shield at point 1 as is indicated by the dashed line, and disconnecting the shield at point 8, as shown in figure 8.36.

*   *   *

In the following section the application of differential amplifiers in measuring set-ups and guarded environments will be discussed in more detail.

8.8   THE DIFFERENTIAL AMPLIFIER AND GUARDING

To discuss the possibilities of differential amplifiers in guarding applications we will repeat some basic concepts which are necessary for a good understanding of the common-mode rejection ratio (CMRR). For a thorough treatment of the differential amplifier reference is made to the relevant textbooks. (See also chapter 4.)

In measuring applications, it very often occurs that differential amplifiers are implemented because this type of amplifier is able to discriminate between a common-mode signal and a differential-mode signal. In general these types of signal are always present. The quality of this kind of amplifier determines in most cases the total quality of the measuring system, because in numerous measuring applications this type of amplifier is used and can be found directly at the input stage of a measuring instrument.

If we consider the different types of active element then a single-input amplifier such as a transistor, with a single-ended output, is a device which cannot be considered as an ideal active element. An ideal active element can be defined as a controllable power source with no limitations in polarity, in voltage and current at both the input and the output. A good example of such an active element is the operational amplifier which is called a nullor in its most ideal form. In a nullor, the input current, i_i, and the input voltage, v_i, by definition are claimed to be zero.

There is some confusion between the different amplifier configurations. For example, sometimes an operational amplifier is called an instrumentation amplifier, but this is just a special configuration in which three operational amplifiers are involved. Such a three-amplifier configuration has improved features with respect to a single operational amplifier: in particular the CMRR and the input impedance are improved. An operational amplifier can be considered as just a building block which can be used either as a differential amplifier or as a single active element, depending on how a few external passive components are connected. For our purposes it will be sufficient to present only some basic concepts which describe the quality of the differential amplifier. Consider the general schematic layout for a grounded circuit as illustrated in figure 8.37.

In this figure the following quantities can be defined. V_s represents the signal source with source impedances R_{s1} and R_{s2}, and R_3 is the amplifier's input resistance. The leakage resistors of the amplifier from input to output ground are represented by R_4 and R_5. The common-mode voltages of input and output with respect to ground are V_{c1} and V_{c2} respectively. Finally, R_{c1}, R_{c2} and R_{c3} are the respective transition impedances to be distinguished and are inherently present with the common-mode signal.

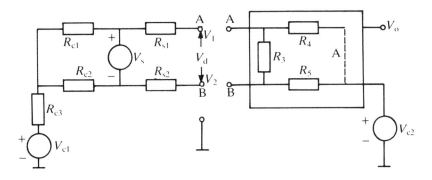

Figure 8.37 General layout of a grounded source circuit (assume $V_{c1} > V_{c2}$).

In the same way as in equations (8.7) and (8.8), the differential-mode and the common-mode voltages can be defined. With different indexes we define for the differential-mode voltage

$$V_d = V_1 - V_2 \tag{8.12}$$

and for the common-mode voltage at the amplifier's input side,

$$V_c = \tfrac{1}{2}(V_1 + V_2). \tag{8.13}$$

Suppose the input resistance is infinite; then no input current will flow and the following equation holds:

$$V_s = V_d = V_1 - V_2 = 0. \tag{8.14}$$

The average value of the input voltage at the amplifier's input side equals

$$\tfrac{1}{2}(V_1 + V_2) = \tfrac{1}{2}(V_{c1} + V_{c2} + V_s). \tag{8.15}$$

When $V_{c1} = V_{c2}$ equation (8.15) yields

$$\tfrac{1}{2}(V_1 + V_2) = \tfrac{1}{2}V_s + V_{c1} = \tfrac{1}{2}V_s + V_{c2}. \tag{8.16}$$

This equation contains the unwanted value V_{c1} or V_{c2}. This need not be a problem, but when for instance $V_{c1} \neq V_{c2}$ then $V_{c1} - V_{c2} \neq 0$ and a current will flow in R_{s1} or R_{s2}.

The following example demonstrates the consequences.

*   *   *

Example 8.7
Suppose a digital voltmeter is applied to execute a measurement on a voltage source delivering a signal output of $V_s = 10$ mV. The required (in)accuracy is 0.1%, which means an allowed $V = 10$ μV as common-mode voltage. Suppose further there is an imbalance involved between R_{s1} and R_{s2} of 1 kΩ; then the input current is

$$I_{in} = \frac{V_{error}}{R_{s1} - R_{s2}} = \frac{V_{error}}{\Delta R_s} = \frac{10 \times 10^{-6}}{1000} = 10 \text{ nA}.$$

If $V_{c1} - V_{c2} = 10$ V then the leakage impedance must exceed

$$R_{leakage} = \frac{10}{10 \times 10^{-9}} = 1000 \text{ M}\Omega = 1 \text{ G}\Omega$$

which is a normal figure of merit for input impedances with respect to ground. When for instance in a 3.5-digit digital voltmeter the lowest digit is 1 μV with a maximum of 1999 mV, this value is too small. Hence, for the desired accuracy an error of 10 μV means that the last two digits are meaningless. This is again an example of how at the input of an amplifier a common-mode voltage is converted into a differential-mode voltage affecting the accuracy to a large extent. The solution is guarding, as will be discussed in the next section.

*   *   *

8.8.1 The common-mode rejection ratio

As claimed already in chapter 3, a differential amplifier can be used to suppress common-mode voltages and to amplify differential-mode voltages. To be able to treat this chapter more independently and to emphasize relevant aspects concerning guarding, we will repeat some characteristics already discussed. The characteristic features of a differential amplifier can be expressed into two figures of merit, of which one is called the discrimination factor F and the other is called the rejection factor H. The rejection factor is the most important one, because it expresses to what extent a common-mode signal can be suppressed by the applied amplifier. Once a common-mode voltage is converted into a differential-mode voltage, any further suppression will fail. The discrimination factor F is just the ratio between the differential-mode gain and the common-mode gain of the same amplifier and can always be improved by a series connection of more stages. For a good amplifier both figures of merit should be high, for instance 120 dB.

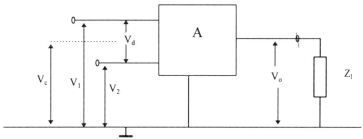

Figure 8.38 A single-ended-output differential amplifier.

To clarify the above statements reference is made to figure 8.38 where a single-ended-output amplifier is depicted. In general the behaviour of a differential amplifier can be described with the following set of equations:

$$V_{od} = A_{dd} V_{id} + A_{cd} V_{ic} \tag{8.17}$$

$$V_{oc} = A_{dc} V_{id} + A_{cc} V_{ic} \tag{8.18}$$

where, when any sign is neglected, A_{dd} is the differential-to-differential signal gain factor, A_{cd} is the common-to-differential signal gain factor, A_{dc} is the differential-to-common signal gain factor, A_{cc} is the common-to-common signal gain factor. The gain factors are defined as

$$A_{dd} = V_{od}/V_{id} \qquad A_{cd} = V_{od}/V_{ic} \qquad A_{dc} = V_{oc}/V_{id} \qquad A_{cc} = V_{oc}/V_{ic} \tag{8.19}$$

with the other quantities kept constant, and where V_{od} and V_{id} respectively are the differential output and input voltages. In the same way, V_{oc} and V_{ic} are the common-mode output and input voltages respectively. Note that the double index in A is usually reserved for double-input/double-output stages, but in this fashion it generalizes the problem.

As can be seen from these equations, four different gain factors can be distinguished in which the indices ascribe the considered gain factor, for instance A_{cd} describes the extent to which a common-mode signal is converted into a differential-mode signal, etc. With the help of these equations, it is a simple affair to define the rejection factor H and the discrimination factor F. The following definition holds for the rejection factor:

$$H \overset{\text{def}}{=} \frac{A_{dd}}{A_{cd}} = \frac{V_{od}/V_{id}}{V_{od}/V_{ic}} = \frac{V_{ic}}{V_{id}}\bigg|_{V_{od}=\text{const}}. \qquad (8.20)$$

This expression can also be written in words as

$$H = \frac{\text{common-mode input voltage}}{\text{differential output voltage caused by common-mode input voltage}}.$$

Usually this quantity is expressed in decibels and is called the common-mode rejection ratio factor, CMRR, and so the manufacturer would claim for instance $20 \log |H| = x$ dB for a given amplifier, where 60 dB is a poor and 120 dB is a good figure of merit. However, an important point must be made here. For a good judgment of the qualifications of an amplifier the measurement conditions should be specified by the manufacturer, such as the applied load and frequency, because they both influence the CMRR. It will be clear that for an ideal amplifier H is infinite. The discrimination factor F is defined as

$$F \overset{\text{def}}{=} \frac{A_{dd}}{A_{cc}} = \frac{V_{od}/V_{id}}{V_{oc}/V_{ic}}. \qquad (8.21)$$

Note that the denominator is the ratio V_{oc}/V_{ic}. We can see from these equations that both H and F can be calculated from measured voltages. As said, H is more important and describes how perfect the amplifier is, or the system is. Very often the causes are symmetry faults as a result of design errors and/or applied technology faults during manufacturing. To improve the CMRR two important measurements can be applied.

1 The input has to be isolated with respect to ground, and hence no ground currents can flow.
2 The amplifier's input must be balanced perfectly.

Isolation gives the best results; balancing gives a further improvement. In the next sections we will see how this can be made effective.

8.8.2 The CMRR in shielding circumstances
So far different possibilities have been discussed for shielding and guarding to reduce the influence of ground loop currents, without looking inside the instrument. However, inside the instrument shielding techniques too can give a large improvement of the CMRR.

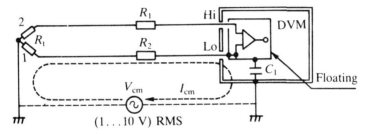

Figure 8.39 Digital voltmeter with floating guard.

Consider for instance an instrument with a metal shield applied inside the instrument as is depicted in figure 8.39. You may think of a DVM in which a metal shield is applied. Very often this shield is referred to as a guard if it is floating on the low-potential side, which here is supposed to be the common.

Although in the DVM, and thus the input amplifier itself, neither the 'Hi' or the 'Lo' terminal of the amplifier is connected to ground, the measuring leads may be grounded at the transducer's side with its symmetrical source resistances.

Although there is no galvanic connection between the amplifier and earth, an alternating current I_{cm} can flow through the stray capacitance C_1 owing to the series mode voltage V_{cm} over R_{t1} and R_2 when a ground loop is made. Bear in mind that for simplicity the input impedance of the amplifier is supposed to be infinite so no input current will flow. In the following example a calculation is executed and, for this purpose, the equivalent electrical circuit for the input is depicted in figure 8.40.

*   *   *

Example 8.8
Suppose the following practical values are given: the power supply frequency is 50 Hz, $C_1 = 10$ nF, $R_{t1} = 1$ kΩ and $V_{cm} = 2$ V; then the input current can be calculated to be

$$I_{cm} = \frac{V_{cm}}{X_{c1} + R_2 + R_{t1}}. \tag{8.22}$$

Substituting the values for X_{c1} it is found that

$$X_{c1} = \frac{1}{2\pi 50 \times 10 \times 10^{-9}} = 318 \text{ k}\Omega$$

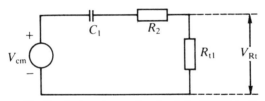

Figure 8.40 Equivalent input circuit of the DVM in figure 8.39.

and this value is in most cases much larger than R_2 and R_{t1}. Hence I_{cm} is found to be

$$I_{cm} = \frac{V_{cm}}{X_{c1}} = \frac{2}{318\,000} = 6.28\,\mu A$$

and thus

$$V_{R_t} = I_{cm}R_t = 6.28 \text{ mV}.$$

Note that the series-mode voltage over R_t is determined by the ratio of R_t and X_{c1} and is directly added to the result of the measurand when AC measurements are performed. If, for instance, this series-mode voltage is compared with the output voltage of a chromel–alumel thermocouple with $40\,\mu V\,K^{-1}$, an unacceptable error is introduced. This error can be reduced by filtering and/or integration techniques, because the magnitude of the error is mainly determined by the magnitude of the parasitic capacitance and the magnitude of the frequency.

In this case the CMRR can be defined as the ratio between the fault potential caused by the common-mode voltage and the magnitude of the common-mode voltage itself, both determined at the input side:

$$CMRR = \frac{\text{present value of } V_{cm}}{\text{fault value caused by } V_{cm}}\Bigg|_{\text{input side}}. \tag{8.23}$$

If this relationship is applied for the above-given values, by substitution we obtain

$$CMRR = \frac{V_{cm}}{F_{fault}} = \frac{2000}{6.28} = 318$$

or in decibels

$$CMRR = 20 \log 318 = 50 \text{ dB}$$

which of course is a rather poor specification.

$$* \quad * \quad *$$

For the CMRR we can conclude that its magnitude is strongly dependent on

1   the applied frequency
2   the value of the parasitic capacitance C
3   the impedance of the transducer

and, what is most important, for an instrument these conditions should be mentioned by the manufacturer when a CMRR is specified. The conditions to be mentioned are standardized by the IEC subcommittee 13B, April 1972, section 23.5.

In the next section we will investigate how the CMRR can be improved.

8.9   IMPROVEMENT OF THE COMMON-MODE REJECTION RATIO

As we have seen in the last section, one way to improve the CMRR is to increase the value of X_{c1}, or conversely, to reduce the parasitic capacitance C_1. To realize a

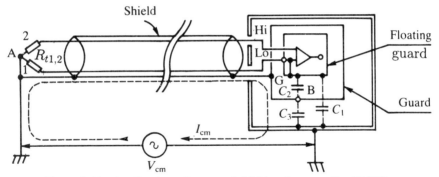

Figure 8.41 Application of a second shield to improve the CMRR.

reduced influence of the parasitic capacitances principles of operation can be applied:

1 the application of a second or floating guard
2 the application of double-shielded cables or active guarding as it is sometimes called.

Both principles will be discussed in detail.

8.9.1 The application of a floating guard

The principle of applying a second or floating guard is illustrated in figure 8.41, which shows a shielded cable of which the shield is connected to the guard at the instrument side and to the ground at the transducer side. Further, in this configuration a cabinet, a guard and a floating housing can be recognized, of which the floating housing is the one placed innermost. In addition to these points, note how the three terminals 'Hi', 'Lo' and 'G' are connected to their housings. If a common-mode voltage V_{cm} is present a current I_{cm} will be produced, but this current now splits up into two parts at point A, at the tip of the transducer. The transducer's impedance is represented by R_t, and R_{t2} and the amplifier's input impedance are supposed to be infinite, so no current will flow in R_{t2}.

The equivalent electrical input circuit is depicted in figure 8.42, in which two current loops have to be distinguished.

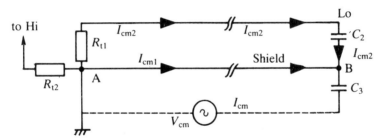

Figure 8.42 Equivalent circuit for figure 8.41 showing the two current loops.

According to Kirchhoff's current law for node B in figure 8.42, the following equation holds:

$$I_{cm} = I_{cm1} + I_{cm2}. \tag{8.24}$$

The current loops are as follows:

1 for I_{cm1} via the cable shield, C_3 and ground, back to V_{cm}, not affecting the result at all
2 for I_{cm2} via R_{t1}, the 'Lo' terminal, C_2, C_3 and ground, back to V_{cm}.

If it is assumed now that R_{t1} has a very small value with respect to X_{c2} then the voltage difference over C_2 will be determined by $I_{cm2}R_t$, independently of the magnitude of X_{c2}. Even if C_2 is of the order of some nanofarads the effective capacitance of C_2 can be of the order of some picofarads because there is almost no potential across the plates of C_2. Hence, the current through C_2 will be very small. Now you can see how the trick works: if a voltage across a capacitor is very small, the capacitor is not effective at all regardless of its magnitude.

The assumption must be made that the ratio between R_{t1} and X_{c2} is very large, or more precisely $|R_t| \ll |X_{c2}|$ and a high impedance must be present between the 'Lo' and the guard. Very often a low source impedance is an advantage, as is the case here. The consequence of this is that, via the shield, C_2 is short circuited and if in this loop, no current due to V_{cm} will flow. Hence, almost all current generated by V_{cm} will flow in the lower loop only.

The effective capacity is strongly reduced by a factor of, say, 1000 which then equals an improvement of another 60 dB. In practice the ultimately effective capacitance between Lo and earth, here denoted by C_1, can have an order of magnitude of 0.5–5 pF, while the capacitance between guard and earth is in the order of some nanofarads. Further, it will be obvious that all three mentioned capacitances are bypassed by the isolation resistances.

A more complete diagram with relevant practical values for the quantities is given in figure 8.43. In the following example it is shown how this can work out. Again the series-mode voltage due the common-mode voltage V_{cm} must be calculated.

*   *   *

Example 8.9
Suppose again the frequency of the mains is 50 Hz and the other values are as depicted in figure 8.43, then the capacitive impedance of X_{c2} is given by

$$X_{c2} = \frac{1}{2\pi 50 \times 700 \times 10^{-12}} \approx 4.5 \text{ M}\Omega.$$

If this value is compared with the isolation resistances R_2 and R_3, then it will be clear that X_{c2} is much smaller than these. Hence we may consider the capacitive impedance only. Suppose further that the input impedance of the amplifier is again infinite and no current flows through C_2, because of an effective short-circuiting of C_2 by R_{t1}, which implies that the voltage difference developed across C_2 is almost

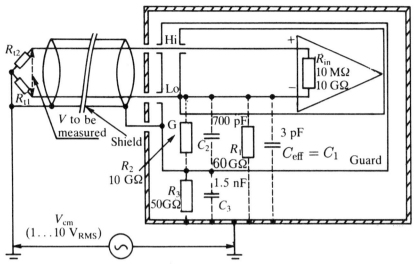

Figure 8.43 Complete diagram of a floating guard configuration. Note that the effective resistance is $R_{eff} = R_1 + R_2$.

zero. The equivalent electrical circuit of figure 8.44 can then be applied to calculate the series-mode voltage V_{R_t}:

$$V_{R_t} = \frac{R_{t1}}{R_{t1} + X_{c1}} V_{cm}. \tag{8.25}$$

If for instance $R_{t1} = 1\,k\Omega$, $C_1 = 3\,pF$ and the common-mode voltage $V_{cm} = 5\,V$, then applying equation (8.25) we find

$$CMRR = 20 \log\left|\frac{V_{cm}}{V_{Rt}}\right| = 20 \log\left|\frac{R_{t1} + X_{c1}}{R_{t1}}\right| = 120\,dB.$$

Of course, the reciprocal value may be calculated, in which case the result will be the same, apart from the minus sign.

Now suppose the shield of the conductors is not connected to the guard; then C_2 is not short circuited by R_t. In this case, in the calculation a current flowing via the parasitic series impedance X_{c2} and X_{c3} should be taken into account. If the values for C_2 and C_3 are substituted, the equivalent parasitic capacitor for C_2 and C_3

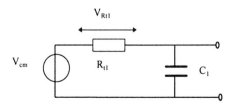

Figure 8.44 Equivalent circuit for calculating the series-mode voltage V_{Rt}.

becomes $C_v = 478$ pF, and if equation (8.25) is applied the CMRR is considerably reduced:

$$\text{CMRR} = 20 \log \frac{1/2\pi 50 \times 478 \times 10^{-12}}{1000} = 76.5 \text{ dB}.$$

For DC measurements, we have to account for the isolation resistances R_2 and R_3 only, because these resistances now determine the total current. Under the condition of DC measurement the CMRR can be calculated as

$$\text{CMRR}|_{DC} = 20 \log \frac{R_2 + R_3}{R_{t1}} \text{ (dB)}. \tag{8.26}$$

Substituting the known values in equation (8.26) yields

$$\text{CMRR}|_{DC} = 20 \log \frac{(10+50)10^9}{1000} = 156.5 \text{ dB}.$$

*   *   *

In the next example it will be shown once more why the CMRR is an important figure of merit.

*   *   *

Example 8.10
Suppose the configuration of figure 8.43 is applied in a digital voltmeter set-up and an inaccuracy of 0.001% is claimed at a resolution of 1 μV. For the lowest measuring range of, say, 1 V the claimed accuracy means a value of 10 μV. Suppose further a series-mode voltage of 50 μV is calculated; then you will understand that the last two digits make no sense at all. If a high accuracy combined with a high resolution is needed, a floating guard system provides the right solution. Of course the costs of such an expensive system must be compared with the requirements when low-level measurements have to be executed.

*   *   *

8.9.2   Active guarding or the application of double-shielded cables
The configuration used to realized active guarding is illustrated in figure 8.45. In this circuit a buffer amplifier or so-called voltage follower is applied. The probe cable is a double-shielded one of which the inner shield is connected to the output of the

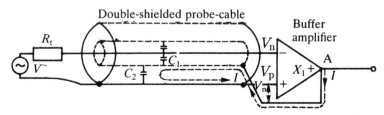

Figure 8.45 Principle of active guarding by applying a double-shielded cable. Note also that $V_n = V_p$.

buffer amplifier and the outer shield is connected to the non-inverting input terminal of the amplifier. As well as these connections, one of the signal source terminals is connected to the outer shield. The result is that the voltage of the inner floating shield equals exactly the voltage of the signal source. The result is a totally ineffective parasitic capacitor C_1 between source terminal and inner floating shield.

The remaining capacitive load is formed by C_2, but any current flowing through C_2 is supplied by the output of the voltage follower and will not result in a current through R_t, which might cause a substantial series-mode voltage error. The source voltage is present between the non-inverting terminal and the inner shield, and hence over the plates of the parasitic capacitor C_2. Basically the same principle is applied as discussed in the previous example, because a parasitic capacitor is made ineffective by a kind of active short-circuiting involving the plates of the capacitor present. Here this short-circuiting is realized by an active element such as a buffer amplifier. Measurement of the signal source voltage is performed by a second amplifier not depicted in figure 8.45.

8.10 WORKED EXAMPLES

In this section we will present two other worked examples illustrating the impact of errors on the CMRR.

$$* \qquad * \qquad *$$

Example 8.11

The configuration in figure 8.46 can be considered as a floating attenuator. This means that the leakage resistances of both input terminals can be considered to be infinite. Then the output voltage V_o due to the source signal V_s can be calculated as

$$V_o = V_s \frac{R_p}{Z_{s1} + Z_{s2} + R_{s3} + R_{s4} + R_p}. \tag{8.27}$$

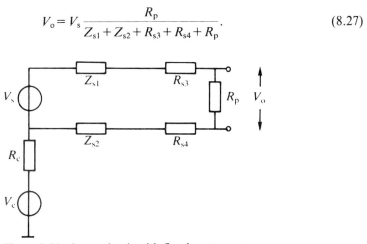

Figure 8.46 Input circuit with floating attenuator.

The transfer function is given by

$$A_{dd} = \frac{V_0}{V_s} = \frac{R_p}{Z_{s1} + Z_{s2} + R_{s3} + R_{s4} + R_p}. \tag{8.28}$$

In this configuration it will be clear that no series-mode voltage or error voltage can be developed by the common-mode voltage V_c; hence $V_{fault} = A_{cd}V_c$ is zero and by definition for the CMRR we find

$$\text{CMRR} = \frac{A_{dd}}{A_{cd}} = \frac{V_{cm}}{V_{fault}} \Rightarrow \infty.$$

Sometimes the reciprocal value is calculated; then the result is $1/H = 0$.

$$* \quad * \quad *$$

Example 8.12
Assume a digital voltmeter with a floating guard as shown in figure 8.43 has the following specifications. The display contains $4\frac{1}{2}$ digits; this is equivalent to a maximum range of 19 999. The minimum range is 20 mV and the resolution is 1 μV. The input impedances are as follows: high–low, 1 GΩ‖500 pF in the 20 mV range; low–guard 10 GΩ‖700 pF; guard–ground 50 GΩ‖1.5 nF. When the imbalance in the input circuit equals the standardized value of 1 kΩ, the CMRR equals 150 dB at DC and 120 dB at 50 Hz AC.

A measurement is performed on a voltage source with an intrinsic impedance R_s of 100 Ω in one lead; the other lead is supposed to be 0 Ω. The source voltage is 100μV which should be measured with an accuracy of 1%.

Demanding the specified accuracy, we will investigate how frequency and a wrong connection of the guard influence the CMRR, and therefore we need to calculate the following cases:

1   the maximum allowed common-mode voltage at an interfering frequency of 20 Hz when the guard is properly used
2   the maximum allowed common-mode voltage with an interfering frequency of 20 Hz when an error is made connecting the guard to the low.

This can be achieved by the following means. Consider the equivalent circuit for the input measurement configuration in figure 8.47. The guard is properly used, which means that it is

1   connected to the screen surrounding the input leads
2   via the screen connected to the common-mode voltage V_{cm} at the transducer side, and hence to earth, in order to make C_2 ineffective by short-circuiting C_2 with R_s.

In this circuit the imbalance is illustrated with 1 kΩ drawn in the Hi terminal lead. Again no current is presumed to flow because of the high input impedance of the amplifier. This means that any error voltage will be developed across the signal source resistance R_s of 100 Ω. The series-mode voltage V_{cm} is converted into a differential-mode (error) voltage.

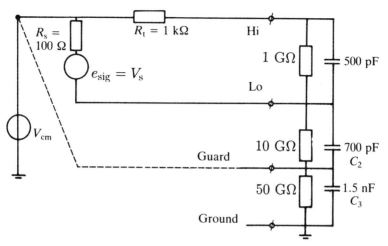

Figure 8.47 Equivalent measuring input circuit of the DVM when the guard is properly used. Note that the guard is connected to V_{cm} at the transducer's side (dashed lines). R_t is specified by the manufacturer.

It has been shown previously that the CMRR can be expressed as a measure of imbalance present in the input circuit. Note that the 1 kΩ imbalance is used as a standardized value at which the CMRR should be specified by the manufacturer. Once the CMRR is specified, the resulting effective impedance Z of the measuring system can be derived from the given impedance ratio. This procedure can be expressed as

$$\text{CMRR}|_{50\text{ Hz}} = 20 \log \frac{R_t + Z}{R_t}. \tag{8.29}$$

Here also the condition holds that $R_t \ll |Z|$, and thus equation (8.29) simplifies as

$$\text{CMRR}|_{50\text{ Hz}} = 20 \log \frac{Z}{R_t} = 120 \text{ dB}.$$

Substituting the specified data yields for Z

$$\frac{Z}{R_t} = 10^6\ \Omega \quad \text{or} \quad Z = 10^6 \times R_t = 10^9\ \Omega.$$

Thus this value equals the resulting effective impedance of the input circuit, and hence

$$Z = \frac{1}{2\pi f C_{\text{eff}}} = 10^9\ \Omega \quad \text{or} \quad C_{\text{eff}} = \frac{1}{2\pi 50 \times 10^9} = 3.18 \text{ pF}.$$

Comparing $X_{C_{\text{in}}}$ with $X_{C_{\text{eff}}}$ it can be seen that $X_{C_{\text{in}}} \ll X_{C_{\text{eff}}}$ and therefore $X_{C_{\text{eff}}}$ will dominate the total effective impedance described by Z. Neglecting the current

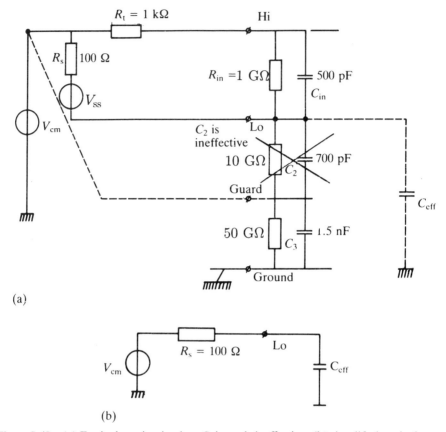

(a)

(b)

Figure 8.48 (a) Equivalent circuit when C_2 is made ineffective; (b) simplified equivalent circuit when the guard is properly used.

through R_{in} of the input amplifier, and keeping in mind that C_2 is made ineffective in figure 8.48(a), figure 8.48(b) represents the remaining equivalent circuit.

Now the effective capacitance C_{eff} is known, the CMRR can be calculated with the source impedance of 100 Ω at the specified frequency of 20 Hz, instead of 50 Hz. To calculate the CMRR the same equation (8.29) as before can be applied; thus with $R_t = R_s = 100$ Ω, we find

$$\text{CMRR}|_{20\text{ Hz}} = 20 \log \frac{Z}{R_t} = 20 \log \frac{1/(2\pi 20 \times 3.2 \times 10^{-12})}{100}$$

$$= 20 \log 25 \times 10^6 = 148 \text{ dB}.$$

A larger CMRR is obtained as can be expected from an applied lower frequency of 20 Hz.

From the specifications we know that the resolution is 1 μV, the source voltage is 100 μV and the required accuracy is 1%; hence the allowable voltage error is 1 μV. On applying a proper guarding the magnitude of the allowed voltage error will again be determined by the ratio of the now present imbalance $R_s = 100\,\Omega$ and the ultimately effective impedance Z, or

$$V_{error} = V_{cm}\frac{R_s}{Z}. \tag{8.30}$$

Rearranging the equation will deliver the first desired answer:

$$V_{cm} = V_{error}\frac{Z}{R_s} = 10^{-6}(25 \times 10^6) = 25\text{ V}.$$

This result equals the maximum allowed common-mode input voltage at an interfering frequency of 20 Hz maintaining the specified accuracy.

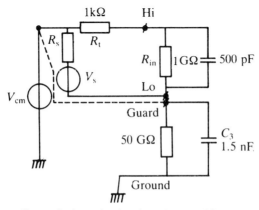

Figure 8.49 The equivalent circuit when the guard is connected to the low.

When the guard is connected to the low, the equivalent circuit of figure 8.49 will illustrate the actual situation.

In this case the remaining effective capacitance C_3 is the capacitance between low and ground and equals 1.5 nF. Again neglecting any input current, the maximum allowable V_{cm} at an interfering frequency of 20 Hz and with the same specified accuracy is

$$V_{cm} = \frac{Z}{R_s}V_{error} = \frac{1/(2\pi20 \times 1.5 \times 10^{-9})}{100} \times 10^{-6} = 53\text{ mV}.$$

It will be clear from this example that a wrong connection of the guard influences accuracy, CMRR or the allowed V_{cm} to a large extent.

*   *   *

In the next section we will present a very real problem that fits perfectly in the discussion of guarding and shielding. This is the formal legislation concerning electromagnetic compatibility which will come into effect January 1996 for the European Community. Also, some existing standards in other non-European countries are given. Every manufacturer, on going to the market, will be confronted with these requirements for his or her electronic products.

8.11 ELECTROMAGNETIC COMPATIBILITY

We have seen already that an increasing number of people and organizations use electronic equipment. In section 8.2.1 we have presented already a collection of interference sources. Basically every electronic system can act as a source of interference for other electronic equipment or as a receiver sensitive for it. In general we are talking about electromagnetic compatibility (EMC). A simple home experiment can illustrate this.

*   *   *

Example 8.13
The interference generated by a personal computer or monitor is easily observed with the help of a simple radio receiver in the vicinity of the PC, when it is slightly mistuned from an arbitrary broadcasting station. Every processing step of the PC can be monitored easily, such as cursor movement, typing, data storing, etc. This interference may not exceed certain predefined limits.

*   *   *

Here follows a rather formal description of the legislation and standards for EMC conformity.

8.11.1 New Approach, Global Approach and CE mark
In the process of removing technical barriers to trade, the European Community has adopted a New Approach to technical harmonization in 1985. This New Approach means that the European Community regulates what is essential for ensuring the necessary level of protection with respect to safety, health and environment. For categories of products the essential requirements are formulated in the directives. The essential requirements are or will be specified in harmonized European technical standards. The European Community has given mandates to CEN, CENELEC and ETSI, the European standardization bodies, to draft and adopt the technical standards. If a product conforms to the relevant standards, it produces a presumption of conformity with the essential requirements.

Since the adoption of the New Approach the Council of the European Community has adopted several directives, as shown in table 8.2.

Other directives will in the near future be adopted, such as

- other medical devices
- *in vitro* diagnostica
- metrology matters
- equipment under pressure
- equipment for use in potentially explosive atmospheres.

Table 8.2 Survey of directives issued by the European Community.

Application area	Directive	Effective
Simple pressure vessels	86/404/EEC	90-07-01
Toy safety	88/378/EEC	90-01-01
Construction products	89/106/EEC	91-07-01
Electromagnetic compatibility	89/336/EEC	92-01-01
Safety of machinery	89/392/EEC	93-01-01
Personal protective equipment	89/686/EEC	92-07-01
Gas appliances	90/396/EEC	92-01-01
Non-automatic weighing instruments	90/384/EEC	93-01-01
Active implantable medical devices	90/385/EEC	93-01-01
Telecommunications terminal equipment	91/263/EEC	92-11-06
Display screen equipment	90/270/EEC	92-12-31

In each directive it is stated when the directive has come into force. The member states must transpose the directives in their own legislation before the date of coming into force.

In the New Approach directives the conformity procedures are defined (procedures to be followed to show that the products conform to the essential requirements). These testing and certification procedures in the specific directives are derived from a modular system as described in the Global Approach (Decision of the Council of the European Community of 13 December 1990). These conformity procedures range from a manufacturer's declaration to assessment of a quality system for the manufacturer and type-testing of the product by a third party, depending on the risks of the product in relation to safety and health. The member states notify bodies to perform the tasks of the third party as described in the directives. These 'notified bodies' must fulfil the minimum criteria of the directives. There is a presumption of conformity with the minimum criteria if the notified body can prove that it satisfies the relevant standard of the EN 45000-series (European accreditation standards). The CE mark is a sign that the product, as declared by the manufacturer, fulfils the essential (minimum) requirements. It is a responsibility of the manufacturer that the products satisfy the essential requirements if these products are brought on the market with the CE mark. All products in the same category will be labelled with the CE mark. So the CE mark is not meant to distinguish in regard to the quality of products of a certain category. Moreover, it remains to be seen whether consumers will have much confidence in the CE mark. Therefore product certification and quality system certification by an independent body for products falling under the New Approach directives or products outside their reach will be needed also in the future. An extra mark may be affixed to the product provided that the mark will not lead to confusion with the CE mark. Products in the range of the New Approach directives may only be brought on the EC market if the products are labelled with the CE mark. This concerns not only manufacturers within the European Community but also manufacturers outside the European Community who want to bring their products on the EC market. The conformity assessment procedures in community legislation as defined by the Global Approach are shown in table 8.3.

Table 8.3 Conformity assessment procedures in community legislation.

A. Internal control of production	B. Type examination					G. Unit verification	H. Full quality assurance EN 29001
Manufacturer Keeps technical documentation at the disposal of national authorities	Manufacturer Submits to notified body Technical documentation Type					Manufacturer Submits technical documentation	Manufacturer Operates an approved quality system (QS) for design
	Notified body Ascertains conformity with essential requirements Carries out tests, if necessary Issues EC type-examination certificate						
Aa. Notified body Intervention							Notified body Carries out surveillance of the QS Verifies conformity of the design Issues EC design examination certificate
A. Manufacturer Declares conformity with essential requirements Affixes CE mark	C. Conformity type Manufacturer Declares conformity with approved type Affixes the CE mark	D. Product quality assurance EN 29002 Manufacturer Operates an approved quality system (QS) for production and testing Declares conformity with approved type Affixes the CE mark	E. Product quality assurance EN 29003 Manufacturer Operates an approved quality system (QS) for inspection and testing Declares conformity with approved type or essential requirements Affixes the CE mark	F. Product verification Manufacturer Declares conformity with approved type or with essential requirements Affixes the CE mark	Manufacturer Submits product Declares conformity Affixes CE mark	Manufacturer Operates an approved QS for production testing Declares conformity Affixes the CE mark	
Aa. Notified body Tests on specific aspects of the product Product checks at random intervals	Notified body Tests on specific aspects of the product Product checks at random intervals	Notified body Approves the QS Carries out surveillance of the QS	Notified body Approves the QS Carries out surveillance of the QS	Notified body Verifies conformity Issues certificate of conformity	Notified body Verifies conformity with essential requirements Issues certificate of conformity	Notified body Carries out surveillance of the QS	

8.11.2 Generic emission standards

In this section a review of the European EMC generic emission standards is given in tables 8.4–8.11.

Table 8.4 Generic emission standards.

EN 50081-1	Generic emission standards Part 1: Domestic, commercial and light industry environment
prEN 50081-2	Generic emission standards Part 2: Industrial environment

Table 8.5 Low-frequency emission standards.

EN 60555-2	Disturbances in supply systems caused by household appliances and similar electric equipment Part 2: harmonics
EN 60555-3	Disturbances in supply systems caused by household appliances and similar electric equipment Part 3: voltage fluctuations

Table 8.6 Radio-frequency emission standards.

EN 55011	Limits and methods of measurement of radio disturbance characteristics of industrial, scientific and medical (ISM) radio-frequency equipment
EN 55013	Limits and methods of measurement of radio disturbance characteristics of broadcast receivers and associated equipment
EN 55014	Limits and methods of measurement of radio disturbance characteristics of household electrical appliances, portable tools and similar electrical apparatus
EN 55015	Limits and methods of measurement of radio disturbance characteristics of fluorescent lamps and luminaries
EN 55022	Limits and methods of measurement of radio disturbance characteristics of information technology equipment

Table 8.7 Emission standards for ISDN terminal equipment.

prENV 550102-1	Electromagnetic compatibility requirements for ISDN terminal equipment Part 1: emission requirements

Table 8.8 Specific emission standard for signalling on installations.

EN 50065-1	Signalling on low-voltage electrical installations in the frequency range 3 kHz to 148.5 kHz Part 1: general requirements, frequency bands and electromagnetic disturbances. Immunity standards

Table 8.9 Immunity standards.

EN 55020	Immunity from radio interference of broadcast receiver and associated products

Table 8.10 Immunity standards for information technology and communication equipment.

prEN 55024-2	Electrostatic discharges
prEN 55024-3	Radiated fields
prEN 55024-4	Fast transient
prEN 55024-5	Surge
prEN 55025-6	Conducted radio interference

Table 8.11 Immunity standards for ISDN terminal equipment.

prENV 550102-2	Electromagnetic compatibility requirements for ISDN terminal equipment Part 2: Immunity requirements

Table 8.12 Survey of electromagnetic compatibility regulations and standards.

Country	Ignition systems	RF equipment including ISM	Household appliances	Radio and TV-video	Fluorescent lamps and luminaries	Solid state controls	IT and EDP equipment
European norm	72/245/EEC	EN55011	EN55014 EN60555-213	EN55013 EN55020	EN55015		EN55022
USA	SAEJ551C	FCC Pt18 FCC MP-5 MDS2010004 NEMA ICS-2 IEEE518-1982 MIL STD461/2	FCC Pt15 ANSIC63-2 ANSIC63-4 NEMA WD2-1970	FCC Pt2 FCC Pt15/C	FCC Pt15J FCC MD-4		FCCPt15J FCC MP-4
Japan	CISPR12 JRTC/MPT	RERART65 JRTC73/74	EA&MCLAW JRTC73/74/75 MPT1970/71	EA&MCLAW JRTC71/74/75 *82 CISPR 13	EA&MCLAW JRTC73/74/75 MPT1970/71		VCCI CISPR22
Canada	SOR75-629 CSA108-4	SOR/75-629 CSA22-1/3/5 CSAC235	CSA108-5-45 CSAC108-5-2	SOR/75-629 SOR/83-352	CSAC108-5	CSA22-4	CSAC108-8
Australia	AS2557	AS2064/2279	AS1044/2279	AS1053	AS2643	AS1054	EN55022
New Zealand	CISPR12 NFFCCOM	RFS49-1	RFS49-1 AS2279 BS5406	RFS49-1 CISPR13	CISPR15		CISPR22
South Africa	R2862-1979 (CISPR 16)	R2862-1979 CISPR11 (CISPR 16)	R2862-1979 (CISPR 16)	R2862-1979 (CISPR 16)	SABS (CISPR 16)	R2862-1979	SABS (CISPR22)
Brazil	Relevant CISPR publications also IEC/C77 standards						

Finally we want to give a table (table 8.12) which outlines some other important directives issued in various countries. For each manufacturer they are also of importance in the design of electronic equipment.

Although much more can be said this will conclude our discussion about guarding and shielding. For more background information reference is made to the literature.

PROBLEMS

8.1 Explain why, in the configuration depicted in figure 8.50, the common-mode voltage V_{cm} is converted into a series-mode voltage.

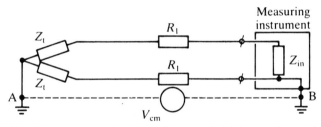

Figure 8.50 A common-mode voltage is converted into a series-mode voltage.

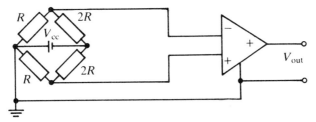

Figure 8.51 Bridge measuring circuit.

8.2 Determine the common-mode voltage at both input terminals of the operational amplifier drawn in figure 8.51.

8.3 Of the measuring circuit illustrated in figure 8.51 the following specifications are known. The supply voltage is $V_{cc} = V$, the CMRR of the operational amplifier for DC is 100 dB and the Wheatstone bridge is in a balanced state. Determine the output voltage V_{out}.

8.4 Determine by which measurement technique the influence of V_{cm} in the configuration of figure 8.52 can be eliminated.

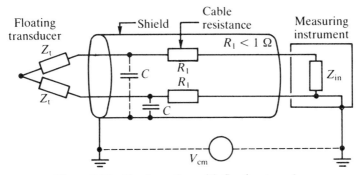

Figure 8.52 Configuration with floating transducer.

8.5 Determine for the configuration depicted in figure 8.53 the CMRR for AC measurements in a country with a mains supply frequency of 60 Hz. The stray capacitance between the low terminal and the cabinet equals 2.6 nF; the transducer impedance is 1 kΩ.

8.6 Determine the CMRR of the earthed attenuator depicted in figure 8.54.

8.7 A digital DC voltmeter has the following specification impedances: high–low, 1 GΩ∥500 pF; low–guard, 10 GΩ∥700 pF; guard–ground, 50 GΩ∥1.5 nF. The common-mode rejection ratio is 130 dB at 50 Hz with 1 kΩ imbalance in the input leads. The series-mode rejection ratio (SMRR) is 60 dB at 50 Hz. A DC measurement is executed

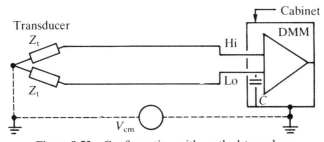

Figure 8.53 Configuration with earthed transducer.

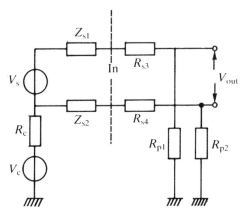

Figure 8.54   An earthed attenuator.

on a source signal of $200\,\mu$V having an imbalanced source resistance of $10\,$kΩ. The following quantities have to be calculated:

1   the error voltage due to a DC common-mode voltage of $100\,$V when (i) the guard is connected to the ground and (ii) the guard is properly used
2   the common-mode rejection ratio in both cases
3   the maximum allowable series-mode voltage with a frequency of $50\,$Hz when an accuracy of 10% is required.
4   the maximum allowable common-mode voltage with a frequency of $50\,$Hz when the guard is properly used and the required inaccuracy again is 10%.

BIBLIOGRAPHY

[1]   Morgan D 1994 *A Handbook for EMC Testing and Measurement* (London: Peregrinus on behalf of the Institution of Electrical Engineers)
[2]   Goedbloed J J 1992 *Electromagnetic Compatibility* (Hemel Hempstead: Prentice-Hall)
[3]   Analog Devices 1988 *Linear Design Seminar*
[4]   Keithley Instruments 1992 *Low Level Measurements*
[5]   Morrison R 1977 *Grounding and Shielding Techniques in Instrumentation* (New York: Wiley)

Nine

NOISE CALCULATIONS

9.1 INTRODUCTION

This chapter is devoted to noise calculations. Noise is one of the most discussed topics in electronics. When very low signal levels are involved noise plays a vital role in signal handling. Noise can have a great variety of causes, but they all have one feature in common with respect to the signal being measured. Noise can have the signal drowned so it cannot be detected any longer. This is mainly determined by the signal-to-noise ratio, but apart from this, the applied measuring technique has an important influence on the threshold value that can be detected. Of course the ultimate limit will be a physical limit and cannot be surpassed.

There is an increasing need to control signal-to-noise ratios because in communication and telecommunication applications the distance to be bridged is still growing. Any increase in distance means a loss in signal-to-noise ratio. Despite an increasing shift to digital signal handling, when very large distances are involved, the noise level will come into scope. However, an even more serious noise problem arises when *biological signal sources* have to be measured. Very often these sources combine a very high source impedance with very low signal level.

A good example of such a source is our heart muscle cells. The electrical phenomenon is due to a flow of K^+, Ca^{2+} or Na^+ ions in the so-called canals in the cell membranes. These cell membranes show an inherently high source impedance with an order of magnitude of gigaohms. The currents flowing through these membranes have an order of magnitude of picoamperes and so the signal power is in the same order as the noise power. Another example you have met already in chapter 8 is the measurement of an electroencephalogram (EEG) in which also very low signal levels (microvolts) have to be measured. Finally, we would like to mention the measurement of magnetic field activities generated by our brains. These are measured with the help of the so-called SQUIDs we met in chapter 6. A Faraday cage is still required here.

This chapter offers you a practical approach with only one objective: to make noise calculations possible with all the relevant concepts required for calculation, without discussing physical noise mechanisms. The following chapter is devoted to

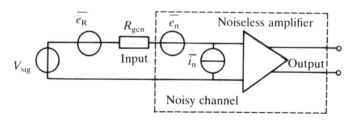

Figure 9.1   The noise characterization of an amplifier.

a more fundamental physical approach and discusses the nature of several noise mechanisms. It can be treated separately or skipped in a first reading.

9.2 NOISE VOLTAGE, NOISE CURRENT AND NOISE FIGURE

The physical aspects of noise are discussed in chapter 10, so in this section we can skip all sorts of terms such as spot noise, thermal noise, broadband noise, shot noise, white noise, fluctuation noise, pink noise, $1/f$ noise, flicker noise, popcorn noise, bipolar spike noise, low noise, no noise and loud noise. This can make it all very confusing but only two things are important here: there is noise and there is signal.

The whole idea is to have the noise level very small compared with the signal level or, stated conversely, a high signal-to-noise ratio (S/N) is always desired. Now it happens that the S/N is related to the *noise figure, noise factor, noise power, noise voltage* and *noise current*. To simplify matters further it also happens that any noisy channel or amplifier can be completely specified for noise in terms of two noise generators $\overline{e_n}$ and $\overline{i_n}$ as is illustrated in figure 9.1. All we really have to understand are the NF, $\overline{e_n}$ and $\overline{i_n}$.

9.2.1 The noise voltage $\overline{e_n}$

The *noise voltage* $\overline{e_n}$, or more precisely the equivalent short-circuit input RMS noise voltage, is defined as that noise voltage which would appear to originate at the input of the noiseless amplifier if the input terminals were shorted. It is expressed in nanovolts per root hertz, at a specified frequency, or in microvolts in a given frequency band.

This noise voltage can be determined by shorting the input terminals, measuring the output RMS noise voltage, dividing by the amplifier gain, and referencing to the input. This explains the term equivalent short-circuit input RMS noise voltage.

In the measurements, an output bandpass filter of known characteristics is used and the measured value is divided by the square root of the bandwidth, $B^{1/2}$, if the data have to be expressed per unit bandwidth or per root hertz. It appears that the level of $\overline{e_n}$ is not a constant over the frequency band but very often increases at lower frequencies as is illustrated in figure 9.2. Very often for every specified amplifier or transistor this type of graph is provided by the manufacturer and is easy to use in calculations as will be seen later on.

9.2.2 The noise current $\overline{i}_n$

The *noise current* $\overline{i}_n$ or, more precisely, the equivalent open-circuit RMS noise current, is that noise which appears to be present at the input of the noiseless amplifier due only to noise currents. It is expressed in picoamperes per root hertz, at a specified frequency, or in nanoamperes in a given frequency band.

This noise current can be measured by shunting a capacitor or resistor across the input terminals in such a way that the noise current will give rise to an additional noise voltage which then equals $\overline{i}_n R_{in}$, or in $\overline{i}_n X_{C_{in}}$, depending on what is connected to the input terminals. The output voltage is measured, divided by the amplifier gain, referenced to the input and the contribution known to be due to the noise voltage $\overline{e}_n$ and the resistor noise is subtracted from the total measured noise. Using a resistor we have the following expression

$$\overline{e^2_{in\ total}} = \overline{e^2_{out}}/G = \overline{i^2_n}R^2_n + \overline{e^2_R} + \overline{e^2_n}. \tag{9.1}$$

If a capacitor is used at the input there is only a noise voltage $\overline{e}_n$ and a noise voltage $\overline{i}_n X_{C_{in}}$, because a capacitor does not generate noise so an $\overline{e}_C$ cannot exist. This yields

$$\overline{e^2_{in\ total}} = \overline{e^2_{out}}/G = \overline{i^2_n}X^2_{C_{in}} + \overline{e^2_n}. \tag{9.2}$$

Again the noise current $\overline{i}_n$ is measured with a bandpass filter of known characteristics and divided by the square root of the bandwidth converting to $pA/\sqrt{B}$ if appropriate. Noise current increases at lower frequencies for operational amplifiers and bipolar transistors, but increases at higher frequencies for field effect transistors. These aspects will be discussed later on in more detail. The current noise behaviour for the same amplifier is depicted in figure 9.2.

9.2.3 The noise figure NF

Finally we have to define the *noise figure*. The noise figure, NF, is defined as the logarithm of the ratio of the input signal to noise ratio and the output signal to noise ratio:

$$NF = 10 \log \frac{(S/N)_{in}}{(S/N)_{out}} \tag{9.3}$$

where S and N are power levels or volt squared levels. It is measured by first determining the S/N at the input with the amplifier disconnected and second determining the S/N at the output with the signal source connected at the amplifier and finally dividing both measured quantities.

Derived from this description we can put the expression for the NF equation (9.3) in a slightly different way:

$$NF = 10 \log \frac{\text{total noise power}}{\text{noise power of the source}}. \tag{9.4}$$

This will be shown in the following section, but as you can see already, the NF expresses a comparison between the total noise power and the noise power of the source itself.

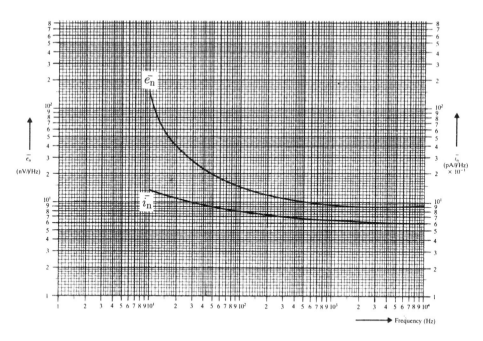

Figure 9.2 Noise voltage and noise current graph for a given operational amplifier as a function of frequency (courtesy National Semiconductor Ltd).

The NF can be calculated in real terms when the values of the source impedances R_{gen} and X_{gen}, as well as frequency, are known. This is because the amplifier's noise current produces a noise voltage source which equals $\overline{i_n}Z_{gen}$ and R_{gen} itself produces a so-called thermal noise voltage. The signal source in figure 9.1 produces some noise; however, V_{sig} is generally considered to be noise free and the input noise of the source is only present as thermal noise of the resistive component R_{gen} of the signal source impedance. As will be shown, this thermal noise is white in nature because it contains a constant noise power density per unit bandwidth. In the next section we will discuss the relationship between the noise voltage source, the noise current source and the noise figure NF.

9.3 THE RELATIONSHIP BETWEEN THE NOISE VOLTAGE, NOISE CURRENT AND NOISE FIGURE

We will examine the relationship between $\overline{e_n}$ and $\overline{i_n}$ at the amplifier's input. When the signal source is connected to the input of the amplifier $\overline{e_n}$ appears in series with the source voltage V_{sig} and $\overline{e_R}$, where $\overline{e_R}$ represents the thermal noise voltage developed across R_{gen}. $\overline{i_n}$ flows through R_{gen} thus producing another noise voltage with magnitude $\overline{i_n}R_{gen}$. It will be clear that this noise voltage is dependent on the value of R_{gen}.

All of these noise voltages add at the input in RMS fashion this means that the square root of the sum of the squares of the noise voltages needs to be considered. If we neglect possible correlations between $\overline{i_n}$ and $\overline{e_n}$ the *total input noise density* $\overline{e_N}$ can be calculated from

$$\overline{e_N^2} = \overline{i_n^2} R_n^2 + \overline{e_R^2} + \overline{e_n^2}. \tag{9.5}$$

So, in general, three different noise sources can be distinguished: the first is delivered by the amplifier only, the second is from the signal source and the third is determined by the product of the noise current of the amplifier and the generator source resistance.

Further examination will reveal the relationship between the different quantities; for this purpose we refer to equation (9.3). Note that this equation is expressed in power densities, and can be rewritten in the following way:

$$NF = 10 \log \frac{S_{in} N_{out}}{S_{out} N_{in}} \tag{9.6}$$

and for N_{out} we can write, when G_p equals the power gain of the amplifier

$$N_{out} = G_p \overline{e_N^2}. \tag{9.7}$$

Substituting equation (9.7) in (9.6) yields

$$NF = 10 \log \frac{S_{in} G_p \overline{e_N^2}}{S_{in} G_p \overline{e_R^2}} \tag{9.8}$$

and this yields for the noise figure

$$NF = 10 \log \frac{\overline{e_N^2}}{\overline{e_R^2}}. \tag{9.9}$$

Compare the result of equation (9.7) with equation (9.3). Equation (9.9) can be written as

$$NF = 10 \log \frac{\overline{e_n^2} + \overline{e_R^2} + \overline{i_n^2} R_{gen}^2}{\overline{e_R^2}} \tag{9.10}$$

or

$$NF = 10 \log \left(1 + \frac{\overline{e_n^2} + \overline{i_n^2} R_{gen}^2}{\overline{e_R^2}} \right). \tag{9.11}$$

From this equation it can be seen that when R_{gen} is small, noise voltages will dominate, and when a large source resistance R_{gen} is involved noise current will become important. A clear advantage accrues here to FET input amplifiers especially at high values of R_{gen} because in an ideal FET there is no $\overline{i_n}$ at all. Note that for a meaningful NF value it must be accompanied by a value for R_{gen} as well as frequency and/or bandwidth.

In the following section a first approach will be given calculating total noise voltage.

9.4 CALCULATING TOTAL NOISE VOLTAGE

As stated already, always keep in mind that we have to reckon with power levels or V^2 levels before the total noise voltage level can be determined, taking the square root. The consequence of this is that every noise term must be squared prior to addition.

Consider now again equation (9.5); then each term for a given amplifier and known signal source must be fixed. Very often the manufacturer provides graphs concerning the noise behaviour of the specified amplifier. Sometimes the total input noise voltage $\overline{e_N}$ versus frequency is given for different values of source resistances

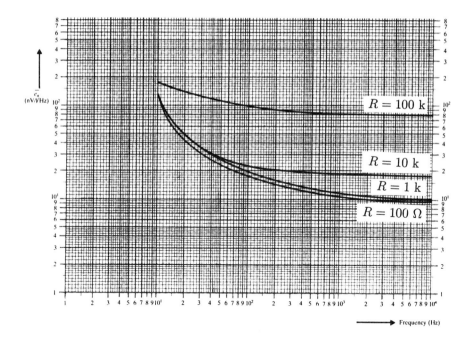

Figure 9.3   Total noise voltage versus frequency with source resistance R_s as parameter (courtesy National Semiconductor Ltd).

$R_s = R_{gen}$ of which figure 9.3 is an example. However, very often the manufacturer provides two separate graphs, one depicting the noise voltage of the amplifier and the other depicting the noise current versus frequency, and occasionally they are combined in one graph. An example of such a graph is figure 9.2. So you have to consider carefully the applied coordinate axis and units. Here are some examples.

*   *   *

Example 9.1

Determine the total equivalent input noise per unit bandwidth for an amplifier operating at 1 kHz from a source resistance of 10 kΩ. The ambient temperature is 300 K. When the manufacturer has provided us with a graph depicting the total noise voltage versus frequency with R_s as parameter we can draw the required value directly from that graph. Assume here that figure 9.3 is applicable; then at 1 kHz with parameter $R_s = 10$ kΩ the total noise voltage has to be read from the graph. This delivers for $\overline{e_N} \approx 17$ nV/$\sqrt{\text{Hz}}$. If the source resistance R_s is not one of the depicted values, an estimation can be made by interpolation. However when this total noise graph is not available you have to perform a little more effort as outlined in the next example.

* * *

Example 9.2

Determine the total equivalent input noise voltage per unit bandwidth for an amplifier operating at 1 kHz from a source resistance of 10 kΩ when the noise data of figure 9.2 are to be applied. The ambient temperature is again 300 K.

The first thing to do is to determine the value of the equivalent noise voltage for R_s; thus $\overline{e_R}$ must be fixed. This can be performed on two different ways.

As mentioned already, the nature of noise generated by R_s is white or thermal noise and can be derived from the following equation, which describes the available thermal noise power at a given temperature and bandwidth of the resistance involved:

$$P = 4kTB \quad (\text{W}) \quad (9.12)$$

where k is Boltzmann's constant (1.38×10^{-23} J K^{-1}), T is the temperature in kelvins and B is the considered bandwidth in hertz. Two source representations can be derived from this equation, one as an equivalent RMS noise voltage source and one as an equivalent RMS noise current source, i.e.

$$P = \frac{\overline{e_{R_s}^2}}{R_s} = \overline{i^2} R_s = 4kTB \quad (\text{W}). \quad (9.13)$$

Then the equivalent voltage power is

$$\overline{e_{R_s}^2} = 4kTBR_s \quad (\text{V}^2) \quad (9.14)$$

and the equivalent current power is

$$\overline{i_{R_s}^2} = \frac{4kTB}{R_s} \quad (\text{A}^2). \quad (9.15)$$

Then the equivalent RMS noise voltage source is

$$\overline{e_{R_s}} = (4kTBR_s)^{1/2} \quad (\text{V}) \quad (9.16)$$

and the equivalent RMS noise current source is

$$\overline{i_{R_s}} = \left(\frac{4kTB}{R_s}\right)^{1/2} \quad (\text{A}). \quad (9.17)$$

This calculation can be executed for a bandwidth of 1 Hz and an ambient temperature of 300 K for the equivalent RMS noise voltage source as well as the equivalent RMS noise current source. The result is depicted in figure 9.4 and is very convenient for noise calculations when the reference temperature is taken as 300 K, which is very usual.

So every time an equivalent RMS noise voltage must be determined you can apply equation (9.12) or you can use figure 9.4 if the referred standard conditions are met.

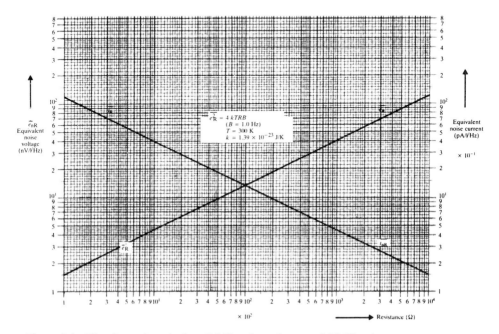

Figure 9.4 The thermal equivalent RMS noise voltage and RMS noise current versus resistance at 300 K and 1 Hz bandwidth (courtesy National Semiconductor Ltd).

If not, then you have to use equation (9.12). To answer the question the following steps have to be taken.

1 Read $\overline{e_R}$ from figure 9.4 at 10 kΩ, the value is 13 nV/$\sqrt{\text{Hz}}$ for 1 Hz bandwidth.
2 Read $\overline{e_n}$ from figure 9.2 at 1 kHz; the value is 9.5 nV/$\sqrt{\text{Hz}}$.
3 Read $\overline{i_n}$ from figure 9.2 at 1 kHz, the value is 0.65 pA/$\sqrt{\text{Hz}}$; then multiply by 10 kΩ to obtain 6.5 nV/$\sqrt{\text{Hz}}$.
4 Square each term individually and substitute into equation (9.5), add all terms and take the square root of the result:

$$\overline{e_N^2} = \overline{i_n^2} R_R^2 + \overline{e_R^2} + \overline{e_n^2}$$

$$= [(9.5)^2 + (13)^2 + (6.5)^2]^{1/2} = \sqrt{301.5} \text{ nV}/\sqrt{\text{Hz}}$$

which slightly differs from the result of example 9.1 owing to reading inaccuracies in the graph.

*   *   *

The result is the total RMS noise voltage at the input in 1 Hz bandwidth at a frequency of 1 kHz. If the total noise voltage is desired in a given bandwidth we must integrate the noise over the specified bandwidth. This is most easily performed in a noise measurement set-up, but may be approximated as follows.

1   If the frequency range of interest is in the flat band between 1 kHz and 10 kHz in figure 9.2 it is simply a matter of multiplying $\overline{e_N}$ by the square root of the bandwidth. In this case with the specified bandwidth range, the total noise voltage is

$$\overline{e_N} = 17.4\sqrt{9000} = 1.65 \ \mu V.$$

2   If the frequency band of interest is not in the flat band of figure 9.2, the band must be broken into sections, calculating the noise in each section, then squaring and multiplying by section bandwidth, summing all sections and finally taking the square root of the sum. This complete procedure is expressed in the following equation where i denotes the total number of subblocks:

$$\overline{e_N} = \left[\overline{e_R^2} B + \sum (\overline{e_n^2} + \overline{i^2} R_{gen}^2)_i B_i \right]^{1/2}. \tag{9.18}$$

For most purposes a subblock of one or two octaves may be sufficient. This procedure is demonstrated in the next example.

*   *   *

Example 9.3
Determine the RMS noise level in the frequency band from 50 Hz to 10 kHz for the amplifier of figure 9.2, operating with a voltage source with an impedance of $R_{gen} = 2 \ k\Omega$. The ambient temperature is 300 K. The easiest way to answer this question is to construct a table.

1   Read $\overline{e_R}$ from figure 9.4 at 2 kΩ. This gives $6 \ nV/\sqrt{Hz}$; square this and multiply by the entire bandwidth. Write the result ($358\,200 \ nV^2$) in the last column of table 9.1.

Table 9.1   Noise calculations for example 9.3.

Frequency band (Hz)	Bandwidth (Hz)	$\overline{e_R^2}$ (nV² Hz⁻¹)	$\overline{e_n^2}$ (nV² Hz⁻¹)	$\overline{i_n^2} R_{gen}^2$ (nV² Hz⁻¹)	Sum × B (nV²)
50–10 000	9950	$6.0^2 = 36$			$36 \times 9950 = 358\,200$
50–100	50		$17.5^2 = 306$	1.7^2	$309 \times 50 = 15\,450$
100–300	200		$12.5^2 = 156$	1.5^2	$158 \times 200 = 31\,600$
300–1000	700		$10.0^2 = 100$	1.3^2	$102 \times 700 = 71\,400$
1000–10 000	9000		$9.0^2 = 81$	1.2^2	$82 \times 9000 = 738\,000$
			Total	$\overline{e_N^2} =$	$1\,214\,650$

2   Read the median value of $\overline{e_n}$ in a relatively small frequency band, say 50–100 Hz, from figure 9.2. This gives $17.5 \ nV/\sqrt{Hz}$; square this and write the result ($306 \ nV^2$) in the table. For the median read the value for frequency 75 Hz.

3   Read the median value of $\bar{i}_n$ in the 50–100 Hz band from figure 9.2 (0.85 pA/ $\sqrt{\text{Hz}}$), multiply by $R_{\text{gen}} = 2\ \text{k}\Omega$, square the result ($1.7^2 = 3\ \text{nV}^2\ \text{Hz}^{-1}$) and enter it in the table.
4   Sum the squared results from steps 2 and 3, multiply by the bandwidth (50 Hz) and enter the result ($309 \times 50 = 15\ 450\ \text{nV}^2$) in the table.
5   Repeat steps 2–4 for band sections of 100 Hz–300 Hz, 300–1000 Hz, and 1– 10 kHz. Note that the median values are taken for frequencies 200, 650 and 5500 Hz respectively.
6   Sum the last column of the table and take the square root to obtain the total RMS noise $\bar{e}_N$ in the 50 Hz–10 kHz band range. The total noise voltage is found to be 1.1 μV.

*   *   *

As you see in this example the contribution of the noise current is very modest because of the relative small value of R_{gen}. Once the total noise voltage is calculated, it is very easy to determine the NF and the signal-to-noise ratio.

9.5   CALCULATING THE NOISE FIGURE AND THE SIGNAL-TO-NOISE RATIO

The noise figure and the signal-to-noise ratio can easily be calculated once the total RMS noise voltage in the required frequency band is determined.

*   *   *

Example 9.4
Applying equation (9.7) for the noise figure and substituting the values of example 9.3, we find

$$\text{NF} = 10\ \log \frac{\overline{e_N^2}}{\overline{e_R^2}}$$

$$= 10\ \log \frac{1\ 214\ 650}{358\ 200}$$

$$= 10\ \log 3.9 = 5.3\ \text{dB}$$

which thus expresses a power ratio between the total present noise power and the noise power developed by the signal source with $R_{\text{gen}} = 2\ \text{k}\Omega$ in the frequency band 50 Hz–10 kHz.

*   *   *

Example 9.5
Determine the S/N for an RMS $V_{\text{sig}} = 4\ \text{mV}$ connected to the input of the amplifier of the previous example for the specified bandwidth and $R_{\text{gen}} = 2\ \text{k}\Omega$. Firstly we have to give here a definition for the signal-to-noise ratio. The signal-to-noise ratio S/N

is defined as the logarithm of the ratio of the *signal source power* and the *total noise power* referenced across the same resistance:

$$S/N = 10 \log \frac{V_{sig}^2}{e_N^2} \tag{9.19}$$

or

$$S/N = 20 \log \frac{V_{sig}}{e_N}. \tag{9.20}$$

Applying equation (9.20) and substituting the calculated value for $\overline{e_N} = 1.1 \ \mu V$ and $V_{sig} = 4 \ mV$ yields

$$S/N = 20 \log \frac{4 \times 10^{-3}}{1.1 \times 10^{-6}} = 20 \log 3.64 \times 10^3$$

$$= 20(\log 10^3 + \log 3.64) = 20(3 + 0.56) = 71.2 \ dB.$$

$$* \quad * \quad *$$

It is also possible to plot NF versus frequency at various values of R_{gen} for any given plot of $\overline{e_n}$ and $\overline{i_n}$. However, no all-purpose conversion plot exists relating NF, $\overline{e_n}$, $\overline{i_n}$, R_{gen} and frequency f. If either $\overline{e_n}$ or $\overline{i_n}$ is neglected, a reference chart can be constructed. An example of such a chart is figure 9.5 when only $\overline{e_n}$ is considered. It

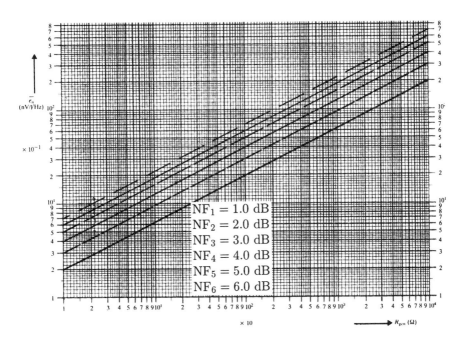

Figure 9.5 Spot noise NF versus R_{gen} when considering only $\overline{e_n}$ and $\overline{e_R}$, and $\overline{i_n}R_{gen}$ may be neglected.

is applicable for most operational amplifiers if R_{gen} is less than about 200 Ω and of course for FETs at any R_{gen}, because no significant $\overline{i_n}$ exists for FETs. However, the actual NF for operational amplifiers with $R_{gen} > 200$ Ω is higher than is indicated on the chart.

The graph of figure 9.5 can be used to find NF if $\overline{e_n}$ and R_{gen} are known, or to find $\overline{e_n}$ if NF and R_{gen} are known. Another possibility is to find maximum allowed R_{gen} for a given maximum NF when $\overline{e_n}$ is known. However, keep in mind that these values are only valid on the condition that $\overline{i_n}$ is negligible, at the specific frequency of interest for the NF and $\overline{e_n}$, and a bandwidth of 1 Hz. If a deviating bandwidth is involved the plot is valid as long as $\overline{e_n}$ is multiplied by $B^{1/2}$.

*   *   *

Example 9.6
Determine the RMS noise voltage $\overline{e_n}$ for a bandwidth of 9000 Hz with $R_{gen} = 1000$ Ω; the NF is 1 dB and the chart of figure 9.5 is applicable.

1   Read at $R_{gen} = 1000$ Ω on the graph of 1 dB $\overline{e_n} = 1.9$ nV/$\sqrt{\text{Hz}}$.
2   Multiply $\overline{e_n} = 1.9$ nV/$\sqrt{\text{Hz}}$ by $\sqrt{9000}$.
3   Then the RMS noise voltage in the specified frequency band is $1.9 \times \sqrt{9000} = 185$ nV $= 0.185\ \mu$V.

*   *   *

In the following section we will discuss the fact that a low NF does not always mean a low noise.

9.6   THE NOISE FIGURE MYTH

The noise figure is easy to calculate because the signal level of the source need not to be known because V_{sig}, denoted by S_{in}, in equation (9.8) drops out. Because NF is so easy to handle you might think that NF is important, but never lose sight of the fact that the S/N_{out} is what remains most important in the final analysis, be it in audio, video, a compact disc player, or digital data systems. You can in fact choose a high R_{gen} to reduce NF to almost being zero if $\overline{i_n}$ is very small.

In this case e_R will be the major source of noise, overshadowing $\overline{e_n}$ completely. The result will be a very low NF but a very low S/N as well because of a very high noise generated by R_{gen}. So do not be misled into thinking that a low NF means a low noise.

A maximum S/N is only achieved when $R_{gen} = 0$. As you know already $\overline{e_R} = (4kTB)^{1/2}$ is strongly temperature dependent, and one way to reduce $\overline{e_R}$ considerably is to cool the signal source with liquid helium which has a temperature of 4 K. This may be costly but in satellite land-communication receivers this technique is applied very often to improve the signal-to-noise ratio to a large extent. Suppose the distance to Uranus has to be bridged, which equals a distance of about 3×10^9 km, then a signal has to travel with the speed of light for $t = 3 \times 10^9/3 \times 10^5 = 10^4$ s which equals 2.78 h before it will reach Earth. Generally, such a distance and time delay will cause a large reduction in the signal-to-noise ratio.

9.6.1 The optimum source resistance

Another term worth considering is the optimum source resistance R_{opt}, which is defined as the value of R_{gen} which produces the lowest NF in a given system at a given frequency. This can be described by the equation

$$R_{opt} = \frac{\overline{e_n}}{\overline{i_n}}. \tag{9.21}$$

This can be proven by differentiating equation (9.10) with respect to R_{gen} and equating it to zero. Note again that this does not mean lowest noise. The derivation is as follows. First conversion of equation (9.10) into a natural logarithm gives

$$NF = \frac{10}{\ln 10} \ln \left(\frac{\overline{e_n^2} + \overline{e_R^2} + \overline{i_n^2} R_{gen}^2}{\overline{e_R^2}} \right). \tag{9.22}$$

Put $R_{gen} = R$, note that equation (9.22) is like

$$NF = 4.34 \ln \left(\frac{T}{N} \right) \tag{9.23}$$

and with $\overline{e_R^2} = 4kTBR$ differentiating equation (9.23) with respect to R gives

$$\partial NF = 4.34 \frac{1}{T/N} d(T/N) = 4.34 \frac{1}{T/N} \frac{N \, dT - T \, dN}{N^2}.$$

This yields

$$\frac{dNF}{dR} = 4.34 \frac{1}{TN} \left[\overline{e_R^2} \left(4kTB + 2R\overline{i_n^2} \right) - \left(\overline{e_R^2} + \overline{e_n^2} + \overline{i_n^2} R^2 \right) 4kTB \right].$$

For a maximum or minimum value this equals zero, yielding

$$4kTBR(4kTB + 2R\overline{i_n^2}) = (\overline{e_R^2} + \overline{e_n^2} + \overline{i_n^2} R^2) 4kTB$$

$$2R^2\overline{i_n^2} = \overline{e_n^2} + \overline{i_n^2} R^2 \tag{9.24}$$

$$R^2\overline{i_n^2} = \overline{e_n^2}$$

and thus

$$R_{opt}^2 = \frac{\overline{e_n^2}}{\overline{i_n^2}} \quad \text{or} \quad R_{opt} = \frac{\overline{e_n}}{\overline{i_n}}. \tag{9.25}$$

If this value for optimum source resistance is substituted in equation (9.11) an expression for optimum NF is found (see problem 9.7). In general, the noise figure is a function of source resistance and operating frequency and plotting this for

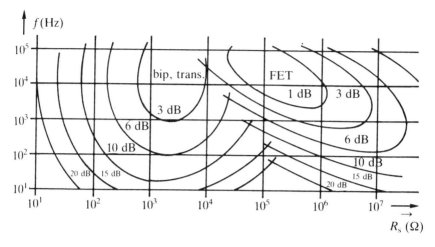

Figure 9.6 Constant-noise-figure contours for a bipolar transistor and junction field effect transistor (courtesy National Semiconductor Ltd).

* * *

constant noise figure the so-called noise figure contours are obtained. This is illustrated in figure 9.6 where for comparison the bipolar and the junction field effect transistor are depicted simultaneously in the same graph. It is interesting to note that a junction field effect transistor shows greater source resistance for the same NF. It is striking and not well understood that the MOSFET shows a much worse noise behaviour.

*   *   *

Example 9.7
Determine R_{gen} for the lowest NF at a frequency of 600 Hz when the operational amplifier of figure 9.2 is applicable.

1   Read from the chart at 600 Hz $\overline{e_n} = 10\ nV/\sqrt{Hz}$.
2   Read from the chart at 600 Hz $\overline{i_n} = 0.66\ pA/\sqrt{Hz}$.
3   Substitute these values in equation (9.25); the result is $R_{opt} = 15.15\ k\Omega$.

Now compare this result with figure 9.3; then you see the following values for $\overline{e_N}$:

1   $\overline{e_N} = 20\ nV/\sqrt{Hz}$ for $R_{gen} \approx 15\ k\Omega$
2   $\overline{e_N} = 10\ nV/\sqrt{Hz}$ for R_{gen} is in the range 0–100 Ω.

*   *   *

From this example a most important conclusion can be drawn: applying R_{opt} does not mean lowest noise *per se*, it only means lowest NF. There is only one condition where applying R_{opt} means lowest noise. Lowest noise, applying $R_{opt} = R_{gen}$, is only guaranteed when $e_R^2 = m R_{gen}$ and this can only be realized when a type of transformer coupling is applied. This will be investigated in more detail in the following section.

9.6.2 The optimum signal-to-noise ratio

Here we will investigate the conditions for R_{gen} on which a maximum signal-to-noise ratio can be realized. The signal-to-noise ratio is already defined in equation (9.19) and can be written as

$$S/N = 10 \log \frac{V_{sig}^2}{B(\overline{i_n^2} R_{gen}^2 + \overline{e_R^2} + \overline{e_n^2})}. \tag{9.26}$$

Again conversion to the natural logarithm and differentiating with respect to R gives the following expression for the differential equation:

$$d(S/N) = \frac{4.34}{TN}(N\,dT - T\,dN) \tag{9.27}$$

where $T = V_{sig}^2$, $N = B(\overline{i_n^2} R_n^2 + \overline{e_R^2} + \overline{e_n^2})$, $R = R_{gen}$, B is the considered bandwidth and $\overline{e_R^2}$ is the noise power of source for 1 Hz bandwidth. The result for the differential quotient is

$$\frac{d(S/N)}{dR} = \frac{4.34}{V_{sig}N}(\overline{i_n^2} R^2 + \overline{e_R^2} + \overline{e_n^2})2\,d(V_{sig})/dR - V_{sig}(4kT + 2\overline{i_n^2} R). \tag{9.28}$$

From this equation, an extreme value for R can be derived when a certain relationship between V_{sig} and R is given. Notice that only the most right-hand term in equation (9.28) is of interest. This means that as long as this term is larger than zero the S/N increases with R and

$$\frac{d(S/N)}{dR} > 0. \tag{9.29}$$

Three conditions for V_{sig} will be considered:

1   $V_{sig}^2 = m^2 R$
2   $V_{sig} = mR$
3   $V_{sig} < mR^{1/2}$.

1 $V_{sig}^2 = m^2 R$

This condition can be realized when a transformer coupling is involved; then

$$dV_{sig}/dR = \frac{m}{2R^{1/2}}. \tag{9.30}$$

Considering only the right-hand term in equation (9.28) and substituting the given relationship for V_{sig} yields

$$\frac{2m}{2R^{1/2}}(\overline{i_n^2} R_n^2 + \overline{e_R^2} + \overline{e_n^2}) > mR^{1/2}(4kT + 2\overline{i_n^2} R). \tag{9.31}$$

Rearranging this equation gives

$$\overline{i_n^2} R_n^2 + \overline{e_R^2}\overline{e_n^2} > 4kTR + \overline{i_n^2} R. \tag{9.32}$$

Note again that $4kTR = \overline{e_R^2}$; thus

$$\overline{e_n^2} > \overline{i_n^2}R^2 \qquad \text{or} \qquad R < \frac{\overline{e_n}}{\overline{i_n}} = R_{opt}. \tag{9.33}$$

Thus the conclusion appears to be that as long as $R < R_{opt}$ the S/N ratio will increase with increasing R_{gen} on the condition that $V_{sig} = m^2R$.

2 $V_{sig} = mR$

In this case the signal level is proportional to R_{gen} and

$$dV_{sig}/dR = m.$$

Again following the same philosophy and substituting now $V_{sig} = mR$ we find for the most right-hand term of equation (9.28)

$$2m(\overline{i_n^2}R_n^2 + \overline{e_R^2} + \overline{e_n^2}) < mR(4kT + 2\overline{i_n^2}R). \tag{9.34}$$

Rearranging this equation gives

$$\overline{e_R^2} + \overline{e_n^2} > 0. \tag{9.35}$$

Thus on the condition that $V_{sig} = mR$ the S/N increases with R_{gen} for any amplifier because equation (9.35) holds always and it makes sense to apply the highest possible practical value for R_{gen}. However, this condition cannot always be fulfilled.

3 $V_{sig} < mR^{1/2}$

Then an optimum may be determined also. If for instance $V_{sig} = mR^{0.4}$ then $dV_{sig}/dR = 0.4mR^{-0.6}$. Substituting in equation (9.28) the given value for V_{sig} gives

$$\frac{0.8m}{R^{0.6}}(\overline{i_n^2}R_n^2 + \overline{e_R^2} + \overline{e_n^2}) > mR^{0.4}(4kT + 2\overline{i_n^2}R)$$

$$0.8(\overline{i_n^2}R_n^2 + \overline{e_R^2} + \overline{e_n^2}) > (4kTR + 2\overline{i_n^2}R^2) \tag{9.36}$$

$$0.8\overline{e_n^2} > 0.2\overline{e_R^2} + 1.2\overline{i_n^2}R^2. \tag{9.37}$$

The conclusion appears to be that the S/N will increase with increasing value of R_{gen} as long as equation (9.37) holds, and thus until

$$\overline{e_n^2} = 0.25\overline{e_R^2} + 1.5\overline{i_n^2}R^2. \tag{9.38}$$

Summarizing this all, it does not make sense to change R_{gen} of an existing signal source in an attempt to make $R_{gen} = R_{opt}$. In particular, do not add series resistance to a source for this purpose. This only makes sense when a transformer coupling is applied manipulating the turns ratio or to redesign R_{gen} to operate with pre-amplifiers where R_{opt} is known.

It does make sense to increase the designed resistance of signal sources to match or to exceed the value of R_{opt} as long as the signal voltage increases with R_{gen}. We will illustrate this with an example.

*   *   *

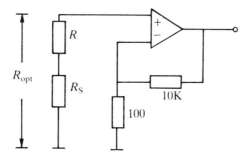

Figure 9.7 Configuration where the optimum source resistance is implemented.

Example 9.8

An amplifier is required to boost a signal coming from a transducer with a resistive impedance $R_{\text{gen}} = 10 \text{ k}\Omega$. The amplifier requirements and specifications are gain $A_V = 100$, frequency range 10 Hz–10 kHz, noise voltage $\overline{e_n} = 12 \text{ nV}/\sqrt{\text{Hz}}$ at 1 kHz and noise current $\overline{i_n} = 0.01 \text{ pA}/\sqrt{\text{Hz}}$ at 1 kHz. The noise characteristics of the amplifier may be considered linear. A non-inverting amplifier configuration will be used. Determine R_{opt} and calculate the NF and the total noise at the input of the amplifier when the transducer is connected to the input without affecting the gain and the noise of the feedback resistors may be neglected if (a) the optimum source resistance is installed or (b) the transducer is directly connected to the amplifier. The optimum source resistance at 1 kHz can be calculated by applying equation (9.21) and substituting the given values:

$$R_{\text{opt}} = \frac{\overline{e_n}}{\overline{i_n}} = \frac{12 \times 10^{-9}}{0.01 \times 10^{-12}} = 1200 \times 10^3 = 1.2 \text{ M}\Omega.$$

In case (a) the circuit involved is depicted in figure 9.7. The total noise $\overline{e_N}$ at the input can be found applying equation (9.5), replacing R by R_{opt}, and taking the square root of equation (9.5). This is done in table 9.2.

1. Read $\overline{e_R} = 130 \text{ nV}/\sqrt{\text{Hz}}$ from figure 9.4, square and put the result into table 9.2.
2. Read $\overline{e_n} = 12 \text{ nV}/\sqrt{\text{Hz}}$, square it and enter into the table,
3. Read $\overline{i_n} = 0.01 \text{ pA}/\sqrt{\text{Hz}}$, multiply with $R_{\text{opt}} = 1200 \text{ k}\Omega$, square the result and enter into the table.
4. Sum the squared results from steps 1, 2 and 3 and multiply the sum by $f = B = 9990 \text{ Hz}$.

Table 9.2 Noise voltage calculations for R_{opt}.

Frequency bandwidth B (Hz)	$\overline{e_R^2}$ (nV^2 Hz^{-1})	$\overline{e_n^2}$ (nV^2 Hz^{-1})	$\overline{i_n^2} R_{\text{gen}}^2$ (nV^2 Hz^{-1})	Sum $\times B$ (nV^2)
9990	16 900			168 831 000
9990		$(12)^2 = 144$	$(0.01 \times 1200)^2 = 144$	2 877 120
				+
		$\overline{e_N^2} =$		171 708 120

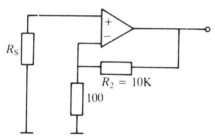

Figure 9.8   Configuration when the transducer is directly connected to the amplifier.

5   Take the square root of the sum; this gives the total noise voltage $\overline{e_N} = 14.2\ \mu V$
    for the specified bandwidth.
6   The NF can be calculated with equation (9.9):

$$NF = 10 \log \frac{\overline{e_N^2}}{\overline{e_R^2}} = 10 \log \frac{171\ 708\ 120}{168\ 831\ 000} = 10 \log 1.017 = 0.06\ dB$$

which is very low indeed.

   In case (b) the current circuit when the transducer is directly connected to the
input of the non-inverting amplifier is depicted in figure 9.8.

1   Follow the same procedure as in case (a), but now substitute the transducer's
    impedance R only. From figure 9.4 $\overline{e_n} = 13.5\ nV/\sqrt{Hz}$. The values are depicted
    in table 9.3.
2   Taking the square root of the sum find $\overline{e_N} = 1805\ nV = 1.8\ \mu V$, which is consider-
    ably smaller than in the case of $R = R_{opt}$. Calculating the NF gives

$$NF = 10 \log \frac{\overline{e_N^2}}{\overline{e_R^2}} = 10 \log \frac{3\ 259\ 337}{1\ 820\ 678} = 10 \log 1.79 = 2.5\ dB.$$

Note that it is not necessary to have multiplied with B to find the NF. For comparison
the results are repeated below:

- case (a), $\overline{e_N} = 14.2\ \mu V$, NF = 0.06 dB
- case (b), $\overline{e_N} = 1.8\ \mu V$, NF = 2.5 dB

*   *   *

Table 9.3   Noise performance calculations for the source directly connected to the amplifier.

Frequency bandwidth B (Hz)	$\overline{e_R^2}$ (nV² Hz⁻¹)	$\overline{e_n^2}$ (nV² Hz⁻¹)	$\overline{i_n^2} R_{gen}^2$ (nV² Hz⁻¹)	Sum × B (nV²)
9990	$(13.5)^2 = 182.25$			1 820 678
9990		$(12)^2 = 144$	$(0.01 \times 10)^2 = 144$	1 438 659
				+
			$\overline{e_N^2} =$	3 259 337

It will be clear that noise figure is only a measurement of the amplifier noise with respect to the source noise. From this we can draw a very important conclusion. Never add a series resistance with the source to improve the NF. The noise figure will improve but the noise performance is degraded considerably. So live with the existing R_{gen} or apply a transformer coupling to make sense. Regarding this aspect sometimes a kind of improvement factor, $I_{impr.}$, is introduced based on the ratio of the matched S/N and the unmatched S/N. In the following example it will be shown that for an operational amplifier with ideal feedback the equivalent input noise is independent of gain.

*   *   *

Example 9.9
Determine the equivalent input noise per unit bandwidth for the feedback configuration shown in figure 9.9. The following specifications are known. The amplifier

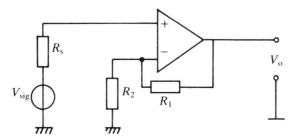

Figure 9.9 Inverting amplifier configuration with ideal feedback.

operates at 1 kHz from a source with a source resistance of 1 kΩ. The values of the feedback resistances are $R_1 = 100$ kΩ and $R_2 = 1$ kΩ. Figure 9.2 represents the noise specifications of the amplifier. Here the thermal noise generated by the feedback resistors has to be taken into account, which can be best performed by moving outside the feedback loop the noise influence of the parallel-connected resistors R_1 and R_2. In addition to this, the noise generated by $i_n(R_1 \| R_2)$ must be added even though the minus input is a virtual ground.

The easiest way to include these effects is to consider $R_1 \| R_2$ in series with R_s. We then obtain the equivalent configuration shown in figure 9.10 and we construct table 9.4 as follows. In section 9.7 we will prove and demonstrate how the position of electrical sources in circuits can be changed.

1 Read $\overline{e_R}$ at 1 kΩ from figure 9.4 (value is 4.55), square it and enter in table 9.4.
2 Read $\overline{e_{R_2 \| R}}$ at 1 kΩ from figure 9.4 (value is 4.55), square it and enter the result in the table.
3 Read $\overline{e_n}$ from figure 9.2 (value is 9.5), square and enter the result in the table.
4 Read $\overline{i_n}$ from figure 9.2 (value is 0.65) multiply by $(R_s + R_1 \| R_2)$, square the result and enter it in the table.
5 Sum all entries; take the square root to obtain the total RMS noise at the input as $\overline{e_N} = 11.5$ nV/$\sqrt{\text{Hz}}$ at 1 kHz in a 1 Hz bandwidth.

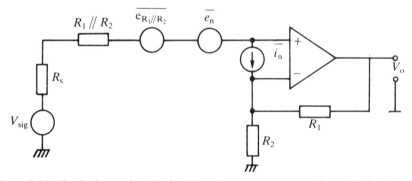

Figure 9.10   Equivalent noise circuit with noise generators moved outside the feedback loop in figure 9.9.

Table 9.4.   Noise calculations for example 9.9.

Source	$\mathrm{nV^2\,Hz^{-1}}$	$\mathrm{nV^2\,Hz^{-1}}$
$\overline{e_R^2}$	$(4.55)^2$	20.7
$\overline{e_{R_1\parallel R_2}^2}$	$(4.55)^2$	20.7
$\overline{e_n^2}$	$(9.5)^2$	90.2
$\overline{i_n^2}(R_s^2 + R_1\parallel R_2)^2$	$(0.65 \times 2.0k)^2$	1.69
	Total	$\overline{e_N^2} = 133.29$

*   *   *

As you see, the gain is not involved in the total noise calculation at the input of the amplifier.

9.7   OTHER NOISE CALCULATING TECHNIQUES

9.7.1   Shifting noise sources

In this section we will present some techniques for manipulating (noise) sources in circuits. This is convenient, because usually we refer noise sources to the input of a circuit. The theorems will be given without proof.

Theorem I
Shifting sources
In a given network the position of voltage sources and current sources may be shifted without affecting any magnitude of currents and/or voltage in that circuit.

This results in the following recipes.

1   *Voltage sources*
It is allowed to shift any arbitrary voltage source through its adjacent node to each of its directly connecting branches, leaving the original position as a short circuit.

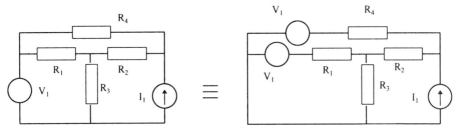

Figure 9.11 Two equivalent circuits after shifting the voltage source.

(Note that this theorem can be proved by applying mesh equations.) In figure 9.11 the voltage source V_1 is shifted through its node towards the adjacent connecting branches short-circuiting the original position. Both circuits are equivalent.

2 Current sources

A current source in a given mesh may be repositioned over its mesh branches leaving the original source position open. This can be proved by applying the current node equation for any node in the circuit. In figure 9.12 the current source is shifted as is indicated.

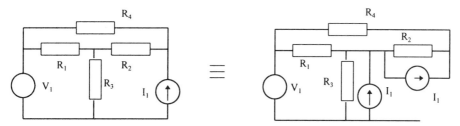

Figure 9.12 Two equivalent circuits after shifting the current source.

Note that this reshuffling of sources may be executed for any type of source in any type of network.

3 Transformers

A voltage source connected to the input of a transformer can be shifted to the secondary winding by taking into account the ratio of the windings. This is illustrated in figure 9.13. In the same fashion we can shift a current source in a transformer as is indicated in figure 9.14.

Figure 9.13 Shifting a voltage source from the primary towards the secondary.

Figure 9.14 Shifting a current source from the primary towards the secondary.

9.7.2 Superposition

Theorem II

In a given linear network each voltage or current equals the sum of the contributions of the present independent sources.

This is illustrated in figure 9.15.

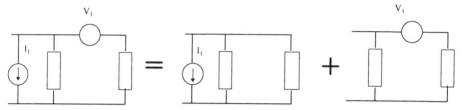

Figure 9.15 The superposition theorem.

9.7.3 Thevenin and Norton

Theorem III

Each one-port network can be replaced by its Thevenin or Norton equivalent.

This is illustrated in figure 9.16.

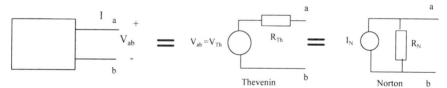

Figure 9.16 Equivalent circuits for a one-port network.

For a two-port network as is illustrated in figure 9.17, this works in a slightly different way and a k matrix should be used. Any two-port can be described with its k matrix as follows:

$$V_i = k_{11}V_o + k_{12}I_o \qquad I_i = k_{21}V_o + k_{22}I_o \qquad (9.39)$$

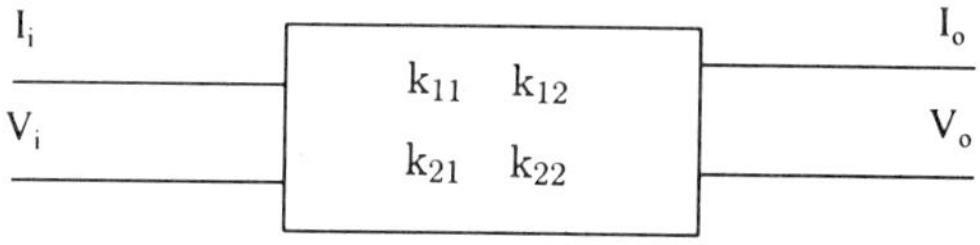

$$
\begin{array}{cc}
k_{11} & k_{12} \\
k_{21} & k_{22}
\end{array}
$$

Figure 9.17 A two-port network described with its k matrix.

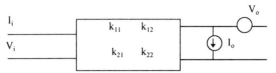

Figure 9.18 Replacement of a two-port with its equivalent noise sources at the output.

where

$$k_{11} = V_i/V_o|_{I_o=0} \quad \text{(open output; no dimension)}$$
$$k_{12} = V_i/I_o|_{V_o=0} \quad \text{(short circuit; transimpedance)}$$
$$k_{21} = I_i/V_o|_{I_o=0} \quad \text{(open output; transconductance)}$$
$$k_{22} = I_i/I_o|_{V_o=0} \quad \text{(short circuit; no dimension)}.$$

Now the noisy two-port can be replaced by a noise-free two-port contributing for all noise sources at the output with one voltage noise source and one current noise source. This is illustrated in figure 9.18. Hence the two noise sources at the output represent all present noise sources. The next step is to recalculate these noise sources with reference to the input. This can be executed with help of the k matrix as is indicated in equation (9.39). We find two voltage noise sources at the input and two current noise sources at the input. This yields the following noise sources as indicated in figure 9.19:

$$V_3 = k_{11}V_o \qquad V_2 = k_{12}I_o$$

and

$$I_2 = k_{22}I_o \qquad I_3 = k_{21}V_o.$$

Finally, we want to end up with one voltage noise source and one current noise source at the input. Therefore we need also the following theorem.

9.7.4 Noise sources in series and in parallel

Theorem IV

1 *Two voltage noise sources in series*
When two voltage noise sources are placed in series then they can be replaced by a corresponding circuitry consisting of one voltage noise source. The following

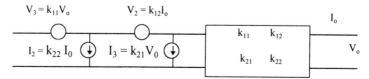

Figure 9.19 The equivalent noise sources referenced to the input.

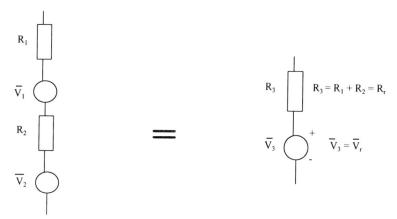

Figure 9.20 Two voltage noise sources in series.

equivalence holds:

$$\overline{V_1^2}=4kTBR_1 \tag{9.40}$$

$$\overline{V_2^2}=4kTBR_2. \tag{9.41}$$

Summing both noise sources gives

$$\overline{V_r^2}=\overline{V_1^2}+\overline{V_2^2}=4kT(R_1+R_2)B \tag{9.42}$$

or

$$\overline{V_r}=(\overline{V_1^2}+\overline{V_2^2})^{1/2}. \tag{9.43}$$

This is depicted in figure 9.20.

2 *Two current noise sources in parallel*
In the same way we have to find the equivalent circuit for two current noise sources in parallel. For a noise power at a given temperature and bandwidth we have again $P=4kTB$ and for the equivalent current noise source we find

$$\overline{i_1^2}=4kTG_1B \qquad \text{and} \qquad \overline{i_2^2}=4kTG_2B. \tag{9.44}$$

Summing both contributions we then find for the total equivalent current noise source

$$\overline{i_r^2}=\overline{i_1^2}+\overline{i_2^2}=4kT(G_1+G_2)B. \tag{9.45}$$

$$\overline{i_r^2} = \overline{i_1^2} + \overline{i_2^2} = 4kT(G_1 + G_2)B$$

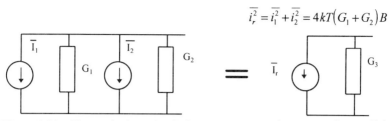

Figure 9.21 The equivalent circuit for two current noise sources in parallel.

This is illustrated in figure 9.21.

Finally, in the case of ideal feedback the input noise sources can be replaced outside the feedback loop without any change, not affecting the noise characteristics of the circuit at all. The S/N ratio remains constant, because when the noise level is reduced the signal level is also reduced.

We will conclude the chapter with three more examples to demonstrate some other techniques of noise calculations.

* * *

9.7.5 Worked examples

Example 9.10

Determine the effective (RMS) noise voltage and noise current for a resistance $R = 1\ \Omega$ at an ambient temperature of 1 K in a bandwidth of 0–10 kHz.

The graph of figure 9.4 will fail because it is constructed for an ambient temperature of 300 K. This means we have to apply the basic formula (9.12). Entering the given values yields

$$P = 4kTB = \frac{\overline{e_R^2}}{R}$$

$$\overline{e_R^2} = 4kTBR = 4 \times 1.38 \times 10^{-23} \times 1 \times 10^4 \times 1 = 5.52 \times 10^{-19}\ \text{V}^2$$

and the equivalent RMS noise voltage is

$$\overline{e_N} = (5.52 \times 10^{-19})^{1/2} = 0.74\ \text{nV}.$$

The RMS noise current is calculated as follows:

$$\overline{i_R^2} = \frac{4kTB}{R} = \frac{4 \times 1.38 \times 10^{-23} \times 10^4}{1} = 5.52 \times 10^{-19}\ \text{A}^2.$$

The equivalent RMS noise current is

$$\overline{i_R} = 0.74\ \text{nA}.$$

* * *

Example 9.11
Determine the equivalent noise voltage source e_{AB} between the points A and B of
the circuit in figure 9.22. In this configuration only one noise voltage source can be

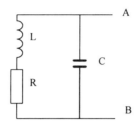

Figure 9.22   Circuit for example 9.11.

distinguished, i.e. R, because, as is known, an ideal coil or capacitor does not generate
any noise contribution. Keep in mind that, here again, power densities must be
considered. Thus we first calculate

$$\overline{e_{AB}^2} = \left| \frac{Z_1}{Z_1 + Z_2} \right|^2 \overline{e_R^2} \tag{9.46}$$

where $Z_1 = 1/j\omega C$ and $Z_2 = R + j\omega L$ and again

$$\overline{e_R^2} = 4kTR.$$

Entering these quantities yields

$$\overline{e_{AB}^2} = \left| \frac{1/j\omega C}{1/j\omega C + R + j\omega L} \right|^2 \overline{e_R^2}. \tag{9.47}$$

multiplying by $j\omega C$ gives

$$\overline{e_{AB}^2} = \left| \frac{1}{1 - \omega^2 LC + j\omega RC} \right|^2 \overline{e_R^2} \tag{9.48}$$

or

$$\overline{e_{AB}^2} = \frac{1}{(1 - \omega^2 LC)^2 + (\omega RC)^2} \overline{e_R^2} \tag{9.49}$$

and finally taking the square root of equation (9.49) gives the required result. Note
that the left-hand factor of e_R^2 is dimensionless.

*   *   *

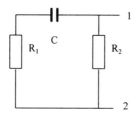

Figure 9.23 Circuit for example 9.12.

Example 9.12

Determine the equivalent noise voltage source $\overline{e_{\mathrm{N}12}}$ for the circuit drawn in figure 9.23. The easiest way to solve this problem is to apply the superposition theorem, taking into account both noise sources separately. This is allowed when it is assumed that there is no correlation of sources. The two noise sources are

$$\overline{e_{R_1}^2} = 4kTR_1 \qquad \overline{e_{R_2}^2} = 4kTR_2. \tag{9.50}$$

To see more clearly what is happening in the calculation the equivalent noise circuit can be split into two circuits just as the superposition theorem dictates. This is drawn in figure 9.24. In figure 9.24(a) the equivalent noise voltage source $\overline{e_{R_1}^2}$ generates a noise power $\overline{e_{12a}^2}$ which is given by

$$\overline{e_{12a}^2} = \frac{R_2^2}{|R_1 + R_2 + 1/j\omega C|^2}\, \overline{e_{R_1}^2}. \tag{9.51}$$

In the same fashion for the circuit of figure 9.24(b) the equivalent noise power at points 1, 2 is

$$\overline{e_{12b}^2} = \frac{|1/j\omega C + R_1|^2}{|R_1 + R_2 + 1/j\omega C|^2}\, \overline{e_{R_2}^2}. \tag{9.52}$$

Summing both expressions gives the required result:

$$\overline{e_{\mathrm{N}}^2} = \overline{e_{12a}^2} + \overline{e_{12b}^2} \tag{9.53}$$

$$= \overline{e_{R_1}^2}\, \frac{(\omega R_2 C)^2}{1 + [(R_1 + R_2)\omega C]^2} + \overline{e_{R_2}^2}\, \frac{1 + (\omega R_1 C)^2}{1 + [(R_1 + R_2)\omega C]^2}. \tag{9.54}$$

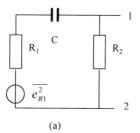

(a)

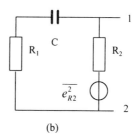

(b)

Figure 9.24 Equivalent noise circuit obtained by applying the superposition theorem to figure 9.23.

Taking the square root of this expression gives the equivalent RMS noise voltage $\overline{e_N}$ seen between the points 1 and 2.

Note that, in figure 9.24(a), R_2 determines finally the fraction of noise power which will appear between the points 1, 2 caused by $e_{R_1}^2$ and, in figure 9.24(b), the term $1/j\omega C + R_1$ determines the fraction of noise power which will finally appear between the points 1 and 2, generated by $e_{R_1}^2$.

*   *   *

This concludes our chapter on noise calculations. In the next chapter we will treat noise in a more physical fashion concerning active and passive components. However, as you have seen it is not necessary to understand all the ins and outs of the physical details to perform noise calculations. Normally it suffices to use the noise specifications delivered by the manufacturer and to use these for the required noise calculations.

PROBLEMS

9.1 Calculate the noise voltage source and the noise current source for the following resistances in the frequency range as specified.

$R = 1\,\text{M}\Omega$, 100–110 kHz, $T = 300\,\text{K}$
$R = 100\,\text{k}\Omega$, 10–11 MHz, $T = 300\,\text{K}$.

9.2 Determine the signal-to-noise ratio for an amplifier with $\overline{e_N} = 17.7\,\text{nV}/\sqrt{\text{Hz}}$ in the bandwidth from 1 kHz to 10 kHz when $e_{\text{sig}} = 100\,\text{mV}$.

9.3 Derive an expression for the equivalent noise voltage source $\overline{e_{12}}$ for the circuit depicted in figure 9.25 at points 1, 2.

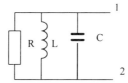

Figure 9.25   Circuit for problem 9.3.

9.4 Derive an expression for the equivalent noise voltage source at terminals 1, 2 for the circuit depicted in figure 9.26.

Figure 9.26   Circuit for problem 9.4.

9.5 Derive an expression for the equivalent noise voltage source at terminals 1, 2 for the circuit depicted in figure 9.27.

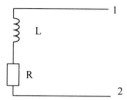

Figure 9.27 Circuit for problem 9.5.

9.6 Calculate the shot noise current and the equivalent thermal noise voltage in a diode with $I_D = 1$ mA, in a bandwidth of 2 MHz and a temperature of 300 K. (See also chapter 10.)

9.7 Derive an expression for the noise figure NF for optimum source resistance.

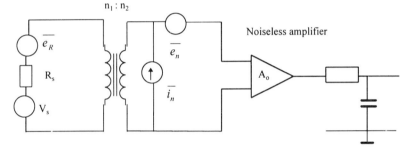

Figure 9.28 Circuit for problem 9.8.

9.8 Figure 9.28 represents a low-noise pre-amplifier with $\overline{e_n} = 10^{-8}$ V/$\sqrt{\text{Hz}}$ and $\overline{i_n} = 0.1$ pA /$\sqrt{\text{Hz}}$. Connected to it is a voltage source with a source resistance $R_S = 20\ \Omega$ via a matching transformer. The transformer may be presumed ideal with a transformer ratio of n_2/n_1. Find (a) the optimum source resistance, (b) the unmatched noise figure without transformer and (c) the matched noise figure with matching transformer.

BIBLIOGRAPHY

[1] Maxwell J 1976 The low noise junction-FET, the noise problem solver *National Semiconductor Handbook*
[2] Sherwin J 1974 Noise specifications, confusing? *National Semiconductor Appl. Note 104*
[3] van der Ziel A 1971 *Noise Source Characterization Measurement* (Englewood Cliffs, NJ: Prentice-Hall)
[4] Gray P R and Meyer R G 1984 *Analog Integrated Circuits* 2nd edn (New York: Wiley) ch 11

Further, all data books from the different manufacturers should be consulted for the latest specifications.

Ten

PHYSICS OF NOISE

10.1 INTRODUCTION

In this chapter, we will continue our discussion of noise by looking at several frequently occurring noise mechanisms. An exhaustive treatment can be found in more theoretical textbooks devoted to noise only. Here we discuss only the most important mechanisms required for the description of noise that is met in commonly used electronic components.

Noise can be defined as a random movement of energy in solid-state materials, conductors and components due to the finite temperature of the molecules of the material. This thermal energy is transported at random by electrons as one of the possible energy carriers, but lattice vibrations can also transport energy at random. For this type of electron transport Ohm's law is no longer valid; the transport of electrons is not subjected to an outside electric field but is caused, for instance, by a thermal activation of the electrons completely at random. This means that the noise mechanisms can be described from a statistical point of view and hence it is a stochastic process. Other mechanisms show a certain relationship with frequency, but very often the absolute temperature is one of the main variables involved. In this chapter three mechanisms will be discussed.

10.2 THERMAL NOISE OR JOHNSON NOISE IN CONDUCTORS

One of the most frequent causes of noise in a conductor is the random movement of electrons due to their finite temperature, resulting in a certain noise power. This noise power, which is known as thermal noise or Johnson noise, can be described with the previously mentioned relation (9.12), and was theoretically derived by Nyquist from a thermodynamic point of view:

$$P = 4kTB$$

where again P (W) is the noise power, k (1.38×10^{-23} J K^{-1}) is Boltzmann's constant, T (K) is the absolute temperature and B (Hz) is the bandwidth. In this equation the frequency bandwidth, the absolute temperature and a physical constant are involved. If this expression is divided by the bandwidth B, a power density per hertz is obtained, which at $T = 300$ K equals

$$P/B = 4kT = 4.14 \times 10^{-21} \text{ W Hz}^{-1}.$$

As you can see from this equation, it expresses a certain amount of available noise power to which a voltage or current source can be assigned to be connected in series or in parallel with a resistor by the known relationships

$$P = \frac{\overline{e_R^2}}{R} \tag{10.1}$$

$$P = \overline{i_R^2} R. \tag{10.2}$$

If this noise power is expressed in an equivalent noise voltage or noise current the following equations hold:

$$\overline{e_R^2} = PR = 4kTRB \quad (\text{V}^2) \tag{10.3}$$

$$\overline{i_R^2} = \frac{P}{R} = \frac{4kTB}{R} \quad (\text{A}^2). \tag{10.4}$$

This is depicted in figure 10.1. Note again that mean squares are involved as a result of statistical considerations. In standardized circumstances an ambient temperature

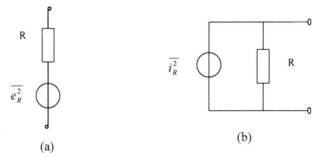

(a)   (b)

Figure 10.1   Equivalent thermal noise voltage and current source circuit of a resistor.

of 300 K is assumed and on this basis the graphs of figure 9.4 are expressed in terms of the root mean square (RMS) of the above expressions. The values are given for a 1 Hz bandwidth which results in units of nanovolts per square root hertz or picoamperes per square root hertz. Because a linear relationship exists between noise power and bandwidth, this type of noise is often called *white noise*.

10.3   SHOT NOISE

In many electronic components we have to deal with flows of charge carriers which cross potential barriers independently of each other. The character of this flow is rather discrete and it is said that these flows show a quantum-like character and show a discontinuous current flow behaviour. This characteristic explains the name 'shot noise'. The current as a function of time is a stochastic function. Good examples are the crossing of holes and electrons over a p–n junction or the emission of electrons in vacuum tubes. For a p–n junction this type of flow of charge carriers is described by the following expression:

$$\overline{i_n^2} = 2qI_D B \tag{10.5}$$

where $\overline{i_n^2}$ (A^2) is the mean square noise current, q the specific charge of an electron (1.6×10^{-19} C), I_D the direct current flowing through the junction and B (Hz) the considered bandwidth. In the same fashion as we have seen for thermal noise, *shot noise* can be represented as a current *density* as follows

$$\overline{i_n} = \left(\frac{\overline{i_n^2}}{B}\right)^{1/2} . \tag{10.6}$$

Here again a linear relationship exists between the noise power and the bandwidth, so this type can be called white noise too. The direct current I_D can be found with the well-known diode equation:

$$I_D = I_0 [\exp(qV_D/kT) - 1] \tag{10.7}$$

where I_D is the direct current in the p–n junction, I_0 is the saturation current, q is the specific charge of an electron, V_D is the voltage over the p–n junction, k is Boltzmann's constant and T is the absolute temperature. Differentiation of this expression yields the differential diode resistance; the result is often represented by r_d to express its differential character. The following equation expresses this relationship between the differential resistance, r_d, and the current I_D:

$$r_d = \frac{kT}{qI_D} \approx \frac{26}{I_D} \tag{10.8}$$

where I_D is in milliamperes. If T equals 300 K then with I_D in milliamperes the rightmost expression is valid and easy to remember. Rearranging equation (10.8) and substituting I_D in equation (10.5) yields for the noise current

$$\overline{i_n^2} = 2q\frac{kTB}{qr_d} = \frac{4kT}{2r_d} B. \tag{10.9}$$

Note that the factor 4 is introduced here in the numerator to obtain again the well known expression $4kTB$. To derive an expression for the equivalent noise voltage $\overline{e_n^2}$, equation (10.9) can first be multiplied by r_d^2; then we find the total noise power. Multiplying this expression by r_d^2 gives

$$\overline{e_n^2} = \frac{4kTB}{2} r_d . \tag{10.10}$$

The factor $\frac{1}{2}$ in (10.10) again is only for convenience to give the well known factor $4kTB$ in our expression.

10.4 FLICKER NOISE

Flicker noise or $1/f$ noise, sometimes called low-frequency noise, excess noise, contact noise, semiconductor noise or pink noise, can be characterized as caused by a flow

Table 10.1 Survey of values for α as a function of source mechanisms.

α	Source mechanism
1	For *pink* noise characterized by an equal power per octave
2	*Red* noise, which can be characterized by the fluctuation in the earth's rotational speed
2.7	For galactic radiation

of charge carriers in a discontinuous medium. Good examples are tubes, carbon composite resistors, diodes, transistors, thermistors, thin-film devices and light sources. Flicker noise is roughly inversely proportional to frequency and starts to be of importance at 1 kHz in both junction FETs and bipolar transistors. As can be seen from the following equation the equivalent noise voltage increases as frequency is decreased. Flicker noise dominates thermal noise at frequencies below 100 Hz and is related to a flow of direct current.

$$\overline{e_n^2} = K I_D^a (1/f)^{\alpha} R^2 B \qquad (10.11)$$

where K is a constant and device dependent, I_D is the direct current, a is a constant ranging from 0.5 to 2, α is a constant (see table 10.1), B is the bandwidth and R is the equivalent resistance through which the current flows. Some authors prefer to describe the noise flicker *current*, in which case equation (10.11) should be divided by the equivalent source resistance R^2 and $\overline{i_n^2}$ will result. The constant α can vary between 0.8 and 1.3 depending on the process to be considered. Some of these processes are mentioned in table 10.1. To give an impression of the level of this type of noise, figure 10.2 shows the relative levels of different types of noise as a function of frequency.

In semiconductors $1/f$ noise is mainly determined by surface properties, and the generation and recombination of charge carriers in surface states. So far we have discussed three different types of physical mechanism of noise sources. In the next

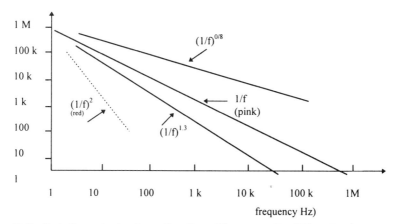

Figure 10.2 Relative noise level as a function of frequency for different noise processes.

section the noise behaviour of three semiconductor components, the bipolar transistor, the field effect transistor and the diode, are discussed, to which all integrated circuits can be related.

10.5 NOISE IN SEMICONDUCTOR COMPONENTS

In this section we will discuss the noise behaviour of the bipolar transistor, the unipolar or the field effect transistor (FET) and the diode.

10.5.1 Noise behaviour of the bipolar transistor
As you have already seen, any active element can be characterized with respect to noise by an equivalent noise voltage source and an equivalent noise current source referred to the input. If a bipolar transistor is considered, the following noise sources can be distinguished:

1 the thermal noise source of the internal base resistor r_{bb},

$$\overline{e_b^2} = 4kTr_{bb}\,B \tag{10.12}$$

2 a shot noise source in the collector current,

$$\overline{i_c^2} = 2qI_C\,B \tag{10.13}$$

3 a shot noise source in the base-emitter junction,

$$\overline{i_b^2} = 2qI_B\,B \tag{10.14}$$

4 a flicker noise component $\overline{i_{bf}}$ in the base emitter region, a term which becomes important when frequencies are below ≈ 100 Hz and will be neglected in our calculations of the equivalent noise sources.

These noise sources are depicted in figure 10.3, and to find the equivalent noise sources this circuit should be equated to an equivalent noise circuit. If the correlation between $\overline{i_c}$ and $\overline{i_b}$ and also the Early effect are neglected and we further suppose that r_{bb} is much smaller than βr_o the circuit for the bipolar transistor can be drawn as is depicted in figure 10.3. In this respect, no correlation means that the noise sources are independent. As said already, this circuit should be equated to an equivalent

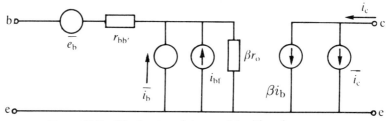

Figure 10.3 Bipolar transistor model with noise sources.

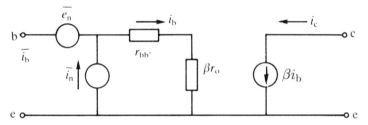

Figure 10.4 The equivalent noise circuit for the bipolar transistor.

circuit which is depicted in figure 10.4 in which $\overline{e_n}$ and $\overline{i_n}$ are drawn. If both circuits are made equivalent with their corresponding input and output terminals, the following equations will hold under the circumstances as stated. First, short-circuiting the input terminals of both circuits and then equating gives

$$\overline{e_b^2} + r_{bb'}^2 \overline{i_b^2} + r_o^2 \overline{i_c^2} = \overline{e_n^2}. \tag{10.15}$$

Note that $\beta i_b = i_c$ and, further, it is assumed that $r_{bb'} \ll \beta r_o$. If in the equivalent circuits, both inputs are opened, then, equating both output currents, this yields the expression

$$\beta^2 \overline{i_b^2} + \overline{i_c^2} = \beta^2 \overline{i_n^2}. \tag{10.16}$$

Because $I_C = \beta I_B$ from equations (10.13) and (10.14) it can be derived that

$$\overline{i_b^2} = \frac{\overline{i_c^2}}{\beta^2}. \tag{10.17}$$

Substituting equation (10.17) in (10.16) yields

$$\beta^2 \overline{i_c^2} + \overline{i_c^2} = \beta^2 \overline{i_n^2} \tag{10.18}$$

or if $\beta \gg 1$

$$\overline{i_c^2} = \frac{\beta^2}{1 + \beta^2} \overline{i_n^2} = \beta \overline{i_n^2} \tag{10.19}$$

and from this expression it can be seen that the equivalent noise current source is approximately given by

$$\overline{i_n^2} = \frac{\overline{i_c^2}}{\beta^2} = \overline{i_b^2}. \tag{10.20}$$

In accordance with (10.13) and (10.14) in this equation the current $\overline{i_c^2}$ and $\overline{i_b^2}$ can be replaced by the respective direct currents I_B or I_C; this yields an expression for the equivalent noise current source referred to the input:

$$\overline{i_n^2} = 2qI_B B = 2q \frac{I_C}{\beta} B. \tag{10.21}$$

Finally, in this expression the DC value I_C can be replaced by kT/qr_o in accordance with equation (10.8):

$$\overline{i_n^2} = \frac{4kT}{2\beta r_0} B. \tag{10.22}$$

If required $\overline{i_n}$ can be found by taking the square root of (10.22).

In the same way, we can find an expression for the equivalent noise voltage source $\overline{e_n}$. This can be performed by substituting equations (10.12), (10.13) and (10.14) in (10.15), which results in

$$\overline{e_n^2} = 4kTr_{bb'}B + r_{bb'}^2 2qI_B B + r_o^2 2qI_C B \tag{10.23}$$

and substituting the corresponding direct current values in accordance with equation (10.8) gives

$$\overline{e_n^2} = \left(4kTr_{bb'} + \frac{4kT}{2\beta r_o} + \frac{4kT}{2r_o} r_o^2\right) B \tag{10.24}$$

and if $2\beta r_o \gg r_{bb'}$ which in most cases may be assumed then

$$\overline{e_n^2} = 4kT(r_{bb'} + \tfrac{1}{2}r_o) B \tag{10.25}$$

and again if required the square root of this expression delivers the equivalent noise voltage source $\overline{e_n}$.

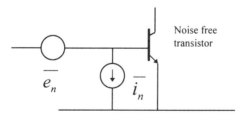

Figure 10.5 The noise-free bipolar transistor model with equivalent noise sources.

The result of this derivation is depicted in figure 10.5 where a noise-free transistor is drawn with its noise sources. For every transistor the manufacturer provides extensive data concerning noise behaviour.

10.5.2 Noise behaviour of the field effect transistor

The same procedure as in the case of the bipolar transistor can be followed to describe the noise behaviour of the FET. Two circuits are to be drawn, one of which depicts the physical noise sources and the second which represents the circuit with the equivalent noise sources. Firstly, the physical noise sources are to be mentioned;

mainly these sources are

1 the thermal noise of the resistive part of the input impedance, which can be expressed by

$$\overline{e_{R_i}^2} = 4kTR_i B \tag{10.26}$$

where R_i represents the input impedance

2 the thermal channel noise source which can be expressed by the equation

$$\overline{i_{ch}^2} = 4kT\tau g_m B \tag{10.27}$$

where τ is a constant related to the operating point and g_m is the transconductance defined as

$$g_m = \frac{\partial I_D}{\partial V_{gs}} = \frac{i_d}{v_{gm}} \tag{10.28}$$

for a given operating point, and $\tau = 1$ below the pinch-off voltage and $\tau = 2/3$ above the pinch-off voltage

3 the gate noise, which can be considered as a shot noise source because of the random behaviour of the current flow in the p–n junction, expressed as

$$\overline{i_{gn}^2} = 2qI_G B \tag{10.29}$$

where I_G is the DC gate current (in MOSFETs the gate current is of the order of nanoamperes and is usually negligible)

4 for low frequencies, the flicker noise current $\overline{i_{df}}$, generated in the drain section, comes into scope (this addition will also be neglected in the calculation of the equivalent noise sources).

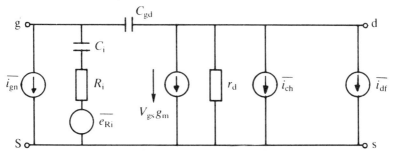

Figure 10.6 The field effect transistor with noise sources.

In figure 10.6 three noise sources are depicted: a noise voltage source in series with the input impedance, a noise current source in parallel with the input terminals and two noise current sources in parallel with the output terminals representing the channel noise and the $1/f$ noise. The equivalent circuit with the equivalent noise sources is depicted in figure 10.7. The magnitude of these noise sources can be determined by equating the equivalent input and output terminals of both circuits with respect to voltage and impedance. Suppose in figure 10.6 the drain source is not loaded and V_{gs} is put equal to $(\overline{e_n^2})^{1/2}$; then, following Kirchhoff's law and

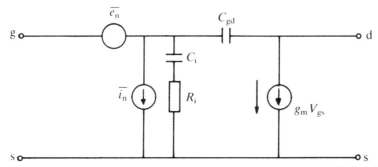

Figure 10.7 The equivalent noise circuit with equivalent noise sources $\overline{i_n}$ and $\overline{e_n}$ of a FET.

neglecting the influence of C_{gd} for node d, we obtain

$$g_m V_{gs} = (\overline{i_{ch}^2})^{1/2}. \tag{10.30}$$

Substituting the assumed value for V_{gs} in (10.30) we find

$$g_m^2 \overline{e_n^2} = \overline{i_{ch}^2}. \tag{10.31}$$

Finally, substituting equation (10.27) for $\overline{i_{ch}^2}$ gives

$$\overline{i_{ch}^2} = g_m^2 \overline{e_n^2} = 4kT\tau g_m B \tag{10.32}$$

and rearranging this to find $\overline{e_n^2}$ yields the equivalent noise voltage source:

$$\overline{e_n^2} = \frac{4kT\tau B}{g_m}. \tag{10.33}$$

Note that the same expression can be found when the input terminals of both circuits are short circuited with no load again.

To find the equivalent noise current source $\overline{i_n}$, V_{gs} for both circuits is calculated. For figure 10.6 we find for V_{gs}^2

$$V_{gs}^2 = \overline{i_{gn}^2} Z_i^2 + \overline{e_{R_i}^2}. \tag{10.34}$$

Note here again that noise power densities are added and that it is assumed that there is no correlation of sources. For figure 10.7 we find

$$V_{gs}^2 = \overline{e_n^2} + \overline{i_n^2} Z_i^2. \tag{10.35}$$

Equating equations (10.34) and (10.35) yields

$$\overline{i_{gn}^2} Z_i^2 + \overline{e_{R_i}^2} = \overline{e_n^2} + \overline{i_n^2} Z_i^2. \tag{10.36}$$

Substituting for $\overline{e_n}$ in equation (10.36) and equation (10.31) and rearranging to make $\overline{i_n}$ explicit we find

$$g_m^2 Z_i^2 \overline{i_n^2} = g_m^2 \overline{e_{R_i}^2} + \overline{i_{ch}^2} + g_m^2 Z_i^2 \overline{i_{gn}^2}. \tag{10.37}$$

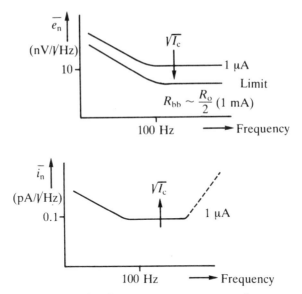

Figure 10.8 Noise voltage and noise current sources versus frequency of the bipolar transistor.

Note the change in sign for $\overline{i_{ch}^2}$ because of the noise power densities. Dividing equation (10.37) by $g_m^2 Z_i^2$ gives

$$\overline{i_n^2} = \frac{\overline{e_{Ri}^2}}{Z_i^2} + \frac{\overline{i_{ch}^2}}{Z_i^2 g_m^2} + \overline{i_{gn}^2}.$$ (10.38)

Finally substituting the corresponding expressions for the different noise sources (10.26), (10.27) and (10.29) delivers the equivalent noise current source at the input:

$$\overline{i_n^2} = \frac{4kTB}{Z_i^2}\left(R_i + \frac{\tau}{g_m}\right) + 2qI_G B$$ (10.39)

where Z_i is the input impedance of the FET and is a frequency-dependent quantity.

In figures 10.8 and 10.9, examples of the noise behaviour characteristics versus frequency are presented for the bipolar and the field effect transistor. The characteristics show the equivalent voltage noise and the equivalent current noise behaviour versus frequency. As said already, these noise measurements can be performed by measuring the output of the amplifier's stage with the input terminals short circuited, giving the noise voltage, or 'open circuited', giving the noise current, and dividing both measured quantities by the gain of the amplifier. As can be seen from these graphs, flicker noise becomes important at lower frequencies.

10.5.3 Noise behaviour of the diode

The equivalent noise circuit for a diode is drawn in figure 10.10. The following physical noise sources are to be found in a diode:

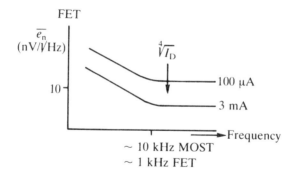

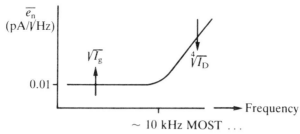

Figure 10.9 Noise voltage and noise current sources versus frequency of the FET.

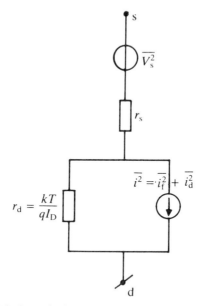

Figure 10.10 Diode equivalent small-signal circuit with noise sources.

1   a thermal equivalent voltage noise source due to the presence of the physical
    resistance r_s of silicon,

$$\overline{v_s^2} = 4kTr_s B \qquad (10.40)$$

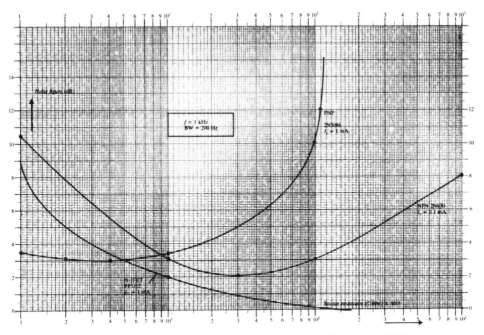

Figure 10.11 Noise comparison of bipolar transistor and junction FET (cour
National Semiconductor Ltd).

2   a shot noise current in the p–n region as a function of the direct current I_D,

$$\overline{i_d^2} = 2qI_D B \tag{10.41}$$

3   a flicker noise current component i_τ, which is only of importance when a low
frequency is present,

$$\overline{i_\tau^2} = K \frac{I_D^a}{f} B. \tag{10.42}$$

To be able to compare some other noise specifications of well known components
two other figures are depicted. Figure 10.11 shows the noise behaviour of a junction
FET and a bipolar transistor. Figure 10.12 shows the noise behaviour of a discrete
junction FET and an operational amplifier.

10.6   NOISE IN SENSORS

Just as we have seen already in chapter 6, a substantial effort is being devoted to
the research and manufacture of semiconductor sensors. Sensors are becoming more
sensitive and noise levels become an important factor to consider when sensitivity
increases. A good example is the use of SQUIDs, which must be cooled in order to
obtain an acceptable noise level. Therefore it makes sense to make a few remarks

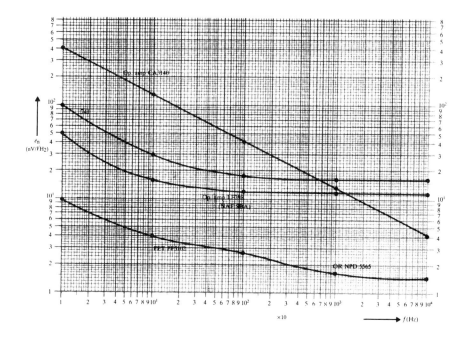

Figure 10.12   Noise comparison of junction FET and an operational amplifier (courtesy National Semiconductor Ltd).

about noise in (semiconductor) sensors. In each sensor one or more sensing elements are used, which can all be traced back to the already-discussed known noise sources. The following sensing elements are frequently used in (semiconductor) sensors:

- resistors
- p–n junctions
- bipolar transistors
- field effect transistors
- thermocouples and thermopiles
- piezo elements
- capacitors
- coils.

For all these elements, a noise discussion can be presented, except for coils and capacitors, because these elements generate no noise. However, this makes no sense where we have done so already. When resistors are used we have to deal with thermal noise and equivalent noise voltage sources and current sources can be introduced. In sensing elements also shot noise can be distinguished. Photon noise may also be considered and it is due to the fluctuation in the number of the quanta or in the power reaching the detector itself. For instance, this can be present in p–n junctions used as CCDs. The fluctuation in the power emitted by a unit-area black body at temperature T can play a role and is described by

$$S_p(f) = 8\sigma k T^5 \tag{10.43}$$

where σ is the Stephan–Boltzmann constant and k is again the Boltzmann constant. The power spectral density for the fluctuation in the number of emitted photons is proportional with

$$S_n(f) \propto T^3. \tag{10.44}$$

Finally, we also meet the already-described $1/f$ noise in sensors, which is of importance when low frequencies are involved. Further, which noise source will dominate will depend on the type of sensing element, the frequency and the temperature. However, as we have seen already, temperature is often one of the main influences in the physical description of noise sources and its influence can be reduced strongly by applying cryogenic techniques. So, for sensors reference can be made partly to noise sources that we have met already and partly to noise sources that are specific to sensors only, such as when photon noise in involved. For more details reference is made to the literature.

BIBLIOGRAPHY

[1] Bordoni F and D'Amico A 1990 Noise in sensors *Sensors Actuators* A **21–23** 17–24
[2] Gray P R and Meyer R G 1984 *Analog Integrated Circuits* 2nd edn, (New York: Wiley) ch 11
[3] Van Putten A F P *et al* 1996 Silicon thermal anemometry, developments and applications *Meas. Sci. Technol.* **7** to appear

Eleven

INTERFACING TO SENSORS

11.1 INTRODUCTION

In this chapter we will discuss aspects of interfacing related to sensors. For interfacing sensors a distinction must be made between self-generating and modulating transducers. Modulating transducers require a type of biasing circuitry, because an auxiliary energy source is needed for this type of sensor to become operational. For self-generating input transducers no auxiliary energy source is needed. They can be connected directly to an appropriate amplifier. In most cases the input signal is very weak and amplification is necessary. Another family of transducers uses the principle of *resonating transducers*, for which special excitation and detection circuitry is needed. This has been discussed already in section 6.8.1.

Interfacing can be defined as the physical connection of different systems in accordance with well defined protocols concerning the physical voltage levels, frequency differences, software protocols and so on. It is observed here that over 13 different definitions for interfacing are known. There exists no conformity on this definition. Everywhere two different domains meet each other, whether they differ in geometry, in type of signal, in type of energy, frequency, phase or (digital) code, a type of interface can be designed to glue them together. Even a dictionary can be considered as a type of interface between two languages.

For modulating transducers a bridge configuration is often applied. This is less applicable where self-generating transducers are involved such as solar cells, thermocouples and so on. Also, it makes no sense for microphones. One of the main reasons to use a bridge configuration is to get rid of the common-mode (supply) voltage. We can pose the following important questions.

1 If a bridge configuration can be applied, what is the bridge output voltage as a function of an induced arbitrary physical variation?
2 If no bridge configuration can be applied which other configuration can be used?
3 What is the type of applied amplifier configuration?
4 Is any kind of filtering required?
5 What is the sensor's noise level?

In the next section we will lay emphasis on bridge configurations.

11.2 SENSORS IN BRIDGE CONFIGURATIONS

We shall consider the most common possibilities for implementing sensor elements in a bridge configuration. Bridge configurations are used with one, two and four sensor elements, and in particular this last of these offers great advantages in integrated circuits. Furthermore, in some measuring systems more than one bridge configuration is encountered, which can offer improved characteristics as we have seen already in chapter 3. To make comparisons between the different configurations possible the easiest way is to calculate the bridge output voltages in the various respective cases.

11.2.1 Bridge configuration with one sensor element
The applied configuration is illustrated in figure 11.1, in which the supply voltage source V_{CC} and one sensor element R_S are depicted. Each time it is supposed that

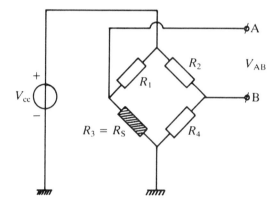

Figure 11.1 Bridge configuration with one sensor element.

the change in physical parameter to be measured results in a change in the resistive sensor element of $R(1+\delta)$. Then solving the bridge equation for V_{AB} and assuming that all bridge resistances are equal at room temperature, we find

$$V_{AB} = V_{CC} \left[\frac{R(1+\delta)}{R + R(1+\delta)} - \frac{R}{2R} \right] \tag{11.1}$$

or

$$V_{AB} = V_{CC} \frac{\delta}{4 + 2\delta} \approx V_{CC} \frac{\delta}{4} \tag{11.2}$$

when δ is positive and $R_S = R_1 = R_2 = R_3 = R_4 = R$ at room temperature, all showing the same temperature coefficient. In this bridge configuration we note again that a constant supply voltage does not guarantee a constant supply current and a constant supply current does not guarantee a constant supply voltage. Also, a constant-temperature configuration does not guarantee constant sensor parameters. The ideal configuration is obtained when a type of feedback is implied for supply current

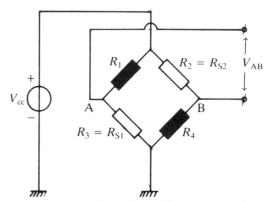

Figure 11.2 Bridge configuration with two sensor elements.

and voltage simultaneously to maintain both quantities constant. (See also section 3.10.3.)

11.2.2 Bridge configuration with two sensor elements

The bridge configuration with two sensor elements is depicted in figure 11.2. When the same assumptions are made as before, we find for the output signal V_{AB}

$$V_{AB} = V_{CC}\left[\frac{R(1+\delta)}{r(2+\delta)} - \frac{R}{R(2+\delta)}\right] \tag{11.3}$$

or

$$V_{AB} = V_{CC}\frac{\delta}{2+\delta} \approx V_{CC}\frac{\delta}{2}. \tag{11.4}$$

As can be seen the bridge output signal is twice that of the configuration with one sensor element. In addition to this difference, in certain applications the output voltage changes sign when the physical parameter changes sign.

*   *   *

Example 11.1
Suppose the sensor elements are strain-sensitive elements and one strain gauge element is subjected to strain and the other to stress; then the output voltage will change sign when strain and stress are reversed for the two strain gauges. This is illustrated in figure 11.3, where a metal bar provided with two strain gauges is sequentially subjected to a force F_1 and a force F_2. So, the force F_1 causes a strain variation converted into a change in resistance of $R(1+\delta)$ in R_{S2} and a stress variation converted into a change in resistance of $R(1-\delta)$ in R_{S1}.

*   *   *

The four-sensor configuration is discussed in the next section.

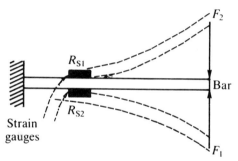

Figure 11.3 Metal bar sequentially subjected to a force F_1 and F_2 generating output voltages with opposite sign.

11.2.3 Bridge configuration with four sensor elements

When four sensor elements are applied in one bridge configuration, it is required that the four sensor elements are subjected to the physical parameter pair-wise with opposite sign. If this condition is not fulfilled a zero output signal will be the result. In most cases this condition can easily be fulfilled even when fully integrated bridge sensors are applied. This four-sensor configuration offers another great advantage, which is more sophisticated than in the case with two sensor elements. As well as the aspect of changing sign when the physical parameter changes sign, when the *direction* of the physical parameter changes the bridge output voltage changes, offering a directional sensitivity. So in this case a *vector* measurement is obtained, which means that the *magnitude* and the *direction* of the physical parameter can be measured. This configuration is depicted in figure 11.4. In this case the output voltage V_{AB} is given by

$$V_{AB} = V_{CC} \left[\frac{R(1+\delta)}{R(1+\delta)+R(1-\delta)} - \frac{R(1-\delta)}{R(1+\delta)+R(1-\delta)} \right] \qquad (11.5)$$

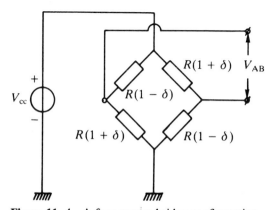

Figure 11. 4 A four-sensor bridge configuration.

or

$$V_{AB} = V_{CC}\,\delta \qquad\qquad (11.6)$$

which is again twice that of the configuration with two sensor elements and four times that of the configuration with one sensor element. Note that in all three discussed configurations the supply voltage is involved in the expression and thus can influence to a large extent the accuracy of the output. A well controlled reference supply voltage always is a prerequisite. Note also that in these configurations the deflection method is used. When a kind of feedback loop is implied the null method can be used and the supply voltage is no longer present in the expression.

*   *   *

Example 11.2
Suppose the four-sensor configuration consists of an integrated circuit with four p-type doped resistive elements with a positive temperature coefficient δ as is depicted in figure 3.28. Suppose further that the bridge is thermally heated by its own dissipation well above ambient temperature; then an air flow will cool down the sensor configuration showing a directional sensitivity. The output signal changes sign when the air flow changes its direction 90° or 180° depending on the applied electronics. This application is called a thermal anemometer (*anemos* is Greek for wind). Based on this principle the alternating direction method (ADM) has been developed as discussed in chapter 7, eliminating offset and drift in vector sensors.

*   *   *

11.2.4   A double-bridge configuration
One application of a configuration with two bridges has already been discussed in section 3.6, but the use of this double-bridge configuration was devoted to biasing, resulting in a constant-voltage and current behaviour in one bridge only. Another application of this configuration is illustrated in figure 11.5. Referring to the same integrated sensor of figure 3.42, by careful inspection of the interconnection pattern it can be seen that both bridges are connected in opposition to each other. When both bridges are series connected to a voltage source and the sensor is again applied as an anemometer, the same airflow will generate two output voltages V_{A1B1} and V_{A2B2} with opposite signs. If a three-amplifier configuration is applied, the output signals will be summed and the common-mode signals will be suppressed to a large extent. High-quality chopper amplifiers are preferred for this configuration. It is interesting that this configuration produces two input signals with opposite signs. However, it can also be shown that the overall signal gain is a factor of $\sqrt{2}$ only and hence is a rather modest improvement.

11.2.5   Biasing bridge circuits
Another application of biasing can be realized with a split power supply for a bridge sensor. The configuration can avoid any on-mode input voltages at the input of the amplifier, keeping these at zero level and hence reducing to a large extent the CMRR

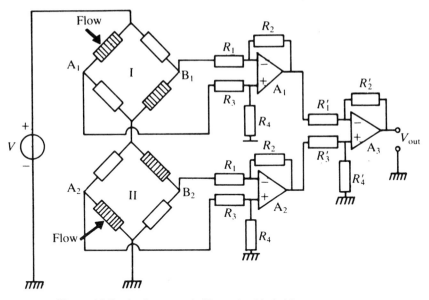

Figure 11.5 An integrated silicon double-bridge anemometer.

requirements. For instance, an amplifier with 80 dB CMRR adds 100 μV error to the output signal per volt of common-mode input voltage. In some cases, the bridge is supplied via a switched power supply to reduce power dissipation. A perfectly symmetrical power supply is then necessary, providing a strongly reduced offset voltage. A schematic layout of such a configuration is depicted in figure 11.6.

Other current biasing circuits use a kind of feedback. An example has been discussed already in section 3.10.3, where a biasing circuit maintaining a constant supply current and supply voltage for the measuring bridge is presented. This is obtained by applying thermal feedback which is possible for transducers integrated on the same substrate. The circuit also provides a constant operational temperature and

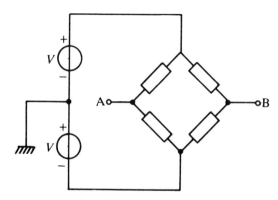

Figure 11.6 A symmetric split power supply to reduce CMRR demands.

consequently constant sensor parameters are guaranteed. Also AC powering (biasing) is used, or a kind of pulse mode powering, offering some other advantages with respect to drift sensitivity and format of output signal. It must be noticed that biasing bridges with a voltage source can guarantee only constant supply voltage and that powering with a current source guarantees constant supply current only.

11.3 BRIDGE AMPLIFIER CONFIGURATIONS

In this section we shall discuss several interfacing connections between the bridge and the applied amplifier configuration. In most cases operational amplifiers are used because of their flexibility as a general building block. In all derivations we assume idealized operational amplifiers, i.e. with infinite open-loop gain, no input currents and hence infinite large input impedance.

11.3.1 Circuit 1

In the circuits of figure 11.7 a standard differential amplifier is connected to the output of the corresponding bridge. The circuit of figure 11.7(a) has one sensor element, and that of figure 11.7(b) has two. The output voltages can be derived from (11.2) and (11.4). For figure 11.7(a) it is found that

$$V_{out} = A V \frac{\delta}{4} \tag{11.7}$$

in which $A = R_2/R_1$ and for figure 11.7(b) with the same expression for A the value is twice that of figure 6.19(a), or

$$V_{out} = A V \frac{\delta}{2}. \tag{11.8}$$

11.3.2 Circuit 2

As can be seen from the expression for the configuration with one sensor element as depicted in figure 11.1 and equation (11.1), the exact result is a non-linear expression in δ. For a bridge configuration with one sensor the behaviour can be linearized if we apply the circuit as depicted in figure 11.8. For this circuit with ideal amplifiers it can be said that node A and node B behave as a virtual common. With this assumption the following current equations for node A and node B will be valid: for node A

$$\frac{V}{R} + \frac{V_{o1}}{R(1+\delta)} = 0 \tag{11.9}$$

and for node B

$$\frac{V}{R} + \frac{V_{o1}}{R} + \frac{V_{out}}{R} = 0. \tag{11.10}$$

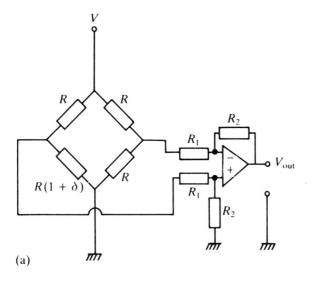

(a)

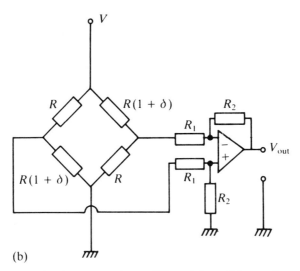

(b)

Figure 11.7 Bridge circuit connected to a differential amplifier configuration: (a) with one sensor element; (b) with two sensor elements.

Solving both equations for V_{out} yields a linear equation in δ:

$$V_{out} = \frac{R_F}{R} \delta V. \tag{11.11}$$

A more accurate and general solution provides the following expression for V_{out}:

$$V_{out} = A\delta V - \delta V + \tfrac{1}{2} A \delta^2 V \tag{11.12}$$

or

$$V_{out} = V(A - 1 + \tfrac{1}{2} A \delta) \delta \tag{11.13}$$

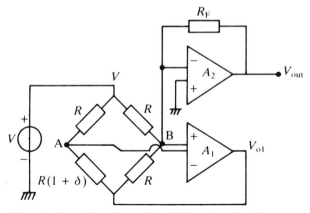

Figure 11.8 Linearized configuration of a bridge circuit with one sensor element applying feedback.

in which $A = R_F/R$. The amplifier A_1 acts as a comparator, maintaining $V_A = V_B$ under all circumstances and at zero level. Amplifier A_2 will deliver the required current via R_F to maintain this equality.

11.3.3 Circuit 3

An alternative feedback circuit is depicted in figure 11.9, in which the sensor element is implemented in the feedback loop. An almost exact solution for the output voltage V_{out} can easily be obtained by considering the amplifiers separately. The result is

$$V_{o1} = -\tfrac{1}{2}V\delta \tag{11.14}$$

and

$$V_{out} = -\frac{R_F}{R_1}V_{o1} = \tfrac{1}{2}V\delta\frac{R_F}{R_1}. \tag{11.15}$$

11.3.4 Circuit 4

In the circuit of figure 11.10 two single-ended amplifiers are used. To reduce the first offset term a matched pair of amplifiers is highly preferred. Chopped amplifiers can

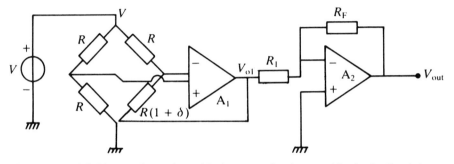

Figure 11.9 A bridge configuration with the sensor implemented in the feedback loop.

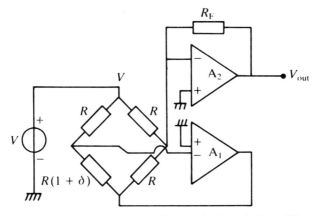

Figure 11.10 Configuration with two single-ended amplifiers.

offer the desired specifications, reducing strongly offset and drift behaviour. The result is

$$V_{\text{out}} = V\delta \frac{R_F}{R}.$$ (11.16)

11.3.5 Circuit 5
In the circuit of figure 11.11 a current-biasing amplifier configuration is implemented. This also reduces non-linearity but instability might easily occur as a result of stray capacitances. Note that two sensors are implied in the bridge configuration. The

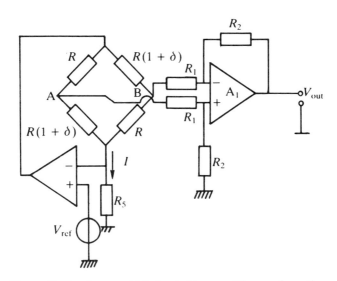

Figure 11.11 A current-biasing bridge amplifier configuration.

expression for the output voltage can be obtained by applying the current node equations for A and B. The result is (see problems 11.11)

$$V_{\text{out}} = A\tfrac{1}{2}(V - V_{\text{ref}})\delta \qquad (11.17)$$

or

$$V_{\text{out}} = A\tfrac{1}{2}(V - IR_5)\delta \qquad (11.18)$$

in which $A = R_2/R_1$ is the gain factor of A_1. In addition to the stability problem the stability of the reference voltage strongly influences the accuracy of this circuit.

11.3.6 Circuit 6

In figure 11.12 a single differential amplifier is applied with which the gain factor can be varied. The disadvantage of this circuit is that any variation in R_1 disturbs the balance of the amplifier's input impedance, resulting in a change in the offset.

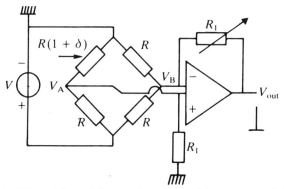

Figure 11.12   A differential amplifier configuration with one sensor bridge element.

Note that the plus and minus signs of the supply voltage V are reversed compared with the usually connected supply circuit. This is performed to obtain a more simplified expression for V_{out}. The expression for V_{out} is not strictly linear but may be approximated with

$$V_{\text{out}} = \frac{R_1^2}{2RR_1 + R^2}\,\delta V \qquad (11.19)$$

where it is assumed that $1 \gg \delta$. A better solution for a variable gain setting is the application of a true instrumentation amplifier as is presented in circuit 7.

11.3.7 Circuit 7

In figure 11.13 a well known instrumentation amplifier configuration is depicted consisting of three amplifiers in which R_1 can be varied to realize a variable gain setting. The matching of the different resistors determines to a large extent the CMRR, and hence 0.1% metal film resistors are preferred. With R_1 the gain can be varied without influencing the balance of the respective input impedances of A_1 and A_2.

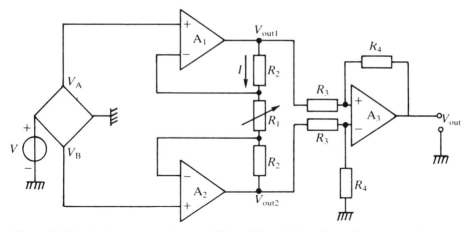

Figure 11.13 An instrumentation amplifier with variable gain setting, connected to a bridge circuit.

The expression for the output voltage is easily found and may be derived calculating the current in R_1 and R_2 twice. We assume again that the amplifiers are ideal, that is to say there are no input currents and the open-loop gain is infinite for all three amplifiers.

$$I = \frac{V_{out} - V_{out2}}{2R_2 + R_1}$$ (11.20)

and

$$I = \frac{V_A - V_B}{R_1}$$ (11.21)

because there is one current only in R_1 and R_2 and V_A and V_B can be found to exist across R_1. Combining both equations yields

$$\frac{V_{out1} - V_{out2}}{2R_2 + R_1} = \frac{V_A - V_B}{R_1}$$ (11.22)

or

$$V_{out1} - V_{out2} = \frac{R_1 + 2R_2}{R_1}(V_A - V_B)$$ (11.23)

and for the third stage it is found that

$$V_{out} = \frac{R_4}{R_3}(V_{out1} - V_{out2}).$$ (11.24)

Substituting equation (11.23) in (11.24) we find

$$V_{out} = \left(1 + \frac{2R_2}{R_1}\right) \frac{R_4}{R_3} (V_A - V_B).$$ (11.25)

The configuration appears to have a very high input impedance and is obtainable as one building block. As discussed in chapter 9, low source impedances produce low noise. It is observed that it is possible to design low source resistances while still retaining high temperature coefficients in integrated circuits. For instance, to this end the aluminium interconnection pattern in the flow sensor chip is made parallel to the sensor area as shown in figure 3.43 and provides a considerable reduction in the source resistance while maintaining a good sensitivity for temperature. For instance, the source resistance of the depicted chip is about 50 Ω at room temperature while the sheet resistance, $R_\square$, is about 5000 Ω/square.

11.4 TELEMETRY

In figure 11.14 a complete telemetry schematic suitable for n transducers and n actuators with remote control provisions for transmitting and monitoring information is shown. In the medical field especially, applications for mobile patients can be found. There is an increasing need for wireless transmitting and monitoring the patient's condition, requiring a very feasible type of interfacing. Also, we can find

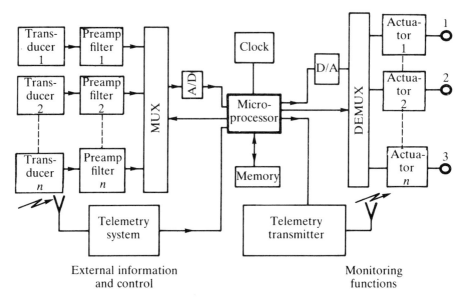

Figure 11.14 Complete interface schematic with telemetry receiver and transmitter for n sensors and n actuators.

this type of application in agriculture. Farmers can monitor their cattle for food consumption and milk production individually by remote control.

11.5 HYBRID INTERFACING CIRCUITS

It will be clear that for sensors many other interfacing circuits exist. One other type of interfacing circuitry will be mentioned here. Several manufacturers have designed integrated hybrids which provide very versatile possibilities for simple interfacing. A good example is Analog Devices' 'Wide Bandwidth Strain Gage Signal Conditioner', AD 1B31. This hybrid circuit provides the following features:

1 convenient offset and span adjustment capability
2 internal half-bridge completion possible
3 remote sensing
4 programmable transducer excitation
5 adjustable low-pass second-order filter.

An accurate reference voltage of 6.8 V is built in. The input impedance is about 1 GΩ. Applications are found in strain, force and pressure transducers. For further details, reference is made to the literature. It must be observed that, as a result of the continuing integration of sensor part and processing electronics, the interfacing circuit has often become an integral part of the sensor itself.

11.6 AUTOZEROING

11.6.1 Autozeroing in electronics

Finally, we will introduce here the concept of *autozeroing* which is another very versatile feature in electronic measurement systems. Each user of an instrument expects the instrument to indicate zero when the input is short circuited. For instance, in a digital voltmeter with a maximum reading of 1999 a zero error of 0.05% of full-scale deflection (FSD) is sufficient to give an error reading of 0001. For optimum accuracy, especially at low values, a zeroing is required and automatic zeroing would be very convenient. This is implemented in most instruments nowadays. As an example, this feature, combined with a dual-slope ADC, is depicted in figure 11.15. Before the real measurement is carried out, the switches S_3, S_4 and S_5 are closed by the logic controller, for example for 50 ms. This grounds the input, giving the integrator a short RC time and connecting the output of the comparator to the capacitor C. This capacitor is now charged by the offset voltages of the amplifier, the integrator and the comparator. When the switches S_3, S_4 and S_5 are opened again, to start the real measurement, the total offset voltage of the circuit, which now equals the zero error, is stored in the capacitor C and as a result the real input voltage is measured correctly.

It is very important to notice that in sensors offset treatment and elimination differ completely with respect to pure electronics circuits, because in sensors an energy conversion between two energy domains is involved. In chapter 7 we have dealt with

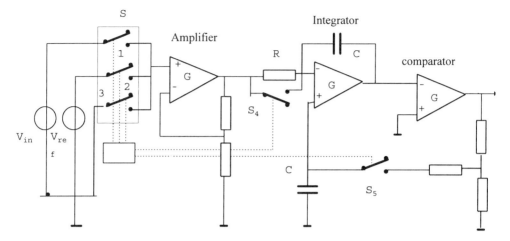

Figure 11.15 Simplified circuit of an autozeroing circuit that is used with a dual-slope ADC.

these aspects in more detail. In particular, the ADM principle concerning full additive offset and drift elimination is discussed in detail.

11.6.2 Autozeroing in transducers

Autozeroing in a transducer differs from autozeroing for pure electronics. Here, a *physical reference* signal is needed which must be connected to the input at the right time. The following statement describes precisely what is needed. The user must know when the measurand is at the reference condition and thereby control the *autoreferencing* at the reference condition, and read or control at the measurement condition. It will be clear that, when physical, pressure, flow, temperature or other physical reference signal is available, a perfect autozeroing can be achieved. Here, a digital solution, with the help of a microprocessor, fits best as a versatile solution. Once such a physical reference is available then all other tricks can be used to eliminate offset and drift.

No discussion is presented about any kind of digital interfacing circuit. The reader is referred to the relevant literature.

PROBLEMS

11.1 Derive expression 11.13. Hint: apply the current node equations for node A and B.

11.2 Prove the expression for the output voltage for figure 11.9. Apply the same hint as before for a general solution.

11.3 Derive the expression for the output voltage for figure 11.10.

11.4 Answer the same question for figure 11.11.

11.5 Answer the same question for figure 11.12.

BIBLIOGRAPHY

[1] Analog Devices 1990 Signal conditioning components *Data Handbook* pp 10–71
[2] Analog Devices 1987 *Linear Design Seminar (1987)*
[3] National Semiconductor 1977 *Pressure Transducer Handbook*
[4] National Semiconductor 1975 *Interface Integrated Circuits*
[5] National Semiconductor 1979 *Interface Databook*
[6] Maxim 1996 *New Releases Databook* vol IV, ch 2
[7] Van Putten A F P, Hitchings D J, Quanjer Ph H 1993 Portable electronic peak flowmeter for improved diagnosis of chest diseases in COPD patients *IEEE/IMTC Conf. Proc. (Irvine, CA, 1993)*

Twelve

ERGONOMICS OR HUMAN ENGINEERING

12.1 INTRODUCTION

This chapter is devoted to one of the most neglected disciplines in measuring techniques. The reason is easy to understand because it is the least possible to quantify and was considered for a long time to be of little importance. However, equipment and systems have grown considerably in complexity and nowadays the relationship between humans and systems very often plays an important and vital role in efficiency, safety, comfort and health.

To understand what is meant by ergonomics the following definition will clarify the implications involved. Ergonomics is defined as the collection of knowledge about skills and limits of the human being and the science of applying this knowledge to equipment and systems to achieve a well-being and functioning of humans in which safety aspects and economical efficiency can play a vital role. The American equivalent for ergonomics is 'human engineering'. The word ergonomics was first introduced by K F H Murrel in 1949 and is derived from the Greek words *ergon* = work and *nomos* = knowledge. The person involved in ergonomics is called a human engineer, for whom the following topics are subjects of investigation:

1 creating good working conditions
2 creating efficient tools, equipment and systems in which a direct relationship with the human being is present
3 the investigation of maximum-load conditions for the human being performing various tasks under static and dynamic environmental conditions
4 the mechanisms with which the required information is perceived and the relevant thresholds for the human being
5 the way in which information is presented to allow unambiguous interpretation.

One of the best examples in which all of these aspects play a vital role is flying an aeroplane. Optimum conditions and circumstances between pilot and machine must be present to fulfill all requirements. A pilot should be in good condition and be able to perform his or her tasks with a machine which is also maintained in good condition. The different flying tasks such as starting, flying, approaching and descending are tasks with different loads for the pilot with possible rapid changes in working circumstances. All equipment, systems and subsystems must be designed in such a fashion that optimum safety, attainability and operability are guaranteed and that the displayed information can be understood without ambiguity.

As will be recognized, in the field of ergonomics the human being is always at the centre of the investigation; this means that the human being is the starting point and functions as the reference to whom all information must be adapted. The consequence of this is that we have to be well aware of the possibilities of our biological senses. This biological equipment will be discussed in the following section.

Ultimately, ergonomics can be defined as the science of finding optimum conditions in the interrelationship between the human being and his or her tools. In principle many disciplines are brought together in ergonomics, such as anatomy, physiology, cybernetics and information theory. Human engineering is not only related to tools such as machines, but may also be applied to the installation of office furniture, shops, schools and so on. As will be clear from the above, human engineering is a multidisciplinary area of science and for a long time it has been more art than science. Consequently, human engineering is difficult to quantify and is therefore initially less attractive for the scientist.

To obtain some feeling for ergonomics we shall discuss some aspects related to electronic systems and equipment only. The chapter is not exhaustive, but its aim is to draw attention to those aspects involved in the design and application of electronic measurement and control systems. Because of their importance, we shall highlight aspects of display design.

12.2 BIOLOGICAL EQUIPMENT OF THE HUMAN BEING

Our biological equipment consists of a number of so-called windows, through which the different physical changes in the outer or inner world can be detected. Each window has a limited spectral width and a limited height of modality. Two of these windows are the most important from a technical point of view, i.e. the *auditory* and the *visual* windows.

12.2.1 The auditory window

The range or width of our auditory frequency window runs from 20 to 20 000 Hz and its modulation height goes from 0 to 140 dB. The resolution of the auditory organ involves 3 Hz below 1000 Hz and 0.3% of the frequency above 1000 Hz. This is the smallest step of distinction we can detect without technical means. Within the auditory window it is curious that nature has provided us with an optimum working area where the sensitivity is the highest. The optimal working area for a hearing organ lies between 300 and 6000 Hz and between 40 and 80 dB, which roughly coincides with the area of speech and most music. This is summarized in table 12.1.

12.2.2 The visual window

The perceivable wavelength of our visual window ranges from 300 to 720 nm and the modulation height varies from 10^{-9} to 10^{9} cd m^{-2}. The resolution for colour

Table 12.1 Review of specifications of the human auditory organ.

Auditory frequency range	20–20 000 Hz
Modulation height	0–140 dB
Resolution below 1000 Hz	3 Hz
Resolution above 1000 Hz	0.3% of frequency
Optimal working modulation area	40–80 dB
Optimal working frequency area	300–6000 Hz

changes varies from 1 to 20 nm and for light intensity changes the resolution must equal 100 cd m^{-2}. This can be expressed in general with the help of Weber's law, which says that two stimuli can be distinguished when there exists a minimum difference in intensity, or

$$\frac{\Delta I}{I} = C \qquad (12.1)$$

where ΔI is the change in intensity, I is absolute intensity and C is a constant dependent on the type of organ. For colours the difference should be in the order of 5%. Within the visual window, the maximum-sensitivity area lies between 450 and 650 nm which is in the yellow to green range and for light intensities between 10 and 100 cd m^{-2}. The candela (cd) is referred to as the standard intensity for a black radiator and proves to be a quite complicated definition seen also in section 6.7. The candela is the light intensity in a perpendicular direction of a surface area of 1/600 000 m^2 of a black body at the congealing temperature of platinum (2043 K) at a pressure of 101 325 Pa. This is summarized in table 12.2.

An illustration of the visual and the auditory windows is given in tables 12.1 and 12.2. Beyond these limits we are not normally able to detect information, and transducers are required to transform the information within our windows. Furthermore, aging plays a role, because with aging our perceptual windows and resolution capability tend to decrease. It is interesting to note that we transform all physical information just to these two types of detectable energy and never for instance to a type of smell or a temperature, because these senses are less developed in human beings and thus less efficient in detecting and presenting information.

Among these two perceptual senses, the visual perception seems to be the most important one. Therefore we will emphasize the way visual information is presented. There is still another reason why measuring instruments must help our biological perception. Perception takes place in the time domain. In order to be perceived the information should not only be equal to or greater than the minimum required size

Table 12.2 Review of the human visual window.

Perceivable wavelength	380–720 nm
Modulation height	10^{-9}–10^9 cd m^{-2}
Resolution for colours	1–20 nm
Resolution for light intensity changes	100 cd m^{-2} or 5%
Maximum sensitivity	450–650 nm (yellow to green)
Maximum sensitivity for light intensity	10–100 cd m^{-2}

and step, but should be presented for some time also, to activate our natural transducers. A smaller change requires a longer duration of the actuating process. The cones and rods of the human retina need some time to react and have a certain time constant. The same reasoning applies for the process of transmitting the pulses across the nerve fibres. In many circumstances we need a transduction not only towards the perceptual area but also towards a perceptual time duration. In a lot of measuring instruments, both mechanisms are implied.

12.3   SOME OTHER ASPECTS OF HUMAN PERCEPTION OF INFORMATION

The artificial aids to natural perception mentioned already indicate the first essential functions of transduction: processing and displaying information. In general they represent a measuring instrument and bridge the gap between the real outer world and the observer to enable the physical changes to be detected in a broader, finer or shorter fashion and on a more reliable and objective basis. This seems to be basic for our curiosity and need to control the outside world and our environment. Three other essential functions must be added to the way we perceive information on higher levels. These functions are transmission, retrieval and processing of the acquired information.

12.3.1   Transmission of information

When transmission of information is involved a transmission channel is always present, in which barriers can be present and in which often noise is added to the signal. Also, it will be clear that we are not able to observe synchronous phenomena occurring at different places without technical means. With well equipped systems and instruments, it is possible to observe a process at large distances, through walls and in places where we are not able normally to be physically present, for instance in space travel or circumstances with extreme conditions.

12.3.2   Retrieval of information

The human brain can receive information in very limited amounts. As time passes the amount and accuracy of information that can be reproduced diminishes. During the first seconds the quantitative losses are huge and over longer periods the losses are also considerable. The human memory and/or brain, from which we can retrieve information in normal circumstances, uses a process of selection, but also proves to deform or sometimes to complete information. So for instance we are able to read a sentence with words missing. What is being perceived and retained differs strongly between individuals and circumstances in which expectation, prejudice and preoccupation can play an important role. This is the main reason why we want to record information.

12.3.3 Processing of information

As indicated already, the human brain is not a passive sensorial mechanism. In principle human perception involves a sequence of selection, deformation and completion in which billions of cells are busy coding the input and processing the received information. Information elements and their functional relations are actively structured into patterns which are easy to remember. A type of hierarchical order can often be recognized in these patterns. There is far from a one-to-one relationship between physical reality, direct perception and the patterns stored in the individual human being.

These aspects stress the fact that making objective statements about phenomena is very difficult, and make it necessary to de-individualize information as far as possible. Here a more philosophical point of view can be made, which says that every observation will influence the process and the information which is retrieved, and is strongly related to the way the information is collected. However, we constantly achieve a consensus about how information can be made more objective. For this purpose, instruments are very valuable: they can process, filter noise, weigh, transform, put into structures and handle information much faster and more reliably than we are able to do.

Throughout history all types of physical and technological principle have been applied to perform measurements. Today these measurements are based on an international system of standard units and measurements in order to have an invariable reference system on which everybody can rely. The SI system is nowadays a well established system of measurements and units. This means that there are two systems or worlds of perception which exist together and can be compared. This is considered further in the next section.

12.4 PERCEPTION

Natural perception or *observation* is human related and thus individualistic, inaccurate, and not reliable, but with guessing capabilities and very inventive in a broad way of perception. *Instrumental perception* is objective, very predictable, constant, highly accurate and reliable but very specific. In a measurement chain the display at the output forms the plane of contact between these two worlds and bridges the gap between. At this point all major aspects of perception come into scope because in displaying information the following requirements have to be considered.

1 It must be possible to detect and scan the information under known conditions.
2 Separate information elements have to be distinguished.
3 It must be possible to decode the presented information.
4 It must be possible to interpret the information involved without ambiguity.
5 It must be possible to remember the displayed information and to take the right actions if required.

The human being can also be considered as a kind of interface between system and outside world. Here it must be borne in mind that a human being is always in the

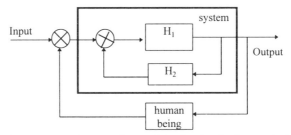

Figure 12.1 The concept of perception and feedback of a human being in a system.

feedback loop of the system and is part of the whole system to be controlled. Ultimately, without this feedback, the system is useless and not determined. Of course in a complex system, for instance a power plant or an aeroplane, a large number of technical feedback loops will be present, but ultimately the human being is always involved in the final feedback loop. This is schematically illustrated in figure 12.1.

The consequence of this is that the industrial designer who is involved in human engineering must be aware that he or she has to translate all the required information in some way on a display which must be interpreted by a human being. This means that the human engineer must be familiar with the laws of human perception. A few examples can illustrate what is meant by these statements. In some cars the speedometer display shows a digital number only and no trend in speed can be detected. Any elementary law of ergonomics fails and in the design stage market trends have been important only. As another example, figure 12.2 shows a digital multimeter where for the same kind of information a combination is shown of an analogue and a digital display even with the possibility of displaying trend charts.

If the display is digital it is very accurate, but that is not always required. If it is only analogue it shows a trend and is likely to be insufficiently accurate. Here only the right combination of analogue and digital display techniques can offer the best solution. Shortcomings with respect to display design, whether visual or spoken, can be very costly in the sense of much loss of performance, dissatisfaction of operators in working circumstances and wrong investments in equipment and manpower. We shall discuss some aspects and trends of human engineering in more detail in the following section. Recently there has been an increasing tendency to present information in spoken commands in cars, aircraft and in an increasing variety of equipment. The voice-controlled computer and typing machine are reality. However, whether these features will become commonly accepted depends on other aspects.

12.5 TRENDS IN THE DEVELOPMENT OF MEASURING SYSTEMS

In this section some trends will be discussed in the development of measuring systems, not only from a viewpoint of human engineering but also to emphasize the impact of electronics on our society in general.

We can recognize an already dramatic increase of the development of robots in manufacturing and control of a great variety of processes. This is mainly due to the

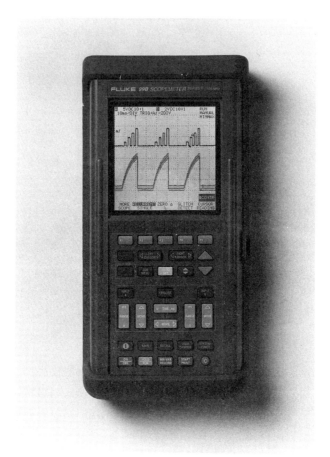

Figure 12.2 Example of a multimeter display with analogue and digital display for the same kind of information (courtesy Fluke Inc. (Netherlands)).

dramatic increase in the application of microelectronics. The impact of this revolution can be treated in the following four points where the definition of a measurement system must be understood in a much broader sense than usual.

1 Miniaturization in electronics is still proceeding rapidly and has caused a qualitative change in information technology. It has become thousands of times smaller, faster, energy saving, more reliable, more complex and cheaper. One of the best examples is the development in memory chips; we are now talking about 1 Gbit DRAM chips. See also chapters 1 and 6.
2 Decentralized measurement and control are now not only to be found in the professional and industrial sectors but are also being rapidly introduced in the consumer and domestic industry. Applications can be found in laboratories, in banking, administration, transport, entertainment, education, etc. The development in display technology for instance has made a big change too; especially

the cathode-ray tube (CRT), the colour plasma screen and the liquid-crystal display have become common in daily instrumentation.

3 Excellent examples can be found in the process industry where one person can monitor and control a complete plant by the applied principle of performing the management task via the CRT monitor which represents in real time all required information in full colour. An overall view and detailed information by selection can be retrieved from the process as required. The way the information is displayed is most vital, and ergonomics and working load for the operator must be considered very carefully. Even safety and health aspects play a role here because CRTs are never completely free of flickering and an uninterrupted workload of 4 h for CRTs is the maximum allowed duration.

4 A very important technological development is still occurring in interface techniques where the connections are made between the inner (systems) and outer world.

In the following section we shall discuss aspects of the observer–display interaction in some detail.

12.6 THE OBSERVER–DISPLAY INTERACTION

In this section we will focus our attention on the observer–display interaction. This means that the process to be considered is at the front end of the measurement system where the human process of interpretation starts. In this process from the very start the human process involves the following steps:

- scanning and detecting one or more quantities
- identifying the different elements
- combining different elements
- interpreting the message
- reacting or memorizing.

For some tasks a passive registration can suffice but sometimes the tasks can be more vigilant and a qualitative interpretation is required. The steps to be investigated can be classified roughly according to the following four questions.

1 *Which type of information is necessary?*
To answer this question a systems analysis concerning the understandability of the data and the way the data will be displayed has to be performed. A clear structure of the data must be developed which must result in a simple coding for the identification and the relationships of the displayed variables. So, it can be necessary to display the data as a function of time, applying for instance the principle of displaying by exception.

2 *For whom is this information destined?*
A clear job description may be required defining task objectives and operating routines. The measurement and control procedures must be adapted to the least competent operator expected. Designing the procedures in case of occurring errors means

that measurements, assessments and rectification of errors can be performed immediately and that if needed an analysis can be executed afterwards. In dangerous situations a display should not only present what is wrong, but preferably also the location and the right action to be executed.

3   *Which general features characterize the required task?*
Here one of the most disturbing characterizations must be recognized if it is involved in a process. A serious amount of monotony in a process can diminish the vigilance considerably, resulting in a loss of quality or even leading to dangerous situations. Diversity of tasks and performances can avoid these problems. The operator must be able to determine his or her own quality checking and working speed. At all times the quality and accuracy of performance must be stressed.

4   *How is the information presented in the available workspace?*
A lot of factors must be recognized which are impossible to quantify and which may only show in the long run, for example, when hours and hours must be spent by a pilot sitting in one place in a cockpit of an aircraft, or in such a case as an operator sitting in front of a console. Here every inconvenience will increase the workload and can cause undesired stress and fatigue especially in the eyes, hands, arms and neck. This is illustrated in the following example.

*   *   *

Example 12.1
An excellent example of a workspace for a helicopter pilot is shown in figure 12.3. This example shows how a multifunctional display and control setting configuration can look. The cockpit is designed in flat-panel technology. Nowadays, this technology

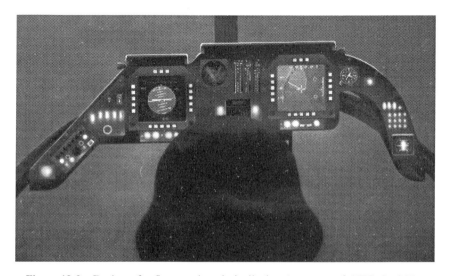

Figure 12.3   Design of a flat-panel cockpit display (courtesy of AEG GmbH).

has become a standard feature in all cockpits and the pilot can decide on which level he or she wants to retrieve information about the functioning of the machine.

*   *   *

The entire workspace must be designed starting from an acceptable position for the eyes and hands. The centre of the visual and manipulative field must be determined and all equipment and most frequently used instruments must be placed in the very middle of the workspace. The less important ones can be placed in the periphery. Any light source must be installed in a way that prevents glare. Very often, coated, tilted glasses and a hood are effective measures. In the following section we will present a number of basic rules of coding of information applicable in the design of displays.

12.7 CODING OF INFORMATION ON DISPLAYS

It is easily recognized that visual perception of information is still the most common way in which information is presented to us. Here, code symmetry plays an important role as the code used to present the information is, ultimately, the visual transducer from instrument to human.

The code can be characterized as a system of agreements and often the user has to learn to understand these codes. Sometimes, this learning process is highly demanding and time consuming, a long time being required to achieve complete comprehension. This is true for example of the CRT oscilloscope; the interpretation of the displayed information depends on different scaling factors and settings.

The human brain prefers the specific language code which already exists in the natural world. If the applied code differs substantially from the preferred code, the learning process is very difficult despite the logic of the applied code. In the design of coding it is always recommended to include some room for error, because for inexperienced users of instruments a so-called half-word may be not enough. The designer of a code can make a choice from several so-called sensory windows, through which the information can be perceived.

A remark must be made here concerning the auditory sense. Audible signals should be restricted to warnings supporting other functions as long as they do not involve human speech. However, the still-increasing possibilities of synthetic speech will undoubtedly offer possibilities which reduce the visual workload. Other possibilities are tactile displays and keyboards originally designed for the blind, but which are still a relatively unknown area in ergonomic design. Until now the visual display is the most common way of presenting detailed information in patterns. A review of coding rules is presented in table 12.3. Table 12.3 lists the most effective information content of different visual coding dimensions for short duration of perception. The column 'Number of steps' shows the number of steps which can be distinguished in the time of signal duration.

The visual channel offers a wide opportunity for coding information according to different dimensions in form, colour, size, resolution, texture, brightness, location, orientation and flicker. The number of steps that can be distinguished coming one

Table 12.3   Review of human visual coding characteristics (courtesy Delft University Press).

Code	Dimension	Number of steps	Bits of information	Signal duration
Digits	—	≥ 100	≥ 6	Short
Letters	—	≥ 1000	≥ 10	Short
Figures	Standard	≥ 15	≥ 3.9	Short
Colour	nm	9	3.1	Very short
Size	cm^2	7	2.8	Very short
Grain	$n\,cm^{-2}$	7	2.8	Short
Brightness	$cd\,m^{-2}$	5	2.3	Very short
Number	n	6	2.7	Long
Length	cm	4	2.0	Long
Angle size	deg	24	4.8	Short
Direction	deg relative	15	3.9	Very short
Flicker	cycles s^{-1}	4	2.0	Very short
Pitch	Hz	5	2.3	Long
Loudness	dB	5	2.3	Short
Smell	?	4	2	Long

after the other in regular order is not the same with all dimensions. Simultaneous signals can be distinguished more easily if they are next to each other than if they are separated in time. Of course the coded information can contain a combination of two or more of the dimensions mentioned in table 12.3. Examples of these combinations are a letters–colour combination, or a digit–size–brightness combination. In such a way the information content can be enhanced but less than might be possible by a simple multiplication of the number of steps of the relevant dimensions which are chosen. A combination of dimensions can even improve the redundancy; for instance a velocity meter can depict all values above 180 km h^{-1} in red as dangerous, and below that value in green as a safe velocity. This table shows also that form, which involves digits, letters and standard figures, has the largest coding capacity, and alphanumeric characters have in their turn a larger capacity than pictograms. Colour is very efficient for a quick search but cannot replace quantitative information. In the following an example is given of information displayed in two different ways for the same process monitored by an operator.

*   *   *

Example 12.2
Suppose the operator watches a machine of which the temperature is 189.75 °C, the oil pressure is 5.78 Pa and 150 s ago the temperature was the same but the oil pressure was 4.78 Pa. This information may be interesting but the operator only wants to know whether the machine is running smoothly, in a safe area, so if the temperature and the oil pressure needles are in the green part of the display the operator receives the same information in a more convenient fashion and with more efficiency.

*   *   *

A basic rule can be abstracted from this example: no more information should be displayed than is strictly required for the process under control. In modern displays the hierarchy on which level information must be presented can be chosen by the operator.

* * *

Example 12.3
Another good example is a wristwatch. For a long time the display was completely analogue. With the rise of modern electronics the wristwatch started to become digital, but it was soon clear that this was not satisfactory at all because too little information is available in a completely digital display. What we see now is that a complete digital display has become obsolete and sometimes both possibilities are combined.

* * *

Table 12.4 provides a survey of the parameters necessary to improve or to guarantee maximum legibility of displays, whether they be CRTs, paper or scale readings. If these guidelines are followed excessive contrasts are avoided and eye fatigue can be prevented, so attention can be maintained for long periods as is often required. In this respect it is noticed that a maximum workload of 4 h uninterrupted labour in front of a CRT is permitted. Up to a reading distance of 6 m the following formula can be applied for the sign height:

$$H = 0.0022D + K_1 + K_2 \qquad (12.2)$$

Table 12.4 Summary of parameters for maximum legibility and display design (courtesy Delft University Press).

Description	Value
Normal reading distance	30–60 cm
Sign height	$\geq 15'$ visual arc or 3.1 to 4.2 with an optimum of 3.7
Width-to-height sign proportions	3:4 and 4:5
Stroke width-to-height	1:8 to 1:6
Spacing between signs	$\frac{1}{2}$ to $\frac{1}{4}$ of sign height
Spacing between words	$\frac{3}{4}$ to 1 of sign height
Interline spacing	$\frac{1}{4}$ to 1 of sign height
Visual field for permanent attention	$10°$–$20°$ of visual arc
Colour wavelength	550 nm (yellow to green)
Background	Dark
Illumination level	80–100 cd m^{-2}
CRT screen illumination level	4–16 cd m^{-2}
Keyboard illumination	500–700 lx
Workspace illumination	≤ 300 lx
Workspace illumination, colour temperature	3000 K

where H (cm) is the sign height, D (cm) is the reading distance, K_1 is a correction factor ranging from 0.15 to 0.65 dependent on reading circumstances such as illumination and K_2 is a correction factor related to the importance of the message and ranges from 0.15 to 0.19.

In many circumstances only capital letters are applied, which can be considered as a mistake. Lower-case letters can more easily be recognized than capitals, so lower-case letters should be preferred for short messages and for body text whenever possible. Also italic letters and letters with serifs are less easily distinguished and therefore should be avoided. If letters are made of a matrix of dots the matrix should preferably be composed of a rectangular pattern of 7×9 dots.

A special field of applications is the design of characters on a CRT display. If less than 10 lines are used for their height the so-called Kell effect can cause blurring and dotting. Pointers with scales are a frequently occurring combination for display design; they are also on CRT displays. The values of the scale should increase in a clockwise direction and from left to right or from bottom to top, dependent on the shape of the display involved.

As said already, in many cases it is sufficient to have only differently coloured regions on a scale, such as green for a safe area and red for a dangerous area. Sometimes more quantitative information is required and a graduation can be applied consisting of three differently sized markers: the headmarker for the decimal intervals, the intermediate markers for the middle of each interval and the minor markers for the smallest units.

Sometimes a percentage scale ranging from 0–100% is very efficient and completely satisfactory. This example of display design, combining analogue and digital information, has already been shown in figure 12.2.

Graphical displays are also used to present information, for instance on a CRT or a printer output on paper. These possibilities offer excellent features to display structures and trends to which our natural perception is most sensitive. Processors are able to calculate a predictive path and so a type of predictive display can be formed. Graphs need to be simple, for instance with vertical bars in a histogram ranging from left to right, whereas the number of curves in one graph should be limited to one or two. The identification of the process depicted in the graphs should be clear, with all the essential units and parameters shown on the graph. These provisions can make a graphical representation very attractive and easy to understand.

When designing a table of numbers a structure with columns proves to be very convenient especially when the decimal points are aligned and long numbers are divided into chunks of three digits. As well as these aspects, tables should be divided into blocks of rows and columns. However, it should always be kept in mind that the information content of a graphical representation is always larger than a numerical representation because a graphical representation is analogue in its very nature.

Last but not least there exists an increasing tendency to apply colours in all types and techniques of display design. This tendency can be seen in CRT design, in software applications and in all types of instrument. Although cost aspects have initially played an important role here, these are becoming less important owing to the improved technology of colour display devices.

PROBLEMS

12.1 Give a definition of ergonomics. What relationship exists between ergonomics and human engineering?

12.2 Which kinds of topic are usual subjects of investigation in human engineering?

12.3 Give a definition of human perception.

12.4 What is meant by a natural transducer?

12.5 Which types of perception can be distinguished? Explain these.

12.6 Which three functions are involved in human perception?

12.7 Discuss the pros and cons of displaying information in digital or analogue form.

12.8 Which kinds of process step have to be distinguished in the observer–display interaction?

12.9 What is meant by coding of information?

12.10 What is the disadvantage of displaying a voltage in a digital fashion only?

BIBLIOGRAPHY

[1] Regtien P L L (ed) 1978 *Modern Electronic Measuring Systems* (Delft: Delft University Press) ch X
[2] Woodson W E 1981 *Human Factors Design Handbook* (London: McGraw-Hill)

APPENDIX 1

Table of frequently used physical constants.

Quantity	Symbol	Value	Unit
Avogadro constant	N_A	602.2045×10^{24}	$kmol^{-1}$
Boltzmann constant	k	$1.380\ 66 \times 10^{-23}$	$J\ K^{-1}$
Elementary charge	q	$1.602\ 18 \times 10^{-19}$	C
Electron rest mass	m_0	9.1095×10^{-31}	kg
Electron volt	eV	$1\ eV = 1.602\ 18 \times 10^{-19}$	J
Gas constant	R	8314.41	$J\ K^{-1}\ kmol^{-1}$
Faraday constant	F	9.6487×10^{4}	$C\ mol^{-1}$
Permeability in vacuum	μ_0	$1.256\ 63 \times 10^{-6}$	$H\ m^{-1}$
Permittivity in vacuum	ε_0	$8.854\ 188 \times 10^{-12}$	$F\ m^{-1}$
Planck constant	h	$6.626\ 17 \times 10^{-34}$	J s
Speed of light	c	$2.997\ 92 \times 10^{8}$	$m\ s^{-1}$
Stefan–Boltzmann constant	σ	56.7032×10^{-9}	$W\ K^{-4}\ m^{-2}$
Standard atmosphere	p	$1.013\ 25 \times 10^{5}$	$Pa = N\ m^{-2}$

INDEX

absolute error, 114, 115
absolute error value, 114, 115
absolute resolution, 114
absolute sensitivity, 114
absolute value of the error, 114
absorption coefficient, 235, 236
absorption of photons, 235
acceleration, 228
acceleration factor, 199
accuracy, 113
acoustic-electric, 258
action potential, 276
activation energy, 199
activation overpotential, 277
active, 11
active guarding, 348, 345
active on-line standby system, 174
active physical quantity, 12
active transducer, 209
actuator, 206
ADC, 79
adder network, 82
additive drift, 286, 304
additive errors, 116, 132
address bus, 139
address (line) bus, 47
ADM, 16, 307
aging, 295
AGND, 327
alias, 29
aliasing error, 30
all-pass filters, 78
alternating direction method, 16, 307
ALU, 147
AMLCD, 244
analogue functions, 67
analogue ground, 327
analogue multiplier, 75
analogue outputs, 46
analogue-to-digital converter, 85
AND function, 170
angle of incident radiation, 235
angle of transmitted radiation, 235

anisotropic, 307, 286
anti-aliasing filter, 44
application software, 47
architecture of bus systems, 47
arithmetic logic unit, 147
arithmetic operations, 146
assembler, 148
asymmetric behaviour, 287
ATE, 47
auditory sense, 429
auditory window, 421
autocorrelation function, 31
automated electronic structures, 39
automated structures, 39
automated test equipment, 47
automatic measurements, 139
automatic nulling, 139
autoreferencing, 418
autozeroing, 417
availability, 166
avalanche breakdown, 243
avalanche effect, 241
avalanche photodiode, 241

balanced attenuator, 332
bandgap, 237, 238
band-pass filters, 78
band-stop filters, 78
bandwidth, 114, 131
barber pole, 271
barrier potential, 237
base, 80
base number, 80
bias-free amplifier, 288
biasing currents, 287
biasing voltages, 287
binary coding, 27
binary multipliers, 83
binary system, 80
biocell membranes, 276
biological equipment, 421
biological signal sources, 361
biological system, 206, 208, 274

biophotonics, 221, 231
biosensor, 220
bipolar differential input stage, 291
bipolar spike noise, 362
bipolar transistor, 289
bit-slice microprocessor, 147
Bode plot, 58, 61
Boltzmann constant, 289, 367
bonding, 230
Boolean algebra, 169
boundary region, 289
branch operations, 146
bridge configuration, 405
broadband noise, 362
buffer amplifier, 348
built-in reliability, 193
built-in test procedures, 139
bulk micromachining, 229
burn-in period, 162, 160
bus-line, 47

cache memory, 143
candela, 236, 422
cantilever, 252
capacitive coupling, 312, 334
capacitive effect, 248, 251
capacitive impedance, 346
capacitive tactile sensor, 256
cardinal measurements, 4
carrier concentration densities, 237
cathode ray tube, 244
CCD, 242
cell membrane, 276
CE-mark, 354
CEN, 354
CENELEC, 354
central processing unit, 139
centralized data-acquisition system, 34
centre of symmetry, 251
channel noise, 397
charge carrier concentration gradient, 237
chemical–dielectric, 280
chemical–electric, 280
chemical–magnetic, 280
chemical–radiant, 280
chopped amplifier, 302
classification of sensors, 206
class 1 clean rooms, 223
clock frequency, 46
CMRR, 114, 121, 311, 342

code, 429
code symmetry, 429
code width, 45
coding, 79, 429
cold on-line standby system, 177
collector, 316
colour, 236
colour displays, 224
colour plasma display panels, 244
combinational circuits, 93
combinational logic, 95
command register, 141
common, 327
common-mode input voltage, 120
common-mode rejection ratio, 114, 121,
    341, 344
common-mode signal, 315
common-mode voltages, 340
common-to-differential signal gain factor,
    341
comparator, 71
compatible materials, 232
compensation techniques, 321
complex calculus, 51
complex frequency response, 25
complex numbers, 51
complexity, 225
component, 164
computer-operated systems, 40
conditional errors, 128, 129
conditional reliability, 173
conductivity, 239
cones, 423
contact noise, 392
contamination, 223
control and timing section, 147
control bus, 47, 139
controller-operated systems, 40
convolution, 19, 29
corona effects, 316
correlation function, 31
Cotton–Mouton, 248, 275
Coulomb interaction, 235
coupling technique, 330
CPU, 139, 144
critical field strength, 268
critical temperature, 268
cross-correlation, 129
cross-effects, 216
cross-sensitivity, 114, 117

CRT, 244, 427
cryogenic techniques, 403
CTD, 241
cumulative distribution, 155
cumulative distribution function, 157
cumulative failure distribution function, 157
Curie and Néel, 275
Curie temperature, 267, 268
current amplifier, 67
current gain factor, 289, 296
current sink, 73
current sources, 73
current-to-voltage converter, 75

DAC, 79
DAQ, 41
DAQ board specifications, 43
data acquisition, 32
data bus, 47, 139
data logging, 39
data transfer operations, 146
De Morgan's theorem, 170
dead zone, 118
decentralized data acquisition, 37
decoder, 91
decomposition technique, 182
deflection method, 104
delay line, 252
Dember effect, 258
demultiplexer, 33, 92
density gradient, 238
depolarization, 276
derating, 196
design failures, 199
design procedures, 193
Destriau, 248
DGND, 327
diamagnetic, 266
dielectric excitation, 253
different ground potentials, 331
differential amplifier, 70
differential diode resistance, 392
differential-mode signal, 315
differential sensitivity, 104
differential-mode input voltage, 120
differential-to-differential signal gain factor, 341
diffraction, 235
diffusion coefficient, 238

digital functions, 91
digital ground, 327
digital I/O, 46
digital micromirror device, 242
digital signal processors, 77
digital-to-analogue converter, 82
DIO, 46
Dirac delta, 18
direct addressing, 144
direct bandgap materials, 237
direct bonding, 230
direct compensation, 85
direct coupling, 331
direct effects, 215
direct indexed, 144
direct bandgap, 237, 238
direct perception, 424
directional sensitivity, 114
discrimination factor, 121
dispersion, 235
displaying by exception, 427
dissatisfaction, 425
distributed multiplexer data acquisition, 37
distribution function, 155
Donnan equilibrium, 276
doping selective etching, 229
double-shielded cables, 345
drift, 114, 303
drift behaviour, 294
drift in sensors, 303
drift nulling, 297
drift-to-signal ratio, 303
driver software, 46
DSP, 77
dual-ported memory, 149
dual-slope integration ADC, 85
dynamic, 73
dynamic devices, 219
dynamic effects, 216
dynamic RAMs, 143
dynamic range, 73, 82, 114, 123
dynamic state description, 216

early failure, 161
earthing, 318
economical reliability, 153
EEG, 313, 361
effective capacitance, 346
effective sampling rate, 45
elasticity, 216

electret microphone, 255
electrical flux, 215
electrically isolated, 43
electrochemical, 280
electroencephalogram, 313
electroluminescence, 248
electromagnetic, 275
electromagnetic compatibility, 112, 354
electron beam, 224
electron excitation, 253, 235
electron–hole pairs, 238, 240
electrostatic excitation, 253
electrostatic fields, 311
electrostatic shield, 327
electrothermal, 266
EMC, 112, 195, 354
EMC conformity, 354
enable input, 93
encapsulation, 230
encoder, 92
energy domains, 207, 211
ENIAC, 226
entropy, 215
environmental circumstances, 196
environmental classes, 197
environmental conditions, 193
EPLD, 101
EPROM, 143
equipment, 163
equipotential, 318
equivalent current power, 367
equivalent noise voltage, 391
equivalent RMS noise current source, 363
equivalent RMS noise voltage source, 362
equivalent voltage power, 367
erasable programmable read-only memory,
    143
ergon, 420
ergonomics, 420
etching, 228
ETSI, 354
Ettinghausen effect, 275
European accreditation standards, 355
European technical standards, 354
excess noise, 392
excitation, 235, 253
excitation signals, 44
expectation, 155
explosive atmospheres, 354
exponential distribution, 155

exponential expansion, 74
exponential failure distribution, 162
extensive quantity, 14, 214
external physical influence, 304
external quantum efficiency, 240

$1/f$ noise, 295, 362, 392
failure probability, 157
failure probability density function, 158
failure rate, 158, 162
failure rate density, 159
failures in time, 158
Faraday, 248
Faraday constant, 276
Faraday effect, 275
Faraday's cage, 361
feedback principle, 65, 130
feedforward coupling, 133
Fermi level, 237
ferrimagnetic, 266
ferromagnetic, 266
ferromagnetic materials, 266
fibre optic sensors, 245
fibre optic thermometry, 246
fibre optical gyros, 246
fibre technology, 245
(field) programmable gate array
    ((F)PGA), 99
(field) programmable logic array
    ((F)PLA), 100, 98
(field) programmable logic sequencer
    ((F)PLS), 102
field-effect transistor, 289
filtering, 44
filters, 77
first-order circuit, 62
FIT, 158
flat-panel technology, 428
flicker noise, 362, 392
floating attenuator, 349
floating guard, 336, 345
floating-guard configuration, 347
floating shield, 349
fluctuation noise, 362
fluidics, 5
flux of radiation, 236
force of mortality, 159
forced cooling effects, 297
Fourier integral transformation, 29
Fourier series, 22

fractional absolute error, 114, 115
fractional error, 114, 115
fractionary binary DAC, 82
full disjunctive form, 95
functional structure, 12
fuses, 96

GaAs technology, 233
GAL, 101
gallium–arsenic, 232
galvanic cells, 279
galvanic separation, 336
galvanoelectric, 280
gamma failure distribution, 162
gas constant, 276
gas discharge lamp, 316
gate noise, 397
gauge factor, 249
general purpose microprocessor, 147
general purpose registers, 147
generic emission standards, 356
global approach, 354
GND, 327
GPIB, 42
ground, 327, 318
ground loop, 318, 319
ground potential, 327
guarding, 307, 311, 317

HAL, 101
Hall coefficient, 269
Hall constant, 269
Hall effect, 232, 265, 268, 272
Hall mobility, 270
Hall MOSFET, 273
Hall voltage, 269
hard array logic, 100
hazard rate, 159
heating transistor array, 263
hexadecimal system, 80
high-pass filter, 78
Hooke's law, 254
HP-IB, 42
hum, 334
human engineer, 420
human factors, 194
human memory, 423
human retina, 422
humidity, 295
humidity sensors, 278

hydrogen-sensitive MOSFET, 279
hysteresis, 114, 118

I/O interface, 41
I/O ports, 141
identification unit, 206
IEC 625, 41
IEEE 488, 41
ignition equipment, 316
illuminance, 236
immediate addressing, 144
immunity standards, 357
implied addressing, 144
impulse function, 18, 22
impulse response, 18
in vitro diagnostica, 354
inaccuracy, 113, 114
incandescence, 248, 266
independent errors, 116
index of refraction, 235
indirect bandgap, 237
infant mortality, 160
information, 2, 205, 427
injection laser, 245
input energy domain, 210
input impedances, 315
input noise per unit bandwidth, 367
input offset current, 287
input offset voltage, 287
interfering sources, 287, 315
instruction decoder, 147
instruction register, 147
instrumental perception, 424
integrated magnetic transducers, 273
integrated silicon flow meter, 263
integrator, 72
intensity sensors, 247
intensive quantity, 12, 214
interface, 35
interfacing, 404
interference, 234, 235, 304, 311
interfering frequency, 350
interfering sources, 287, 315
interferometry, 246
internal physical effects, 304
internal quantum efficiency, 240
interval measurement, 4
intrinsic impedance, 350
intrinsic material, 237
intrinsic safety, 228
intrinsic semiconductors, 239, 261

inversion symmetry, 256
ion electrolytic conduction, 278
ion-sensitive field-effect transistor, 279
ionization, 235
irradiance, 236
ISFET, 232, 279
ISO 9000, 165
ISO 9001, 165
ISO 9002, 165
ISO 9003, 165
isolating transformer, 316
isolation breakdown, 334
isolation resistances, 346
isolation transformer, 334
isotropic signal, 286, 307

jamming, 21, 118
Johnson noise, 390
junction field-effect transistor, 290
junction-FET differential input stage, 293

k-matrix, 382
Karnaugh diagram, 96
Kelvin relation, 259
Kerr electro-optic, 248
Kerr magneto-optic, 248
kinematic micromechanical structures, 219
knowledge, 2
Kynar piezo film, 257

ladder network, 83
lamination, 230
Langevin function, 267
Laplace operator, 64
Laplace transform, 20, 55
laser, 248
laser micromachining, 229
lateral photoeffect, 240, 255
lateral photoelectric, 258
lateral photovoltaic, 258
LCA, 101
leakage impedance, 340
learning factor, 196
least significant bit, 81
legibility, 430
life length, 154
lifecycle, 154
LIGA, 229
light emitting diodes, 244
lightning, 323

line widths, 224
linear region, 289
linear systems, 56
linearity, 117
linearization configuration, 412
linearization routines, 45
lock-in amplifier, 306
logarithmic amplifier, 73
logarithmic compression, 125
logic operations, 146
logon, 2
long-term measurements, 309
loop area, 329
Lorentz force, 265
Lossev, 248
loud noise, 362
low noise, 362
low-frequency emission standards, 357
low-frequency noise, 392
low-pass filter, 62, 78
LPCVD, 251
LSB, 81, 82
luminance, 236
luminous energy, 236
luminous flux, 236
luminous intensity, 236

machine code, 147
MAGFET, 272
Maggi–Righi–Leduc, 275
magnetic core moment, 267
magnetic coupling, 331, 333
magnetic dipoles, 267
magnetic excitation, 254
magnetic field, 311, 328
magnetic orbital moment, 267
magnetic permeability, 271
magnetic resistor, 274
magnetic sensitivity factor, 270
magnetic spin moment, 267
magnetobridge, 273
magnetodiode, 272
magnetoresistivity, 270
magnetoresistor, 272
magnetostriction, 258, 275
magnetotransistor, 272
magnetrons, 316
maintenance protocol, 164
manufacturing failures, 200
Markov chain, 186

Markov techniques, 186
maximum legibility, 430
maximum signal-to-noise-ratio, 375
maxterm, 96
MBBA, 243
mean time between failures, 114, 159
mean time to failure, 160
mean time to first failure, 161
mean time to repair, 167
mechanical chopper disc, 306
mechanical shocks, 316
mechanical transducers, 247
mechatronics, 219
medical devices, 354
memory map, 145
memory unit, 139, 142
metal–electrolyte junctions, 276
metal-oxide junctions, 278
metric, 2
metric information, 2
metrology matters, 354
metrons, 2
micro-tuning forks, 252
microcontrollers, 147
micromachining, 219
micromechanics, 218
microprocessor fundamentals, 139
microprocessor-based systems, 139
microresonator, 252, 255
microsystem, 218
microwave coupling, 331
military standard class, 195
Miller indices, 211
minimal cut-set technique, 185
minterm, 96
MIS structure, 241
mission availability, 169
mnemonics, 147
mobility, 239
modulating, 306
modulating energy domain, 210
modulating transducer, 210, 209, 404
moduli, 215, 216
modulus of elasticity, 253
mole of electrons, 276
momentary availability, 168
Moore's law, 226
MOS field-effect transistor, 290, 302
MOSFET differential input stage, 293
MOSFET switches, 302

most significant bit, 81
MSB, 81, 82
MTBF, 114, 159
MTTF, 160
MTTFF, 161
MTTR, 167
multimode, 246
multiple-card system, 327
multiple-input/output configuration, 33
multiplexer, 33, 92
multiplexing, 33, 92
multiplicative drift, 304
multiplicative errors, 116, 304
multiplying errors, 116

natural perception, 424
Néel temperature, 266
negative temperature coefficient, 261
Nernst effect, 266
Nernst equation, 276
nerve fibres, 423
network reduction method, 180
New Approach, 354
New Approach directives, 355
NF, 365
no noise, 362
noise, 117, 304, 390
noise current, 362, 363
noise factor, 362
noise figure, 362, 363, 364, 370, 372
noise flicker current, 393
noise generators, 362
noise power, 362, 390
noise power density, 364, 390
noise sources in parallel, 383
noise sources in series, 383
noise voltage, 362
noise-to-signal ratio, 303
nominal measurement, 4
nomos, 420
non-inverting amplifier, 70
non-linearity, 114, 118
non-polarizable electrode, 277
non-repairable systems, 160
normal failure distribution, 162
normal operating region, 161
Norton, 382
null voltage shift, 219
nulling element, 103
nullor, 68, 339

Nyquist criterion, 67
Nyquist frequency, 30

observation, 424
octal system, 80
offset, 114, 119, 304
offset error, 69
offset error changes, 304
offset in transducers, 294
offset nulling, 293, 297
offset symmetry, 291
offset-to-signal ratio, 303
Ohm's law, 217
one-port network, 382
opcode, 148
operand, 148
operational amplifier, 67
optical heating excitation, 254
optical semiconductor, 240
optimum source resistance, 373
optocouplers, 44
optocoupling, 331
optoelectronic magnetic sensor, 274
optoelectronics, 220
OR array, 97
order of the Markov chain, 186
ordinal type of measurement, 4
output buffer, 142
output enable, 142
output energy domain, 210
output impedance, 136
output resolution, 46
overpotential, 277
oxygen sensor, 279

p-domain, 57
paging, 144
PAL, 101
parallel active on-line system, 175
parallel converter, 91, 84
parallel k out of n system, 179
paramagnetic, 266
parasitic capacitance, 313
parasitic current, 324, 333
parasitic series impedance, 347
part, 163
partial fraction method, 58
passive liquid crystal displays, 243
passive physical quantity, 14
passive quantity, 11

passive registration, 427
passive transducer, 210
path-tracing technique, 182
PCB, 326
peak detector, 77
PEEL, 101
Peltier coefficient, 259
Peltier effect, 217, 261, 259
Peltier elements, 265
penetration depth, 329
perception, 423
perceptual knowledge, 2
perfect non-polarizable electrode, 277
perfect polarizable electrodes, 277
periodic signals, 22
peripheral, 141
permeability of vacuum, 267
permittivity, 251, 216
permittivity of vacuum, 251
phase sensors, 246
phasor, 53
phonon drag, 259
photochemical, 248
photoconductivity, 243
photoconductor, 241
photocurrent, 240
photodielectric, 248
photodiode, 241
photodiode arrays, 242
photoelastic, 258
photoelectric, 248
photointegrated circuit, 242
photolithographic, 224
photoluminescence, 248
photomagnetoelectric, 248, 275
photometric, 235
photon energy, 234
photon noise, 402
photonics, 221
photons, 228
photoresistor, 241
photosynthesis, 233
phototransistor, 241
photovoltaic, 243
photovoltaic effect, 242
physical changes, 295
physical effects, 233
physical reality, 424
physical reference, 418
PI-FET, 257

pictograms, 431
piezoelectric, 256
piezoelectric effect, 248
piezoelectric excitation, 254
piezoelectric polymer film, 257
piezoelectricity, 251
piezojunction, 254
piezojunction effect, 249
piezooptic, 258
piezoresistance, 258
piezoresistivity effect, 249
piezoresistor, 256, 254
piezotransistor, 254
pinch-off voltage, 290
PIN diode, 241
pink noise, 362, 392
Planck's constant, 234
plasma addressed LCD, 244
PLC, 40
PLD, 40
plug-in cards, 40
plug-in data acquisition, 43
pn junction, 239
pn luminescence, 248
Pockel's effect, 248
Poisson constant, 249
Poisson distribution, 156
polar form, 53
polar plot, 58, 59
polarization, 235, 277
polarized light, 235
pole–zero plots, 63
polymers, 278
popcorn noise, 362
posterior technique, 129
potentiometric voltages, 277
power consumption, 226
power gain, 365
power supply rejection ratio, 114, 121
power supply variations, 114, 121
power–delay product, 227
preamplifier with adjustable gain, 69
predictive display, 432
price per gate, 230
prior technique, 129
probability distribution, 155
probability function, 155
processing system, 220
program counter, 147
programmable array logic, 101

programmable logic controllers, 40
programmable logic device, 40, 91, 93
programmable n-input AND gate, 96
programmable read only memory
   (PROM), 99, 98
programmer-operated systems, 39
PSD, 233
PSRR, 114, 121, 295
PTAT bandgap reference, 73
PTC, 261
pulse amplitude modulation, 27
pulse position modulation, 27
pulse width modulation, 27
pulse-shaped signals, 316
PVDF, 257
pyroelectric, 257, 266

qualitative description, 3
qualitative interpretation, 3, 427
quality, 154
quality classes, 196
quality factor, 195
quantitative process, 3
quantization uncertainty, 85
QW laser, 233

radiance, 236
radiant energy, 236
radiant excitance, 236
radiant intensity, 236
radiation heating, 248
radio-frequency emission standards, 357
radioluminescence, 248
radiometric, 235
radiometry, 236
radix, 80
RAM, 143
ramp function, 21, 22
random access memory, 143
random failures, 200
range, 45, 123
range of power, 228
ratio measurements, 4
Rayleigh failure distribution, 162
Rayleigh mode, 251
read only memory, 143
redundancy, 1, 174
reference source code, 79
reflectance, 240
refraction, 235, 234

register indirect indexed, 144
rejection factor H, 121, 342
relative addressing, 144
relative non-linearity, 118
relative permittivity, 251
relative resolutions, 114
relative sensitivity, 104, 114
reliability, 114, 126, 154, 227
reliability priority, 194
remnant magnetism, 267
repair rate, 167
repeatability, 219
reproducibility, 219
resistive heating excitation, 254
resistive pressure transducer, 255
resolution, 45, 104, 114, 348, 421, 422
resonating frequency, 253
resonating transducers, 404
response analysis, 17
response time, 219
retina, 423
reverse-biased diode, 301
rf coupling, 331
Righi–Leduc, 275
rigidity, 215
rise time, 219
RMS, 17
RMS noise current, 363
RMS noise voltage, 362
rods, 423
ROM, 143
root loci, 66
RS-232 (V24), 41
RTD, 264
R/W memory, 143

sacrificial layers, 229
safety earth, 317
safety guard, 334
safety voltage, 317
sample and hold, 27
sampled signals, 26
sampling, 26
sampling rate, 45
saturation, 118
SAW, 248, 251
SAW sensors, 252, 280
scaling errors, 304
Schottky barrier diode, 241
Seebeck coefficient, 258

Seebeck effect, 218, 258, 212, 263, 316
selective information, 1, 2
selector lines, 93
self-generating transducers, 209, 404
semantic, 2
semiconductor noise, 394
sensitivity, 116, 214, 422
sensor, 206, 219
sensor classification, 206, 246
sensorial mechanism, 424
sensory windows, 429
sequential circuits, 94
sequential logic, 101
sequential tests, 201
serial DA converters, 84
series mode voltage, 320, 323
series mode voltage error, 319, 349
series system, 172
settling time, 46, 219
Shannon, 2
shielding, 307, 311, 321
shielding cabinet, 321
shifting noise sources, 380
shifting sources, 380
shot noise, 362, 391
sigma–delta modulator ADC, 87
signal source power, 371
signal structure, 12, 16
signal-to-noise ratio, 362, 370, 372, 375
signals with a unique character, 17
single mode, 246
slew rate, 46, 70, 114, 119
SLM, 233
smart nose transducer, 279
smart sensor, 218
S/N, 362, 375, 370, 372
Snell's law, 235
sources of interference, 315
space angle, 236
spatial structures, 12, 32
specific conductivity, 217
specific probability density function, 30
specific resistivity, 217
specific thermal conductivity, 297
spectral density, 403
speed of light, 234
speed of propagation, 228
spot noise, 362
SQUID, 274, 361
stability, 219

stack, 146
standard deviation, 155
standard half-cell voltages, 276
standard normal form, 95
star structure, 47
state description, 213
state diagram, 187
static devices, 219
static RAMs, 143
statistical process control, 223
status register, 141
steady-state availability, 168
steady-state description, 214
step function, 21
step stimulus, 21
steradian, 236
stiffness, 215
stochastic errors, 129
stochastic signals, 30
stochastic variables, 155
stochastic variations, 295
strain gauge signal conditioner, 417
stray capacitance, 335
strong inversion, 289
structural information, 2
successive approximation ADC, 90
Suhl effect, 275
superconductivity, 268, 274
superposition, 382
superposition theorem, 382, 387
surface acoustic wave filters, 251
surface acoustic waves, 248, 251
surface micromachining, 219
survival function, 156
switched capacitor filters, 78
switches, 316
synchronized detectors, 25
system, 163
systematic errors, 128

technical reliability, 153
technical systems, 208
telemetry, 38, 416
temperature coefficient 219, 260, 298
temperature drift, 298
temperature integrated sensor, 262
temperature range, 114
temperature rejection ratio, 121
texture, 429
TFT, 244

thermal EMF, 297
thermal expansion, 216
thermal feedback, 107
thermal matrix, 263
thermal noise, 367, 362, 390
thermal noise power, 367, 390
thermal noise voltage, 364
thermal transport, 259
thermistors, 261, 264
thermochemical, 266, 280
thermoconductivity, 266
thermocouple, 258, 344
thermocouple effects, 297
thermocouple materials, 265
thermodielectric, 266
thermodiffusion, 259
thermodiode, 262
thermoelastic, 266
thermoelectric, 266
thermoluminescence, 266
thermopiles, 263
thermoresistance, 258, 260
thermoresistor, 262
Thevenin, 382
thin-film transistor technology, 244
Thomson coefficient, 259
Thomson effect, 259
three-state buffers, 143
threshold voltage, 290
time analysis, 17
time multiplexing, 33
time-invariant, 19
timing I/O, 46
tolerances, 111
total input noise density, 365
total input noise voltage, 366
total noise power, 363
total noise voltage, 366
total noise voltage level, 366
total RMS noise voltage, 369
traducere, 206
transconductance amplifier, 67
transducers, 3, 205, 206
transfer function, 25, 57
transformer coupling, 333, 379
transient, 18
transient response, 19
transient term, 168
transimpedance amplifier, 67
transistor pressure transducer, 255

transition, 187
transition impedance, 312, 318, 339
transverse photoelectric effect, 240
triboelectric, 258
TRR, 121
two-port network, 382

uncertainty, 114
unilateral Laplace transform, 20
unit step function, 22
unreliability, 157
UPLD, 93
useful life period, 161
user-programmable logic devices, 93

V24, 41
valence band, 237
variance, 155
vector measurement, 407
vector sensor, 307
VFC, 89
vibrations, 295
virtual instruments, 40
visual perception, 422
visual window, 422
voice-controlled, 425
Volta effect, 280
voltage amplifier, 67

voltage error, 319, 349
voltage feedback, 136
voltage gradient, 217
voltage power supply rejection ratio, 295
voltage-to-frequency conversion, 85
voltage-to-frequency converter (VFC), 89
voltage-to-time conversion, 85
VXI/VME bus, 41, 42

wafer size, 225, 226
wave number, 237
weak inversion, 289
wear-out period, 161
Weber's law, 422
Weibull failure distribution, 162
Weiss field, 267
Westcott theorem, 135
wet etching, 228
Wheatstone bridge, 104
white noise, 31, 362, 390
wireless data transmission, 38

x-ray beam, 224

Zener voltage, 262
zero-offset voltage, 219
zero-signal reference potential, 322
zero-state response, 20